Bioinformatics: Design, Sequencing and Gene Expression

Bioinformatics: Design, Sequencing and Gene Expression

Editor: Daniel McGuire

R CALLISTO REFERENCE

www.callistoreference.com

Callisto Reference,
118-35 Queens Blvd., Suite 400,
Forest Hills, NY 11375, USA

Visit us on the World Wide Web at:
www.callistoreference.com

ISBN: 978-1-64116-171-8 (Hardback)

Cataloging-in-Publication Data

Bioinformatics : design, sequencing and gene expression / edited by Daniel McGuire.
    p. cm.
Includes bibliographical references and index.
ISBN 978-1-64116-171-8
1. Bioinformatics. 2. Sequence alignment (Bioinformatics). 3. Gene expression.
I. McGuire, Daniel.
QH324.2 .B56 2019
570.2--dc23

Table of Contents

Preface

Bioinformatics is an interdisciplinary field of science, which combines biology, mathematics, computer science and statistics, for the development of methods and software tools to understand biological data. These may include the interpretation of the genetic basis of disease, unique adaptations, differences between populations, etc. It also aids in the understanding of the organizational principles between nucleic acids and protein sequences. Techniques such as image and signal processing facilitate the extraction of useful results from large amounts of data. Besides these, computational techniques of pattern recognition, visualization, data mining, etc. are used in bioinformatics. Research in this field explores the areas of drug design and discovery, gene finding, sequence alignment, prediction of gene expression and protein-protein interactions, among others. This book is a compilation of chapters that discuss the most vital concepts and emerging trends in the field of bioinformatics. The various advancements in this field are glanced at and their applications as well as ramifications are looked at in detail. This book is a vital tool for all researching and studying this field.

This book unites the global concepts and researches in an organized manner for a comprehensive understanding of the subject. It is a ripe text for all researchers, students, scientists or anyone else who is interested in acquiring a better knowledge of this dynamic field.

I extend my sincere thanks to the contributors for such eloquent research chapters. Finally, I thank my family for being a source of support and help.

Editor

GATK hard filtering: tunable parameters to improve variant calling for next generation sequencing targeted gene panel data

Simona De Summa[1†], Giovanni Malerba[2*†], Rosamaria Pinto[1], Antonio Mori[2], Vladan Mijatovic[2] and Stefania Tommasi[1]

Abstract

Background: NGS technology represents a powerful alternative to the standard Sanger sequencing in the context of clinical setting. The proprietary software that are generally used for variant calling often depend on preset parameters that may not fit in a satisfactory manner for different genes.
GATK, which is widely used in the academic world, is rich in parameters for variant calling. However the self-adjusting parameter calibration of GATK requires data from a large number of exomes. When these are not available, which is the standard condition of a diagnostic laboratory, the parameters must be set by the operator (hard filtering). The aim of the present paper was to set up a procedure to assess the best parameters to be used in the hard filtering of GATK. This was pursued by using classification trees on true and false variants from simulated sequences of a real dataset data.

Results: We simulated two datasets, with different coverages, including all the sequence alterations identified in a real dataset according to their observed frequencies. Simulated sequences were aligned with standard protocols and then regression trees were built up to identify the most reliable parameters and cutoff values to discriminate true and false variant calls. Moreover, we analyzed flanking sequences of region presenting a high rate of false positive calls observing that such sequences present a low complexity make up.

Conclusions: Our results showed that GATK hard filtering parameter values can be tailored through a simulation study based-on the DNA region of interest to ameliorate the accuracy of the variant calling.

Keywords: NGS, Variant calling, Variant filtering, Targeted gene panel, SNV, Indel

Background

In the last decade, sequencing technologies, the so-called next generation sequencing (NGS), have delivered a step change in the ability to sequence genome, leading to a state of permanent evolution.

NGS platforms allow to detect mutations significantly reducing time and costs [1, 2]. Ion Torrent Personal Genome Machine (PGM) started to be distributed in 2011 [3] and thus to be used for the identification of genetic variants associated to human diseases [4, 5]. Indel detection, in particular, in homopolymer region have a high positive rate [6, 7] which have to be lowered for clinical applications [8, 9]. Thus the major challenge in NGS regards the correct manipulation of output data [10] assembling appropriate pipeline, including aligner and variant caller. Diverse algorithms for alignment have been compared in many studies [11, 12]. Caboche et al [13] compared different mapping algorithms with Ion Torrent data, in terms of computational requirement, mapper robustness, ability to map reads in repeated

* Correspondence: giovanni.malerba@univr.it
[†]Equal contributors
[2]Department of Neuroscience, Biomedicine and Movement Sciences, Section of Biology and Genetics, University of Verona, Strada Le Grazie 8, 37135 Verona, Italy
Full list of author information is available at the end of the article

regions and behavior with mutated reference genome. They were able to optimize a benchmark procedure from whole genome sequencing of small genomes, highlighting the importance to evaluate mappers or to optimize parameters of a chosen mapper for a specific application. Moreover, variant calling pipelines have been compared in relation to different applications and platforms. Yeo et al. [14] optimized an indel detection workflow for BRCA1/2 PGM sequencing panel. They compared the proprietary software Torrent Suite and two open source variant callers, GATK and SAMtools. Their results showed that SNV detection was less problematic than indel identification using Torrent Suite. Moreover, they demonstrated how the combination of BWA or TMAP mappers and SAMtools is able to improve indel detection.

As demonstrated by these studies, the bioinformatic challenge on NGS data and, in particular, Ion Torrent data from targeted sequencing requires a lot of efforts in order to correctly identify the best analysis pipeline.

GATK is a well-known toolbox for NGS data analysis. Variant Quality Score Recalibration (VQSR) step generate an adaptive model based on metrics, such as strand bias, from true variants. Thus it could be possible to calculate if a variant is true or false. However, this step could be used only for whole genome data or for dataset including more than 30 exomes. For targeted gene panels, GATK's Best Practices suggest to set up hard filters specific for the study. In the present study, we compared variant calling results of GATK pipeline including the use of hard filtering, suggested by GATK's Best Practices, and the proprietary Torrent Suite Variant Caller regarding a custom panel including 11 genes. Then, we focused on two simulated datasets (100 replicates for each dataset), with high and low coverage, and then we processed the raw variants called by GATK to set parameters of quality in order to increase the number of true variants and decrease the number of false variants.

Methods
Real dataset
A dataset of 26 metastastic melanoma formalin-fixed paraffin embedded (FFPE) samples was studied. Exonic regions of a panel of 11 genes were sequenced with Ion Torrent PGM. Sequencing data were analyzed using 2 standard pipelines including either the Torrent Suite (TMAP 4.0.6 aligner and Torrent Variant Caller version 4.2-18 (TVC) with the parameters "somatic" and "high stringency" switched on) or the GATK suite for few variants (bwa aligner and GATK programs; see GATK pipeline section for details).

Simulated datasets
The above reported panel of 11 genes was devised to identify sequence variants in the coding regions. When

the present study was conceived we prepared a catalog of known variants and their allele frequency using the information from the real 26 sequenced individuals. The simulation-based study was then set up using a reference dataset of 26 individuals with randomly assigned variants according to the catalog of known variants. The simulated genotypic profiles of each individual were recorded. For every sample it was prepared by simulation a file (fasta format) containing the amplicon sequences that were modified to introduce the assigned variants, repeated 2 times (since humans are diploids) in both the forward and in the reverse sequence. The dataset of 26 fasta files was then processed with the ART simulator [15] to generate files (FASTQ format) similar to those produced by a sequencer of new generation, mimicking its features and biases. For every sample the ART simulator was launched twice under the hypothesis of a sequencing depth of 20x (low coverage, LC) and 100x (high coverage, HC), respectively. The 26 files were then processed using a standard pipeline for variant calling that includes the aligner BWA and the GATK suite of programs (see GATK pipeline section). All the still unfiltered variants were tagged as true (actually present in the fasta file of the simulated subject) or false (not present in the fasta file and therefore representing the product of an erroneous call by the bioinformatic variant calling process) variants. Thus one hundred independent datasets of 26 individuals were simulated for a total of 5200 simulated samples. By this approach we tried to simulate several times the most likely scenario (number of samples per dataset, amplicons used, selected genes, reported variants) resembling to the real one.

GATK pipeline
We followed the Toolkit for Genome Analysis (GATK, https://software.broadinstitute.org/gatk/) [16] recommendations of DNAseq best practices for calling variants. Hence the following software were used: BWA-mem (http://bio-bwa.sourceforge.net/) for sequence alignment [17] and GATK 3.4 software for the later steps. In more detail, sequences underwent the following steps: 1) alignment to the human genome reference version hg19, 2) realignment around Indels, 3) base recalibration and 4) variant discovery (using the haplotype caller function in ERC mode) without been marked for duplicates.

The discovered variants were hard filtered after having selected the rules to setup the filters from the classification trees as described in the "Filters for the variant calling" and "Classification trees" sections.

Filters for the variant calling
Hard filtering evaluated 7 standard GATK filters (BaseQRankSum, ClippingRankSum, DP, MQ, GQ, MQRankSum, ReadPosRankSum, see Additional file 1 for a

description) and 3 filters (FS, ADT, ADTL) that were not present in the standard GATK vcf output files. FS is the p value from the contingency table of the number of reads calling the alleles at the variant site on either the DNA strands (forward and reverse). ADT and ADTL evaluate imbalances in calling the reference and the alternative allele (ADT), also depending on the amount of reads that map on the variant locus (ADTL). ADT and ADTL are defined as follows:

$$ADT = \frac{|(AD1-AD2)|}{AD1 + AD2}$$

$$ADTL = \log_{10}(AD1 + AD2)^* ADT$$

where AD1 and AD2 are the number of unfiltered reads calling the reference and alternate allele, respectively.

Descriptive statistics were performed using R 3.2.3 and the kruskal.test function, and ROCR library version 1.0-7.

Classification trees

Every filter was included into classification trees to target the filter rules that better discriminate the true variants (listed by GATK and also present in the simulated sample) from false variants (proposed by GATK but not present simulated in the sample under examination). These rules have been then used for the hard filtering. Eight classification trees were generated to investigate separately SNV and INDEL, homozygotes and heterozygotes (as shown by GATK) under a simulated coverage of 20x or 100x. For every tree we extracted the selected filters and their threshold values in order to use them in the phase of hard-filtering. We extracted the filters that were listed starting from the root node of the classification tree to the next 2 consecutive daughter nodes (for a maximum of 7 filters). The filters selection for each classification tree was made as follows. Comparing every daughter node with the root node we targeted the nodes that 1) contained at least 10% of the true calls and 2) where the ratio of the number of true calls and false calls was greater than 3. To minimize the number and complexity of the filter rules we then considered for each targeted node l (lower node) the upper connected node u (upper node) closer to the root node. We then selected the node u instead of the node l when 1) the node l contained less than 80% of the true calls or 2) when containing more than 70% of false calls of node u (i.e. reduction of at least 30% of false calls), respectively. Once the relevant nodes were selected we extracted the filter rules starting from the root node toward each of the finally selected daughter nodes.

Classification trees were produced using the library rpart (version 4.1-9) of the R package (version3.1.2)

Results

Results in the real dataset: GATK vs TVC

In the first part of this study we looked at the results obtained using 2 different standard pipelines on the same set of sequences. TVC is the software use for targeted sequencing in bundle with the Ion Torrent sequencer. GATK is considered the "gold standard" in managing large NGS data (i.e. exomes and genomes) and can be used for targeted sequencing. For this reason, we compared TVC calls with those produced by GATK 3.4. TVC called 399 variants in the entire dataset, 73 of which were shared with GATK that detected 83 SNVs. Then we performed the VCF files, which are the output files of both TVC and GATK, focusing on some Parameters Of sequencing Quality. In particular, DP (Coverage) and AF (Allele Frequency) tags were shared by VCF outputs. TVC calls showed mean, DP and AF values of 1658.13 and 0.15, respectively. The 73 SNVs called by both TVC and GATK showed higher mean. This observation could suggest that shared SNVs are true positive calls, having higher coverage and quality by depth values. It is noteworthy that mean POQ values of the variants called only by GATK were similar to TVC calls (DP: 1795.33; AF: 0.16). SNVs identified only by TVC showed lower mean DP and AF values than shared variations (663.75, 0.08, respectively). Such results suggest that SNVs called by TVC may be enriched of many false positives. Indel calling is a highly debated problem when we refer to Ion Torrent data. In a very preliminary way, we evaluated mean DP and AF values of 107 Indels identified by GATK, which displayed similar values to those of shared SNVs (3814.58, 0.16, respectively). No indel was detected by TVC.

High-coverage and low-coverage simulated datasets: descriptive statistics

To better explore GATK variant calling and to try to tune the hard filtering parameters (filters), we performed a simulation-based study, as described in the "Methods" section. In the High-Coverage (HC) dataset, GATK identified 91115 unfiltered SNVs and 81640 unfiltered Indels. It is noteworthy that 98.49% and only 21.23% of SNVs and Indels were true variants, respectively. In the Low-Coverage (LC) dataset, 113246 and 88145 unfiltered SNVs and unfiltered Indels were respectively called. As expected, we found that the percentage of true variants was lower in the LC dataset (84.95% for SNVs and 8.68% for Indels) (Table 1).

We observed that 3.9% of SNVs and 53.8% of Indels were homozygous in the HC dataset whereas 14.8% of SNVs and 51.9% of Indels were homozygous in the LC dataset.

We investigated the distribution of the values of the individual GATK filters, namely BaseQRankSum,

Table 1 Overall GATK unfiltered alterations identified in HC and LC dataset (100 replicates of a dataset of 26 individuals and 11 genes)

	TV	FV
HC dataset		
Indels	17,359	64,281
SNVs	89,743	1,372
LC dataset		
Indels	7,656	80,489
SNVs	96,203	17,043

TV true variants, *FV* false variants

ReadPosRankSum, ClippingRankSum, DP, MQ, MQRankSum, and GQ (see Additional file 1 for details) between true and false variants. Table 2 reports the descriptive statistics for the HC dataset and shows that BaseQRankSum, ReadPosRankSum and DP display a statistically significant difference both in SNVs and Indels. MQRS showed a statistically significant difference in Indels only.

Table 3 reports the descriptive statistics for the LC dataset and shows a statistically significant difference for all the GATK filters with the exception of MQ both in SNVs and Indels subsets. Of note that GQ always (= the median value of each of the 100 replicates) presented a value equal to 99 for the true variants. It is intriguing that in the case of SNV subset, DP presents the lowest values in the TVs compared to FVs even the difference is relatively small.

The performance of all individual filters to discriminate between true and false variants was summarized by estimating the area under the ROC curve (AUC). Table 4 reports that ADT and GQ showed the best performance. Additionally every filter showed a better performance for SNVs rather than Indels (Additional file 2).

It could be noticed that correctly called alterations showed a higher coverage than false variants, highlighting the importance of this parameter.

Classification trees

We performed the analyses of either SNVs or Indels subsets stratified by genotype, in HC and LC datasets. Classification trees (Additional file 3) allowed to set a series of filter rules for each of the 2 type of sequence alteration. Table 5 shows the parameters and threshold values to be used for hard-filtering been extracted from the classification trees.

Notably, the classification tree did not select any reliable filter for homozygous Indels in the LC dataset (Table 6).

We then explored the sequence of the flanking regions of each type of alterations, in particular we observed that short homopolymeric strings are recurrent and therefore partly responsible for false positive calls (Table 7, Additional file 3).

Discussion

It is well known that there are many technical challenges involved in getting an accurate variant calling procedure of NGS data including the bioinformatic analysis. A number of tools based on complex statistical models has been developed but many concerns related to their performance remain still open. Since the number of the called variants varies from software to software, typically more than one computer program is then used. If the variant is actually called by all the programs then its support increases. However, the problem occurs when the variant is called only by some programs, raising the suspicion that it is not true. NGS is now applied in many fields. We were interested in studying the case of targeted sequencing of small set of genes when using a common NGS platform such as Ion Torrent. When analyzing a few variants (as in the case of a panel of genes rather than an exome) the GATK guides suggest to use filters that must be set by the user (hard-filtering) rather than the adaptive filtering that needs a high number of variants to work properly. Under these conditions

Table 2 Descriptive statistics of GATK filters in the HC dataset, stratifying calls by type (SNV/Indels). Data are displayed as mean ± sd

	SNVs			Indels		
	TV mean ± sd	FV mean ± sd	p-value	TV mean ± sd	FV mean ± sd	p-value
BQRS	0.11 ± 0.03	-0.6 ± 0.5	<0.0001	0.28 ± 0.12	0.15 ± 0.05	<0.0001
RPRS	-0.067 ± 0.05	0.05 ± 0.4	0.0009	-0.74 ± 0.22	0.23 ± 0.05	<0.0001
CRS	0.0007 ± 0.02	-0.009 ± 0.29	0.72	0.001 ± 0.07	0.007 ± 0.03	0.6
DP	96.61 ± 0.58	49.25 ± 5.06	<0.0001	109.4 ± 9.57	96.01 ± 0.1	<0.0001
MQ	60 ± 0	59.99 ± 0.07	-	60 ± 0	60 ± 0	-
MQRS	-0.03 ± 0.02	-0.05 ± 0.28	0.3	-0.02 ± 0.09	-0.21 ± 0.04	<0.0001
GQ	99 ± 0	79.15 ± 12.06	-	99 ± 0	73.16 ± 2.7	-

The mean value is the mean value of the median value from each of the 100 replicates
BQRS BaseQRankSum, *RPRS* ReadPosRankSum, *CRS* ClippingRankSum, *DP* depth of coverage, *MQ* MappingQuality, *MQRS* MappingQualityRankSum, *GQ* genotype quality, *TV* true variants, *FV* false variants

Table 3 Descriptive statistics of GATK filters in the LC dataset, stratifying calls by type (SNV/Indels). Data are displayed as mean ± sd

	SNVs			Indels		
	TV mean ± sd	FV mean ± sd	p-value	TV mean ± sd	FV mean ± sd	p-value
BQRS	0.02 ± 0.02	-0.27 ± 0.12	<0.0001	0.16 ± 0.16	0.01 ± 0.03	<0.0001
RPRS	-0.19 ± 0.03	-0.31 ± 1.1	<0.0001	-1.29 ± 0.3	0.04 ± 0.04	<0.0001
CRS	-0.02 ± 0.02	-0.05 ± 0.07	<0.0001	-0.004 ± 0.12	-0.04 ± 0.01	0.001
DP	19.97 ± 0.17	22.72 ± 1.3	<0.0001	21.83 ± 1.97	20.24 ± 0.42	<0.0001
MQ	60 ± 0	60 ± 0	-	60 ± 0	60 ± 0	-
MQRS	-0.06 ± 0.01	-0.1 ± 0.08	<0.0001	-0.15 ± 0.15	-0.1 ± 0.04	0.03
GQ	99 ± 0	20.04 ± 2.9	-	99 ± 0	17.94 ± 0.92	-

The mean value is the mean value of the median value from each of the 100 replicates

BQRS BaseQRankSum, *RPRS* ReadPosRankSum, *CRS* ClippingRankSum, *DP* depth of coverage, *MQ* MappingQuality, *MQRS* MappingQualityRankSum, *GQ* genotype quality, *TV* true

Table 4 Perfomance of the individual filters evaluated to discriminate between true and false variants by the AUC values from ROC curve, grouped by type of variants (SNV or Indel) and status of the genotype call (homozygote or heterozygote) according to the depth of sequencing (LC or HC dataset)

	SNV		Indel	
	Homo	Het	Homo	Het
HC dataset				
BQRS	0.73	0.53	0.5	0.53
RPRS	0.57	0.61	0.52	0.68
CRS	0.5	0.51	0.53	0.5
DP	0.79	0.8	0.76	0.6
MQ	0.55	0.5	0.6	0.63
MQRS	0.52	0.53	0.58	0.58
GQ	0.65	0.95	0.77	0.77
ADT	0.96	0.8	0.56	0.94
ADTL	0.8	0.77	0.72	0.94
FS	0.51	0.62	0.51	0.54
LC dataset				
BQRS	0.58	0.65	0.5	0.52
RPRS	0.53	0.5	0.54	0.74
CRS	0.52	0.5	0.51	0.51
DP	0.63	0.67	0.52	0.62
MQ	0.51	0.54	0.54	0.52
MQRS	0.52	0.52	0.51	0.5
GQ	0.79	0.99	0.53	0.97
ADT	0.98	0.99	0.5	0.92
ADTL	0.67	0.98	0.54	0.98
FS	0.5	0.54	0.5	0.54

(which are very common since many laboratories developed their own panel of specific genes to study the association with a specific phenotype) it becomes important to tailor the filters to call the true variants on the specific design. It is likely that different panels of genes and even different designs for the same panel of genes require a different setup of filters. We therefore tried to explore through a simulation-based study the outcomes that the pipeline for the variant calling may encounter. We were interested in the study of a specific scenario made up a group of individuals with a specific sequencing design. We measured the performance inthe calling true variants for each of the filters that can be set when working with hardfiltering. Hence we used classification trees on a large data simulated dataset of true and false variants.

In the present paper, we studied several standard and non standard GATK filters to be used for hardfiltering in the context of a targeted gene panel sequencing. Firstly, we analyzed a real dataset coming from the sequencing of an Ion Torrent targeted gene panel observing a high discrepancy between TVC and GATK, particularly for Indels, suggesting that such type variants are even difficult to be detected by the present bionformatic tools. In fact the importance to define a "gold standard" dataset to test variant calling methods is a very hot topic. Recently, "synthetic" matched tumor/normal samples was created for comparing performances of popular variant callers in detection of "somatic" SNVs. However, even if they had the advantage to refer to NIST-GIAB [18] as gold standard, authors could not discriminate "somatic" SNVs from germline background, an important issue when studying tumors, and moreover the batch that they purchased was not the same used for NIST-GIAB [19]. In the present study, we decided to simulate two datasets, each with a different coverage and carrying alterations found in real data. Notwithstanding this investigation did not simulate tumor

Table 5 Parameters and their thresholds selected by regression trees

Sequencing depth	Variant type	Genotype by GATK	Filter rule
20x	SNV	homozygous	ADT > =0.98
20x	SNV	heterozygous	ADT < 0.55
20x	INDEL	homozygous	N/A (*)
20x	INDEL	heterozygous	ADT < 0.26 & GQ > =98.5 & DP > =23.5 & MQ > =59.5
			ADT < 0.26 & DP > 23.5 & ReadPosRankSum < -1.55
100x	SNV	homozygous	ADT > =0.96
100x	SNV	heterozygous	GQ > =68.5
100x	INDEL	homozygous	ADTL > =5.08
100x	INDEL	heterozygous	ADT < 0.15 & MQ > =59.91 & GQ > =98.5

(*): no reliable filters were selected by classification trees

Table 6 Results by the application of selection parameters and their thresholds on simulated datasets

	TV N (%)	FV N (%)	Variant selected by hard filtering %
HC dataset			
Homo SNVs			
Overall	2,382 (66.6)	1,195 (33.4)	93.9
Selected	2,238 (98.6)	31 (1.3)	
Het SNVs			
Overall	87,361 (99.8)	177 (0.2)	98.6
Selected	86,166 (99.9)	24 (0.03)	
Homo indels			
Overall	54 (0.12)	43,871 (99.8)	27.7
Selected	15 (75)	5 (25)	
Het indels			
Overall	17.305 (45.8)	20,410 (54.1)	84.6
Selected	14,646 (94)	935 (6)	
LC dataset			
Homo SNVs			
Overall	2,084 (12.38)	14,721 (87.6)	96.9
Selected	2,020 (92.2)	171 (7.8)	
Het SNVs			
Overall	95,119 (97.6)	2,322 (2.3)	99.4
Selected	94,602 (99.9)	80 (0.08)	
Homo indels			
Overall	154 (0.4)	45623 (99.6)	100
Selected	154 (0.4)	45623 (99.6)	
Het indels			
Overall	7,502 (17.6)	34,889 (82.3)	43
Selected	3,226 (99.1)	27 (0.8)	

% have to be intended as the percentage of unfiltered variants for "overall"calls and as the percentage of alterations which were not filtered out in the hard filtering process for "selected"calls; % of selection indicates the amount of variants selected from the total callset. *TV* true variants, *FV* false variants

GATK hard filtering: tunable parameters to improve variant calling for next generation sequencing targeted...

7

Table 7 Homopolymeric sequences flanking false positive variants

	Chr	Position	Flanking sequence	N° of occurrences
HC dataset				
Homo SNVs	chr10	131565164	CCGGT**T**GGGGA	77
	chr3	178921420	GGACT**G**TTTTT	73
Het SNVs	chr13	48919347	TAAAC**A**TTTTA	63
	chr3	178937372	CTTGG**T**AAAAG	9
Homo Indels	chr4	55602995	AGAGC**C**AAAAA	1842
	chr10	89693016	AAGTT**A**TTTTT	1802
Het Indels	chr13	48955363	AGTTA**C**TTTTT	2175
	chr3	178941853	CTATC**C**TTTTT	1678
LC dataset				
Homo SNVs	chr2	204736165	GGGTT**G**TTTTT	334
	chr13	48954225	GGTAA**A**TTTTT	241
Het SNVs	chr7	140534584	AAACA**G**AAAAA	32
	chr13	48955464	CTTTG**A**TTTTT	20
Homo Indels	chr7	140481508	AACAG**T**AAAAA	1153
	chr7	140481513	TAAAA**A**AGTCA	1084
Het Indels	chr3	69915434	TAAAG**G**AAAAA	1202
	chr10	89693016	AAGTT**A**TTTTT	1107

Variant locus is on the 6th nucletide (bold) of the 11 nucleotide string (flanking sequence)

heterogeneity, GATK variant calling was tested both in a relative high coverage and low coverage conditions.

Recently, Vanni et al [20] highlighted the discrepancy between TVC and GATK, which was also observed in our study, excluding indel calls from comparison. They considered Phred score ranging 5–30 to mark low-quality variants. Our results show that such approach could not be enough to have a high quality GATK call set. In detail, we evidenced that different parameters could be tuned depending on type of mutations and genotypes suggested. In a previous study, authors focused on the detection of parameters that could allow to improve indel detection [7]. They focused on two parameters regarding the frequency of reference and alternate alleles and the variance of the width of inserted/deleted sequences. The first parameters is similar to our ADT and ADTL filters, which were involved in the step of selection for the reduction of false positives. They observed that the numbers of false positive regarded in particular indel in homopolymeric regions. In a similar way we observed that flanking regions, were homopolymeric in a high number of false positive calls. It is important to reduce errors in these regions because they occur in genomic regions where the occurrence of true alterations is also higher [21]. Variant

calling of TVC is improving but Indels are still a problem and thus parallel pipeline with opportune set up and filters could be helpful to solve this question, with particular attention on type of platform used for sequencing and on type of design (e.g., exome, targeted gene panel). Carson et al [22] recently demonstrated how DP and GQ filters could be able to enhance sensitivity and specificity in whole exome sequencing data. They tested different thresholds and showed that over a certain threshold accuracy reached a plateau and notably they demonstrated that VQSR is not enough to improve variant calling. Indeed, they concluded that, also when VQSR could be applied, opportune hard filtering strategies need to be set up. Our intent was not to target the precise hard filter parameter values and our results have to be intended as suggestion in handling data coming from targeted gene panel sequencing in order to optimize GATK variant calling outputs.

We observed that hard filtering was able to reduce the number of unfiltered false positives, with a different efficiency between SNVs (higher) and Indels (lower). True Indels were hard to be filtered and the performance of filtering was generally lower than in the case of the SNVs (i.e. high loss of true variants and high abundance of false variants; see Table 6). We also observed that the very majority to the unfiltered Indels called at the homozygous status is represented by false variants. Hence there are some regions of the DNA reference sequence that are prone to be recognized as carrying Indels by the bioinformatic pipeline. Our preliminary investigations show that these regions often contain low complexity sequences (for instance a short sequence made up of the same base). A good strategy would be to train in advance the program that operates in the indel parameter recalibration phase to recognize these regions but this hypothesis needs to be investigated in more detail. The reader should note that even if some NGS technologies are known to read with difficulty the DNA regions having low complexity, we are here asserting that the call errors of the variants in the low complexity regions are due to the bioinformatic analysis and not to sequencing errors as we worked with data produced by the simulator and not by a NGS sequencer. So, in the real world, regions with low complexity sequences are doubly condemned to possible abundant errors due to both sequencing and the following bioinformatics analysis.

Some results were quite different from the expected. In particular, we observed that they were detected (unfiltered and then filtered) more SNVs in the case of low coverage (see Table 6, Het SNVs) than in high coverage. Of note that in such cases the rate of false variant is very small even for the unfiltered variants. However, the rate of false variant is about 10 folds greater in the LC dataset. Therefore we hypothesize that variant calling process is more sensitive because of less specific when analyzing dataset with a low coverage.

In general terms, as already known, the Indels are more difficult to be analyzed than the SNVs and a deeper sequencing helps to improve the performance of filtering the true variants. However, such a performance could vary since complexity of the sequence changes along the sequence itself. In the case of targeted sequencing, our suggestion is to study in advance the region that we will be sequenced in order to evaluate the performance of the variant calling procedure over such regions in order to figure out the most problematic areas to be treated with caution. We also suggest the use of simulations based on the specific target region which can help to calibrate the filters for the specific problem. We argue that it could be useful to set specific filters for different regions and for different known variants.

Conclusions

The results of our study showed that filters could be correctly tuned according to coverage and type of alterations. Moreover, it could be useful to test by appropriate simulations the design of amplicon gene panels to gain a priori knowledge of the possible issues in variant calling by GATK.

Abbreviations
SNV: Single nucleotide variants

Acknowledgment
Fondazione Cariverona

Funding
Publication of this article was funded by Italian Ministry of Health (Ricerca Corrente 2016).

Authors' contributions
SDS, GM: conceived and designed the study; RP: experiments for real dataset data; AM, VM: bioinformatic analyses; ST: revision of the manuscript. All authors have read and approved the final manuscript.

Competing interests
The authors declare that they have no competing interests.

About this supplement
This article has been published as part of BMC Bioinformatics Volume 18 Supplement 5, 2017: Selected works from the Joint 15th Network Tools and Applications in Biology International Workshop and 11th Integrative Bioinformatics International Symposium (NETTAB / IB 2015). The full contents of the supplement are available online at https://bmcbioinformatics.biomedcentral.com/articles/supplements/volume-18-supplement-5.

Author details
[1]IRCCS-Istituto Tumori "Giovanni Paolo II", Molecular Genetics Laboratory, viale Orazio Flacco, 65, 70124 Bari, Italy. [2]Department of Neuroscience, Biomedicine and Movement Sciences, Section of Biology and Genetics, University of Verona, Strada Le Grazie 8, 37135 Verona, Italy.

References
1. Chan M, Ji SM, Yeo ZX, Gan L, Yap E, Yap YS, Ng R, Tan PH, Ho GH, Ang P, Lee ASG. Development of a next-generation sequencing method for BRCA mutation screening: a comparison between a high-throughput and a benchtop platform. J MolDiagn. 2012;14:602–12.
2. Costa JL, Sousa S, Justino A, Kay T, Fernandes S, Cirnes L, Schmitt F, Machado JC. Nonoptical massive parallel DNA sequencing of BRCA1 and BRCA2 genes in a diagnostic setting. Hum Mutat. 2013;34:629–35.
3. Rothberg JM, Hinz W, Rearick TM, Schultz J, Mileski W, Davey M, Leamon JH, Johnson K, Milgrew MJ, Edwards M, Hoon J, Simons JF, Marran D, Myers JW, Davidson JF, Branting A, Nobile JR, Puc BP, Light D, Clark TA, Huber M, Branciforte JT, Stoner IB, Cawley SE, Lyons M, Fu Y, Homer N, Sedova M, Miao X, Reed B, et al. An integrated semiconductor device enabling non-optical genome sequencing. Nature. 2011;475:348–52.
4. Hadd AG, Houghton J, Choudhary A, Sah S, Chen L, Marko AC, Sanford T, Buddavarapu K, Krosting J, Garmire L, Wylie D, Shinde R, Beaudenon S, Alexander EK, Mambo E, Adai AT, Latham GJ. Targeted, high-depth, next-generation sequencing of cancer genes in formalin-fixed, paraffin-embedded and fine-needle aspiration tumor specimens. J MolDiagn. 2013;15:234–47.
5. Yousem SA, Dacic S, Nikiforov YE, Nikiforova M. Pulmonary Langerhans cell histiocytosis: profiling of multifocal tumors using next-generation sequencing identifies concordant occurrence of BRAF V600E mutations. Chest. 2013;143:1679–84.
6. Jünemann S, Sedlazeck FJ, Prior K, Albersmeier A, John U, Kalinowski J, Mellmann A, Goesmann A, von Haeseler A, Stoye J, Harmsen D. Updating benchtop sequencing performance comparison. Nat Biotechnol. 2013;31:294–6.
7. Yeo ZX, Chan M, Yap YS, Ang P, Rozen S, Lee ASG. Improving indel detection specificity of the Ion Torrent PGM benchtop sequencer. PLoS One. 2012;7, e45798.
8. Elliott AM, Radecki J, Moghis B, Li X, Kammesheidt A. Rapid detection of the ACMG/ACOG-recommended 23 CFTR disease-causing mutations using ion torrent semiconductor sequencing. JBiomol Tech. 2012;23:24–30.
9. Bragg LM, Stone G, Butler MK, Hugenholtz P, Tyson GW. Shining a light on dark sequencing: characterising errors in Ion Torrent PGM data. PLoSComputBiol. 2013;9:e1003031.
10. Nielsen R, Paul JS, Albrechtsen A, Song YS. Genotype and SNP calling from next-generation sequencing data. Nat Rev Genet. 2011;12:443–51.
11. Ruffalo M, LaFramboise T, Koyutürk M. Comparative analysis of algorithms for next-generation sequencing read alignment. Bioinformatics. 2011;27:2790–6.
12. Pattnaik S, Vaidyanathan S, Pooja DG, Deepak S, Panda B. Customisation of the exome data analysis pipeline using a combinatorial approach. PLoS One. 2012;7, e30080.
13. Caboche S, Audebert C, Lemoine Y, Hot D. Comparison of mapping algorithms used in high-throughput sequencing: application to Ion Torrent data. BMC Genomics. 2014;15:264.
14. Yeo ZX, Wong JCL, Rozen SG, Lee ASG. Evaluation and optimisation of indel detection workflows for ion torrent sequencing of the BRCA1 and BRCA2 genes. BMC Genomics. 2014;15:516.
15. Huang W, Li L, Myers JR, Marth GT. ART: a next-generation sequencing read simulator. Bioinformatics. 2012;28(4):593–4.
16. McKenna A, Hanna M, Banks E, Sivachenko A, Cibulskis K, Kernytsky A, et al. The Genome Analysis Toolkit: a MapReduce framework for analyzing next-generation DNA sequencing data. Genomeresearch. 2010;20(9):1297–303. doi:10.1101/gr.107524.110.
17. Li H, Durbin R. Fast and accurate short read alignment with Burrows-Wheeler transform. Bioinformatics. 2009;25(14):1754–60. doi:10.1093/bioinformatics/btp324.
18. Zook JM, Chapman B, Wang J, Mittelman D, Hofmann O, Hide W, Salit M. Integrating human sequence data sets provides a resource of benchmark SNP and indel genotype calls. Nat Biotechnol. 2014;32:246–51.
19. Xu H, DiCarlo J, Satya RV, Peng Q, Wang Y. Comparison of somatic mutation calling methods in amplicon and whole exome sequence data. BMC Genomics. 2014;15:244.
20. Vanni I, Coco S, Truini A, Rusmini M, Dal Bello MG, Alama A, Banelli B, Mora M, Rijavec E, Barletta G, Genova C, Biello F, Maggioni C, Grossi F. Next-Generation Sequencing Workflow for NSCLC Critical Samples Using a Targeted Sequencing Approach by Ion Torrent PGM™ Platform. Int J MolSci. 2015;16:28765–82.
21. Albers CA, Lunter G, MacArthur DG, McVean G, Ouwehand WH, Durbin R. Dindel: accurate indel calls from short-read data. Genome Res. 2011;21:961–73.

2

In silico approach to designing rational metagenomic libraries for functional studies

Anna Kusnezowa and Lars I. Leichert*

Abstract

Background: With the development of Next Generation Sequencing technologies, the number of predicted proteins from entire (meta-) genomes has risen exponentially. While for some of these sequences protein functions can be inferred from homology, an experimental characterization is still a requirement for the determination of protein function. However, functional characterization of proteins cannot keep pace with our capabilities to generate more and more sequence data.

Results: Here, we present an approach to reduce the number of proteins from entire (meta-) genomes to a reasonably small number for further experimental characterization without loss of important information. About 6.1 million predicted proteins from the Global Ocean Sampling Expedition Metagenome project were distributed into classes based either on homology to existing hidden markov models (HMMs) of known families, or de novo by assessment of pairwise similarity. 5.1 million of these proteins could be classified in this way, yielding 18,437 families. For 4,129 protein families, which did not match existing HMMs from databases, we could create novel HMMs. For each family, we then selected a representative protein, which showed the closest homology to all other proteins in this family. We then selected representatives of four families based on their homology to known and well-characterized lipases. From these four synthesized genes, we could obtain the novel esterase/lipase GOS54, validating our approach.

Conclusions: Using an in silico approach, we were able improve the success rate of functional screening and make entire (meta-) genomes amenable for biochemical characterization.

Keywords: Functional metagenomics, Global ocean sampling project, GOS, Lipase, Protein function

Background

Modern sequencing technology allows for fast and relatively inexpensive sequencing of large amounts of DNA. Whole genome sequencing of genomes of single microorganisms or even whole microbial communities are now state-of-the-art and commercially available. Especially the sequencing of microbial communities, in which the isolation of a single organism is no longer necessary, significantly expanded the number of known proteins in public databases.

The global ocean sampling project (GOS), still one of the largest metagenomic projects to date, was initiated

by the J. Craig Venter Institute in 2007 [1]. When the ~6.1 million protein sequences from the GOS dataset were published, it more than doubled the number of known proteins in public databases at that time [2]. Nonetheless, already before the first metagenomic datasets were published, the functional analysis of proteins could not keep up with the speed, with which new gene sequences were discovered. Thus, the functions of most proteins have been, and still are, predicted based on their homology to a much smaller number of well-characterized proteins. Therefore, the function of a major part of all proteins in large data repositories, such as NCBI or EMBL is still unknown (e.g. more than 75% of sequences in Trembl [3]). In some cases, depending on the taxonomic origin, up to 80% of the gene functions of a given organism cannot be inferred from

* Correspondence: lars.leichert@ruhr-uni-bochum.de
Institute of Biochemistry and Pathobiochemistry – Microbial Biochemistry, Ruhr University Bochum, Universitätsstr. 150, 44780 Bochum, Germany

homology [4, 5]. The publication of large metagenomic datasets only exacerbated this challenge.

Within the vast amount of proteins of unknown or predicted function lies great biotechnological potential. Biocatalysts that could be found among those proteins might help us to move away from a petrol-based economy to a more bio-based economy. However, given the sheer size of our databases, it is challenging to make this in silico knowledge of genetic sequences amenable for functional testing in the lab.

Here we present our approach to tackling this challenge using the GOS metagenomic dataset as an example. We leveraged the available information from public databases to classify all proteins of the GOS dataset with known domains into existing families. We classified the remaining set of proteins de novo and devised an algorithm that selects a representative protein sequence from these families based on a minimal phylogenetic distance (i.e. the closest possible relationship) to all other members. Representatives from 4 families containing predicted lipolytic enzymes were functionally characterized in the lab. One protein, termed GOS54, contains the alpha/beta hydrolase domain PF07859 and showed high lipase/esterase activity when expressed in *Escherichia coli*. We demonstrate that our approach can be used to substantially reduce the number of genes from metagenomic datasets that need to be screened for functions. It thus might accelerate biocatalyst discovery.

Results

The majority of GOS proteins can be classified based on HMM domains from public protein family databases

The GOS metagenomic project is, to date, still one of the largest publicly available metagenomic datasets. The goal of our study was to make the vast protein sequence diversity contained in this dataset amenable to protein-biochemical functional studies. In a first step, we, therefore, annotated all predicted protein sequences contained in the GOS dataset based on existing HMMs from public protein family databases. We chose two HMM databases: PFAM, a comprehensive protein family database [6] and the complementary, more bacteria-focused TIGRFAMs [7]. HMM searches are fast and can find more distantly related protein family members when compared to standard homology searches such as BLAST [8]. Using an E-value cutoff of $\leq 10^{-5}$, we could classify, based on the combined PFAM and TIGRFAMs HMMs, 4,436,387 of GOS's 6,123,395 protein sequences. The use of a less strict cutoff of $\leq 10^{-3}$ resulted in the additional annotation of less than 4% of proteins in the dataset and an increase in the theoretical number of false positive matches by 2 orders of magnitude. We, therefore, decided to use the more stringent significance threshold for our analysis.

For the purpose of experimental testability, we ultimately wanted to associate each protein to one, and only one class. However, many proteins matched to multiple HMMs with an E-value that passed our significance threshold. Such an overlap may well be significant, if it was either due to a similarity in the HMMs, or due to a frequent co-occurrence of two distinct domains in proteins of the GOS dataset. If either of those two cases were true, we predicted that a majority of proteins that scored above the threshold for a particular HMM should also score above the threshold for one or more of the other HMMs. We, therefore, considered all proteins that were matching a certain HMM above the threshold as a distinct set. As a measure of the co-occurrence of certain HMMs, we then calculated the Jaccard index for each of these HMM-based sets with all other sets. If the Jaccard index of two sets was above 0.75 (i.e. more than 75% of its combined members were present in both sets), we created a new classifier, which combined both HMMs. In this way, we created classifiers, which were either based on one HMM (13,417 classifiers) or on multiple HMMs (890 classifiers, for the distribution of the number of HMMs in these classifiers, see Additional file 1). We then took each individual protein and considered only the most significant HMM match (i.e. the match with the lowest E-value) and assigned it to the classifier that contained this particular HMM. In this way, we could assign 72% of the proteins in the GOS set to 14,307 HMM-based classifiers, which we then considered protein families (Fig. 1a, Additional files 2 and 3).

Almost 40% of GOS proteins that do not contain domains recognized by known protein family HMMs were classified de novo

However, 28% of GOS proteins did not match HMMs from PFAM and TIGRFAMs with a significant e-value $\leq 10^{-5}$ (Fig. 1a). Thus, we clustered these remaining proteins using the Markov Cluster algorithm (MCL) [9]. To obtain MCL-based classes comparable to the classes we annotated using known HMM domains, we adjusted the so-called inflation parameter of this software, which basically adjusts the diversity of the clusters created. Based on a test clustering of ~15% of our HMM annotated proteins we found that an inflation parameter of 1.1 resulted in the largest overlap of MCL-based classes with the existing HMM-based classes and thus this value was used in our subsequent analysis (Fig. 1b).

We then used the MCL algorithm on all proteins from the GOS dataset that could not be annotated using HMMs of known domains. Thus, we distributed these proteins into 148,013 classes containing between 1 and 17,282 sequences (Additional file 4). Theoretically, non-coding DNA can randomly result in open reading frames of significant size. To rule out that we included protein-

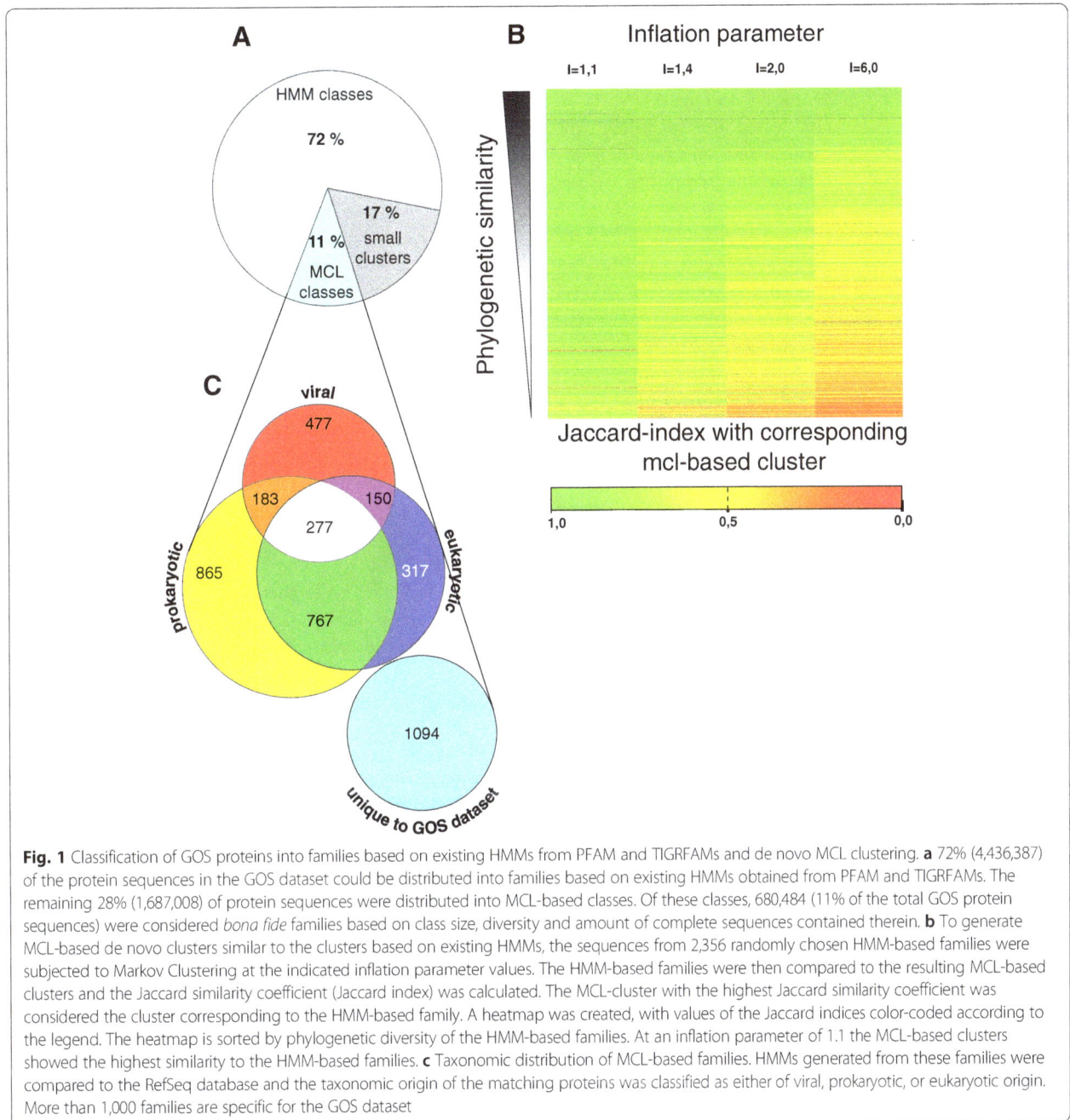

Fig. 1 Classification of GOS proteins into families based on existing HMMs from PFAM and TIGRFAMs and de novo MCL clustering. **a** 72% (4,436,387) of the protein sequences in the GOS dataset could be distributed into families based on existing HMMs obtained from PFAM and TIGRFAMs. The remaining 28% (1,687,008) of protein sequences were distributed into MCL-based classes. Of these classes, 680,484 (11% of the total GOS protein sequences) were considered *bona fide* families based on class size, diversity and amount of complete sequences contained therein. **b** To generate MCL-based de novo clusters similar to the clusters based on existing HMMs, the sequences from 2,356 randomly chosen HMM-based families were subjected to Markov Clustering at the indicated inflation parameter values. The HMM-based families were then compared to the resulting MCL-based clusters and the Jaccard similarity coefficient (Jaccard index) was calculated. The MCL-cluster with the highest Jaccard similarity coefficient was considered the cluster corresponding to the HMM-based family. A heatmap was created, with values of the Jaccard indices color-coded according to the legend. The heatmap is sorted by phylogenetic diversity of the HMM-based families. At an inflation parameter of 1.1 the MCL-based clusters showed the highest similarity to the HMM-based families. **c** Taxonomic distribution of MCL-based families. HMMs generated from these families were compared to the RefSeq database and the taxonomic origin of the matching proteins was classified as either of viral, prokaryotic, or eukaryotic origin. More than 1,000 families are specific for the GOS dataset

sequences derived from these "random" open reading frames in our de novo classes, we only considered classes with more than 30 members and no less than 10% complete protein sequences (i.e. sequences derived from DNA sequences with an unambiguous start and stop codon) for further analysis. We argued that the presence of 30 homologous open reading frames excludes that these sequences are random, non-coding DNA. Within this set, we then tested the phylogenetic significance based on multiple sequence alignments. These alignments were generated with MAFFT using default parameters [10] and optimized with MaxAlign [11]. Classes with more than 80% gaps were then excluded, since such large numbers of gaps typically only occur in alignments of proteins with very low to no similarity [11, 12]. For the remaining classes, which we considered *bona fide* families (Additional file 5), we generated Hidden Markov Models. For this purpose, seed alignments were created using MAFFT's *G-INI* strategy with gap regions removed by Gblocks [13]. Low-quality alignments that could not be improved by Gblocks running with relaxed parameters (similar to the parameters described in [14]) were rejected. Amino acids were considered "conserved" if they were present in at least 50% of the sequences at any given position and gaps were only considered if present in more than 50% of the sequences. Multiple sequence alignments fulfilling these criteria were modified by Gblocks as described and were then used as input for HMMER 3.0 to create HMMs [8] (Additional file 6). We then tested the specificity and sensitivity of these HMMs with a test set containing the sequences used to build the HMM (as true positives), and with 80,506 mammalian and plant proteins, matched in size distribution to the (exclusively microbial) GOS protein set (Additional file 7). We then obtained the number of false negatives (i.e. all unmatched proteins from the set of proteins used to build the HMM) and estimated the upper limit of false positives (i.e. all matched mammalian and plant proteins) to calculate the F_1 score, a measurement of HMM performance [15].

This resulted in a library of 4,130 HMMs with an F_1 score ≥ 0.5, which contain the information of 680,484 proteins in total (Additional file 8). To test if the high F_1 scores are the result of overfitting of the data, we tested our approach with a training set derived from the largest of our de novo classes. To this end, we randomly selected two thirds of the sequences contained in FUME-FAM002132, which contains 1177 sequences in total and built an HMM from this subset. We then tested the selectivity and specificity of this HMM against the complete set of proteins in FUMEFAM002132. Repeating this approach 10 times, we consistently achieved an F_1 score of 1.

We then used this set of HMMs and searched for potential matches in the NCBI RefSeq database, a set of curated prokaryotic, viral, and eukaryotic sequence datasets [16, 17]. By this approach, we found that 477 and 865 of our HMMs were specific for viral or prokaryotic organisms, respectively. 1,094 HMMs did not produce any match in any of these datasets, suggesting that they are specific for the ocean metagenome (Fig. 1c, Additional file 8).

For each family, we determined one representative sequence

With 14,307 families based on HMMs from public databases and 4,130 novel families, we had the GOS proteins subdivided into 18,437 families in total. To be able to test these families for their biochemical function, we decided to define one protein from each family, which best represents this family. We selected this representative based on its similarity to all other proteins within its family. For this purpose, we first generated a guide tree using MAFFT [10]. We then calculated the sum of the distances in this tree of each individual member to all other members of the class. We reasoned that the member with the minimal sum of distances is the most closely related member to all other members and, therefore, best represents its class (see Fig. 2a-c for a

schematic overview of this procedure). Because the representative should be testable in the lab, we considered only complete sequences (i.e. sequences derived from DNA sequences with an unambiguous start and stop codon). To remove bias introduced in the tree building, we randomly created subsets containing 90% of all members and recalculated the guide tree and the associated distances for these subsets 100 times. The member selected most often was then defined as the representative for this family. In this way, we could define one representative for all families containing at least one complete protein, giving us a total of 9,771 representatives (Fig. 2d, Additional file 9).

Representatives of lipolytic GOS-proteins were tested for activity in the lab

To test the validity of those representatives, we decided to test the representatives of families matching HMMs of well-characterized lipolytic protein families. Lipolytic enzymes such as lipases and esterases constitute an important group of biocatalysts for biotechnological applications [18]. We therefore identified carboxylic ester hydrolases family (EC 3.1.1.-) proteins from the Uniprot database with the highest possible annotation score of 5. After excluding enzymes from potentially pathogenic organisms, we singled out 7 individual proteins, which we then could match, based on homology to four of our families (Table 1).

We synthesized codon-optimized genes for the representatives of these 4 families and cloned them into an IPTG-inducible *Escherichia coli* expression vector. We then transformed these plasmids into *E. coli* and screened for lipolytic activity using plate-based activity assays. As a positive control, we decided to use LipA, a well-characterized lipase from *Bacillus subtilis*. One of these proteins, GOS54 showed activity on a tributyrin plate, producing a clear halo around the clone, similar to the LipA positive control, indicating the ability of this enzyme to degrade triglycerides containing short-chained fatty acids (Fig. 3a). A triolein-based plate assay showed also activity, albeit to a lesser extent (Fig. 3b). We could verify this activity and GOS54's preference for short-chained fatty acid esters in an activity assay using p-NP esters of butyrate and palmitate as substrates (Fig. 3c - f). This indicated to us that GOS54 is indeed a lipolytic enzyme and that we can use our representative approach to determine the function of protein classes in the GOS dataset.

Discussion

The functional annotation of proteins from metagenomic datasets is challenging. One approach is the distribution of proteins into families based on homology to already known protein families. Using this approach,

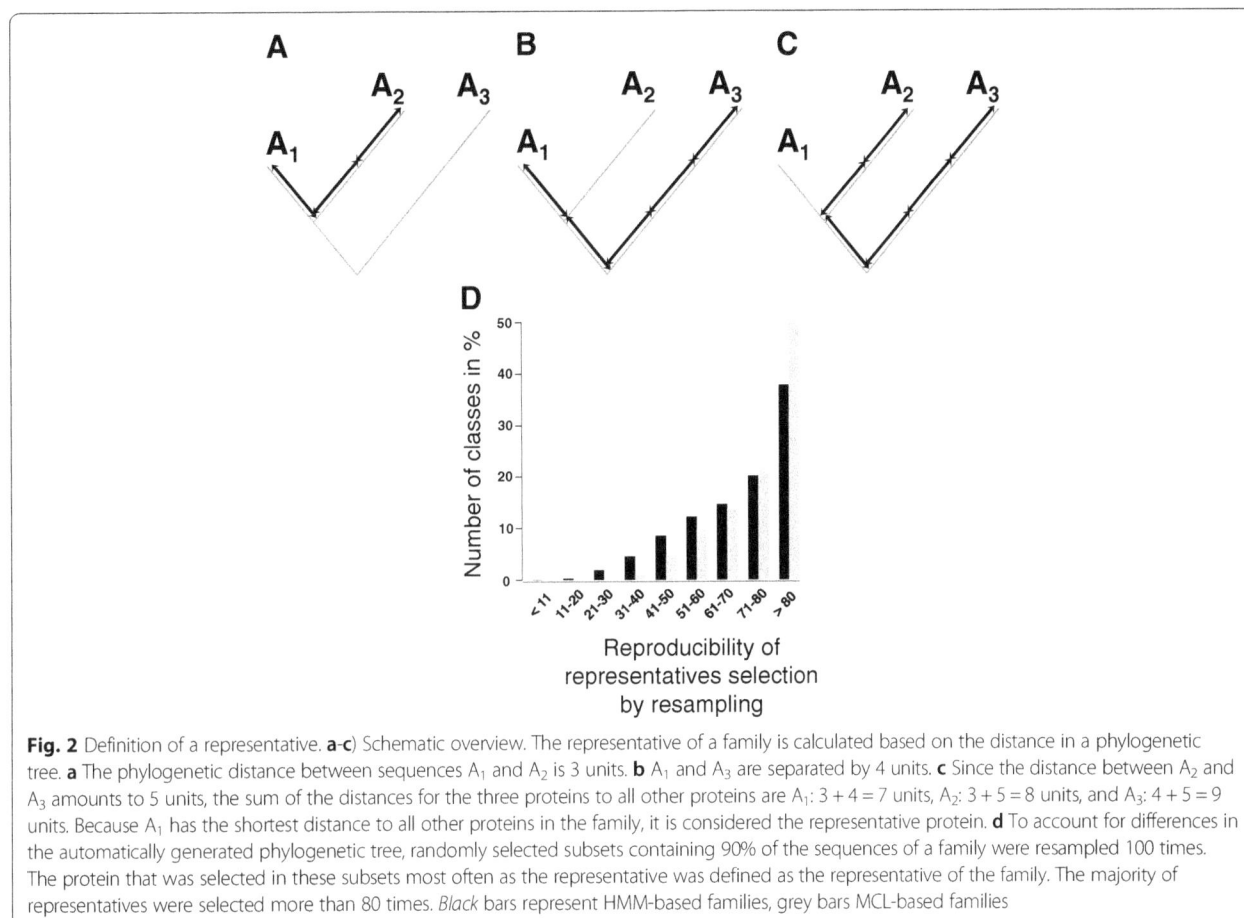

Fig. 2 Definition of a representative. **a-c**) Schematic overview. The representative of a family is calculated based on the distance in a phylogenetic tree. **a** The phylogenetic distance between sequences A_1 and A_2 is 3 units. **b** A_1 and A_3 are separated by 4 units. **c** Since the distance between A_2 and A_3 amounts to 5 units, the sum of the distances for the three proteins to all other proteins are A_1: $3 + 4 = 7$ units, A_2: $3 + 5 = 8$ units, and A_3: $4 + 5 = 9$ units. Because A_1 has the shortest distance to all other proteins in the family, it is considered the representative protein. **d** To account for differences in the automatically generated phylogenetic tree, randomly selected subsets containing 90% of the sequences of a family were resampled 100 times. The protein that was selected in these subsets most often as the representative was defined as the representative of the family. The majority of representatives were selected more than 80 times. *Black* bars represent HMM-based families, grey bars MCL-based families

truly new protein families cannot be discovered. The discovery of novel families unique to the tested dataset can be achieved by a de novo definition of protein families. A de novo definition is typically based on homology of proteins in the data set and computationally substantially slower than an approach based on known HMMs. We therefore decided to use a hybrid approach, assigning families in the GOS metagenomic dataset based on HMMs of known protein families, where possible (see Fig. 4 for a schematic overview). From the remaining proteins that did not contain any known HMMs, we defined protein families using a markov clustering approach [9]. Categorization of the GOS dataset based on HMMs and using the Cd-hit algorithm [19] has been successfully performed before [1], but the resulting data is currently unavailable. The use of the more precise MCL algorithm allowed us to create HMMs based on a substantial number of our de novo-defined families.

Based on the assumption that high homology correlates with an identical function, it should be possible to test just one protein of a family to deduce the function of all members of this family. To define a protein that represents a

Table 1 Representatives of families matching HMMs of well-characterized lipolytic proteins

Family/Representative	Members with complete sequence	Associated HMM from PFAM	Well-characterized enzymes from UniProt with accession number
FUMEFAM011958/**GOS54**	383	Alpha/beta hydrolase fold PF07859	Acetyl esterase EcE *Escherichia coli* P23872
FUMEFAM010194/**GOS55**	376	GDSL-like Lipase/Acylhydrolase family PF13472	Arylesterase *Streptomyces coelicolor* Q9S2A5, Lipase *Streptomyces rimosus* Q93MW7
FUMEFAM018084/**GOS88**	22	Alpha/beta hydrolase fold PF00561	Pimeloyl-[acyl-carrier protein] methyl ester esterase P13001
FUMEFAM012527/**GOS89**	7	PB011927	Thermostable organic solvent tolerant lipase *Bacillus sp.* Q5U780 (EC 3.1.1.3)

Reviewed amino acid sequences with the maximal annotation score of 5 and bacterial origin were downloaded from UniProt [27]. Based on matching HMMs from Pfam (Release 27.0) we determined the 4 protein families with the highest homology to the well-characterized proteins from UniProt

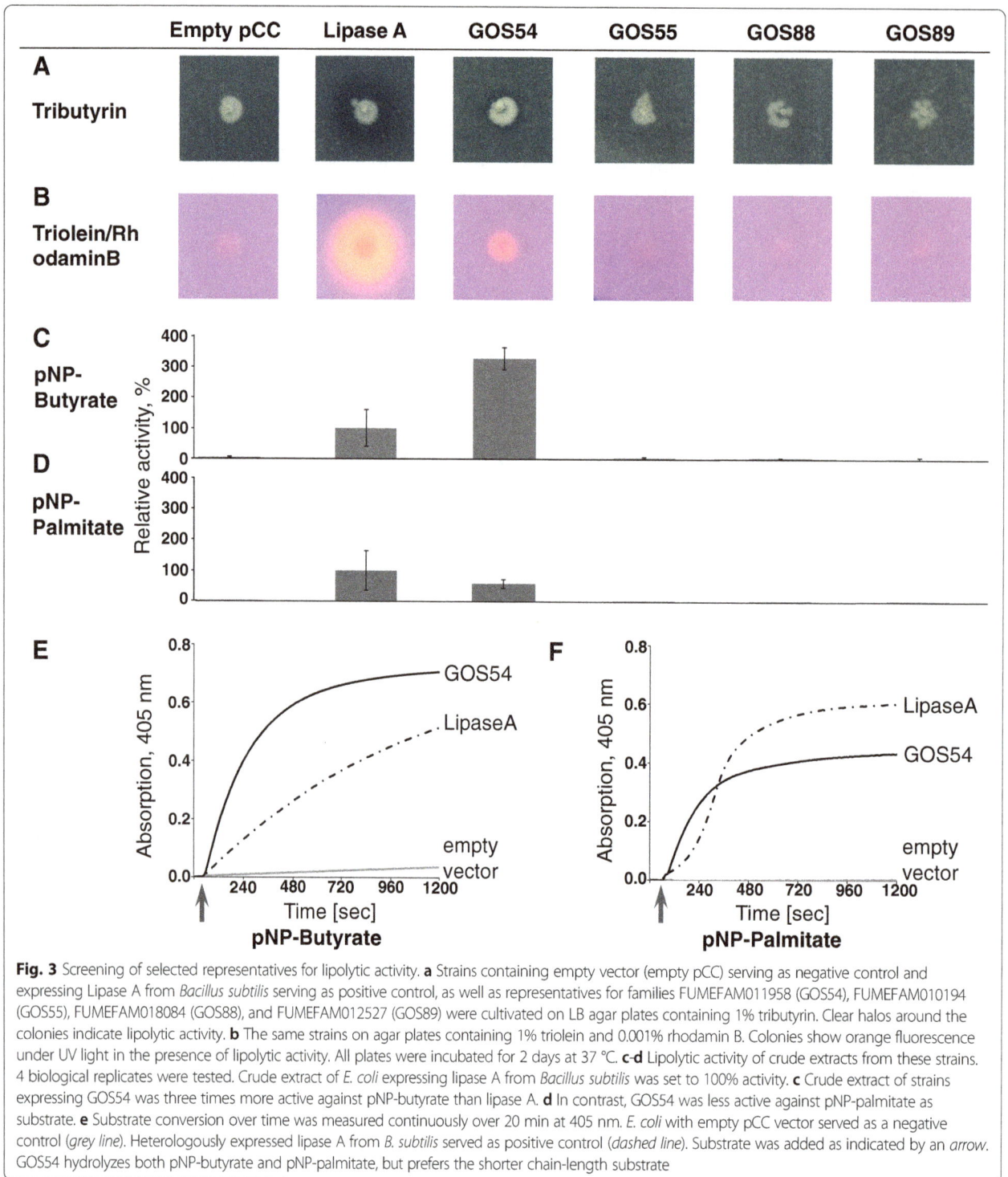

Fig. 3 Screening of selected representatives for lipolytic activity. **a** Strains containing empty vector (empty pCC) serving as negative control and expressing Lipase A from *Bacillus subtilis* serving as positive control, as well as representatives for families FUMEFAM011958 (GOS54), FUMEFAM010194 (GOS55), FUMEFAM018084 (GOS88), and FUMEFAM012527 (GOS89) were cultivated on LB agar plates containing 1% tributyrin. Clear halos around the colonies indicate lipolytic activity. **b** The same strains on agar plates containing 1% triolein and 0.001% rhodamin B. Colonies show orange fluorescence under UV light in the presence of lipolytic activity. All plates were incubated for 2 days at 37 °C. **c-d** Lipolytic activity of crude extracts from these strains. 4 biological replicates were tested. Crude extract of *E. coli* expressing lipase A from *Bacillus subtilis* was set to 100% activity. **c** Crude extract of strains expressing GOS54 was three times more active against pNP-butyrate than lipase A. **d** In contrast, GOS54 was less active against pNP-palmitate as substrate. **e** Substrate conversion over time was measured continuously over 20 min at 405 nm. *E. coli* with empty pCC vector served as a negative control (*grey line*). Heterologously expressed lipase A from *B. subtilis* served as positive control (*dashed line*). Substrate was added as indicated by an *arrow*. GOS54 hydrolyzes both pNP-butyrate and pNP-palmitate, but prefers the shorter chain-length substrate

family in the best way possible, we devised an algorithm that determines the protein with the closest phylogenetic relationship to all other proteins in the family. This determination is based on a phylogenetic tree in which we determine the phylogenetic distance between all members and select the member with the shortest sum of distances to all other members. A similar approach has been used in the COMBREX Project [20]. In this way, we could reduce 4,969,723 proteins that were assigned to families to 9,771 representatives. 1,153,672 proteins are not represented by a representative, either because they were not assigned to a family or because their family contained only incomplete protein sequences. The families, their members and representatives are summarized in Additional file 10.

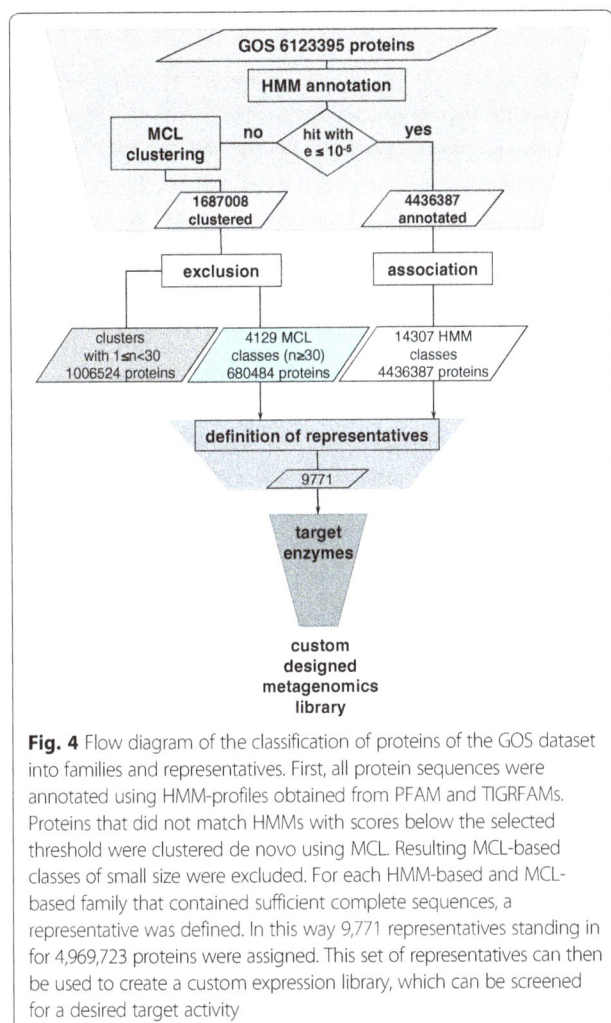

Fig. 4 Flow diagram of the classification of proteins of the GOS dataset into families and representatives. First, all protein sequences were annotated using HMM-profiles obtained from PFAM and TIGRFAMs. Proteins that did not match HMMs with scores below the selected threshold were clustered de novo using MCL. Resulting MCL-based classes of small size were excluded. For each HMM-based and MCL-based family that contained sufficient complete sequences, a representative was defined. In this way 9,771 representatives standing in for 4,969,723 proteins were assigned. This set of representatives can then be used to create a custom expression library, which can be screened for a desired target activity

As a proof of concept of the representative approach we then tested representatives of four families for lipase activity. We selected these representatives based on their homology to well-characterized lipases, because these lipid-degrading enzymes are one of the most commonly used biocatalysts [21]. One representative showed esterase/lipase activity, attesting the usefulness of the representatives approach.

Conclusion

Here we present a workflow to categorize large metagenomic datasets into protein families. Proteins homologous to known protein families are categorized based on publicly available HMMs. The residual proteins, which do not show homology to known protein families could be categorized de novo. We devised a new algorithm to select one representative from each protein family, which can then be functionally tested in the wet lab. Using representatives from lipolytic families we could verify our approach, discovering the novel esterase GOS54.

Methods
GOS dataset

All 6,123,395 hypothetical protein sequences from the GOS dataset were obtained from the NCBI BioProject 13694. Corresponding scaffolds were downloaded from GeneBank (ftp.ncbi.nih.gov/genbank/wgs/gbcon[33–108].seq). This information and the associated annotations were stored in a PostgreSQL database (version 9.0.5) on a Mac mini server (8 × 2 GHz Intel Core i7, 8 GB 1333 MHz DDR3, running Mac OS X Server Lion 10.7.5, Apple, Cupertino, CA). Based on the information contained in the scaffolds, proteins derived from DNA sequences with an unambiguous start and stop codon were defined as "complete". All other sequences and sequences for which no scaffold information was obtainable were considered "incomplete".

HMM profile-based annotation

The GOS data set was annotated using predefined HMMs from PFAM and TIGRFAMs by HMMER 3.0 [8]. The reporting threshold was set to 10^{-5}, the output retrieved in table form, all other parameters remained at the default settings.

The 34,833 HMMs from Pfam A and B (Release 27.0) and the 4,424 HMMs from TIGRFAMs (Release 14.0) were obtained from ftp://ftp.ebi.ac.uk/pub/databases/Pfam/, and ftp://ftp.jcvi.org/pub/data/TIGRFAMs/, respectively.

Family assignment based on HMM annotation

It is possible that individual sequences from the GOS dataset could match multiple HMMs above the reporting threshold. However, we wanted each sequence to be assigned to one and only one protein family. We thus compared the sets of sequences that matched any given HMM with all the other sets of sequences matching the other HMMs by calculating their Jaccard similarity coefficient (Jaccard index). If two sets of sequences matching different HMMs reached a Jaccard similarity coefficient ≥ 0.75, they were combined into one classifier. Multiple HMM-profiles were combined transitively. After all classifiers were defined in this way, each individual sequence was assigned to the classifier, which contained the HMM that it matched with the lowest (i.e. best) threshold. The set of proteins matching a given classifier was defined as protein family and assigned a FUMEFAM number. The alignment of all members of HMM-based families with > 30 members was then optimized using MaxAlign [11] with the command perl -w maxalign.pl -d -f = [$PATH]$I $I.

All-vs-all BLASTP

To reduce computing time for the blastp program, sequence redundancies were first removed using Cd-hit [19]

with the following parameters: cd-hit -i lacking_hmm_-seq_14.fasta -o lacking_gos90_14 -c 0.9 -n 5 -M 0 -T 0.

NCBI BLAST 2.2.28+ version was used for a calculation of E-values needed for MCL clustering. To accelerate pairwise comparison, sequences were organized as a BLAST database with a hash index. Protein sequences were split into files with ~10,000 sequences. The BLASTP-based comparison was performed in multi-threads mode on a 32 Core AMD Opteron 6274, 2.2 GHz machine, containing 128 Gb DIMM DDR3 RAM, and 800 Gb hard disk space, running Ubuntu (GNU/Linux 3.13.0–95-generic × 86_64). The e-value was set to 10^{-5}, output was set to "table" and the -parse_deflines and -show_GIs parameters were set to "true".

Markov clustering

Markov clustering of proteins was performed using the MCL program (version 12–068) according to the protocol by Van Dongen and coworkers [9]. To find an optimized value for the Inflation parameter, 648,901 GOS sequences from 2,356 (~15%) randomly selected HMM-based families were clustered by MCL at values between 1.1, and 6.0. The HMM-based families were then compared to the resulting MCL-based clusters by calculating the Jaccard index. The MCL-based cluster that matched with the highest given Jaccard index was considered the corresponding cluster to that HMM-based family. The highest number of classes which had a Jaccard index > 0.5 with their corresponding cluster was obtained using the inflation parameter 1.1 (see Fig. 1d). We thus used the inflation parameter i = 1.1, other parameters set were: -stream-mirror; -stream-neg-log10; -stream-tf'ceil (200)'. After the Markov clustering, sequences previously removed using the Cd-hit algorithm were added back to the corresponding cluster, in order to account for all sequences found in the GOS dataset.

Creation of HMMs and protein families based on MCL clustering

MCL generated 145,314 clusters. To define families, small clusters containing less than 30 protein sequences were removed. To assess the quality of the clusters, alignments were created from the remaining clusters using MAFFT with the default parameters [10]. These alignments were improved by removing sequences that create significant gaps using MaxAlign [11] with the command perl -w maxalign.pl -d -f = [$PATH]$I $I. Clusters, which still contained more than 80% of gaps after that procedure were removed. The number of gaps was calculated using alistat from the biosquid package biosquid_1.9 g + cvs20050121-2_i386 [8]. Clusters that contained less than 10% of

complete sequences were also removed. To create a seed alignment for HMM creation, the alignment of the cluster members was further optimized with MAFFT [10] using mafft –reorder –bl 62 –op 2.73 –maxiterate 1000 –globalpair. Poorly aligned regions were removed by Gblocks 0.91b, using Gblocks -t = p -b1 = [half of the numbers of sequences in the alignment] -b2 = [half of the numbers of sequences in the alignment] -b3 = 256 -b4 = 2 -b5 = a -e = .sto. If Gblocks did not identify conserved blocks, the clusters were dismissed. Based on these improved seed alignments, Hidden Markov Models were created using the hmmbuild utility from HMMER 3.0: hmmbuild –fragthresh 1.0 -n [NAME] –o [NAME].-out -O [NAME].alig [NAME].hmm [NAME].stockholm. The new HMMs were validated on a test set. In this test set all members from the cluster were added to 80,506 mammalian and plant proteins from RefSeq (see Additional file 7). The HMMs were validated against this test set with

hmmsearch –incE 0.001 -E 0.00001 –tblout [].out [NAME].hmm [NAME]_test_set.fasta.

We then calculated the F1-score for each HMM using the formula

$$F_1 = \frac{2 \cdot TP}{(2 \cdot TP + FP + FN)}$$

where

TP (true positive) = number of family members that matched the HMM;

FP (false positive) = number of RefSeq-based mammalian and plant proteins that matched the HMM. Given that proteins in the GOS dataset are of prokaryotic origin, we argued that matched mammalian and plant proteins would give us an upper limit of false positives;

FN (false negative) = number of family members which did not match the HMM.

Clusters with an F1-score > 0.5 were considered *bona fide* protein families and were assigned a FUMEFAM number.

Definition of representative family members

In a given family, we defined the relationship distance between each protein pair (i, j) as the sum of the length of all branches directly connecting the pair in a phylogenetic tree. To this end we used the Newick-formatted guide tree generated from MAFFT output (mafft –retree 2 –reorder –6merpair –averagelinkage –treeout) based on the members of the final optimized (where applicable) alignment of the family. This guide tree is build using a modified UPGMA algorithm. We then calculated for each protein in the family the sum of all relationship distances to all other proteins using the formula:

$$Sum\ of\ distances\ (j) = \sum_{i=0}^{n} d(i,j)$$

where

d (i,j): distance between sequence i and sequence j in the phylogenetic tree

n: number of sequences in family

The complete protein with the smallest sum of relationship distances was determined as a possible representative for any given family. This process was repeated 100 times using each time 90% of randomly selected sequences from the family. The protein that was selected as possible representative most often was designated as the representative for the family.

Selection of lipolytic representatives

We selected from the UniProt database bacterial enzymes from the carboxylic ester hydrolases family EC 3.1.1.- with the highest annotation score 5. Enzymes from pathogenic organisms or with oligosaccharides as substrates were excluded manually. The selected proteins were then compared to the PFAM HMMs. The representatives of the families from our database that were classified by the best matching HMMs were retranslated into DNA sequences optimized for *E. coli* codon usage using JCat [22]. These DNA sequences (Table 1) were then synthesized by GeneArt (Thermo Fisher Scientific GENEART GmbH, Regensburg, Germany).

Protein expression in *Escherichia coli*

Genes were cloned into an IPTG inducible expression vector based on pTAC-MAT-Tag-2 (Sigma– Aldrich, St. Louis, MO, USA) termed pCC and transformed into *Escherichia coli* XL1-Blue using a heat-shock transformation protocol [23]. Strains were grown routinely in LB-medium [24] containing 200 μg/ml ampicillin. For strains used in this study see Table 2.

Plate-based esterase and lipase assays

For all activity assays, at least three biological replicates were tested. To test for lipase activity, LB plates containing 1% (w/v) triolein, 0.001% (w/v) Rhodamin B (Sigma-Aldrich, St. Louis, USA), 200 μg/ml ampicillin, and 100 μM IPTG were prepared [25]. A formation of orange fluorescent halos around colonies on these plates is an indicator of lipase activity. For an esterase activity screening, LB agar plates containing 1% (w/v) tributyrin, 200 μg/ml ampicillin, and 100 μM IPTG were prepared. Esterase activity can be identified by clear halos around colonies on these plates. Overnight cultures of strains were adjusted to an OD_{600} of 2. 2 μl of these cultures, corresponding to ~ 8×10^5 cells were spotted onto the lipase and esterase screening plates.

Lipolytic activity assays in crude cell extracts

Cell cultures were grown at 37 °C in LB medium containing 200 μg/ml ampicillin. At an OD_{600} between 0.5 and 0.6 the expression of the plasmid-encoded proteins was induced by addition of 500 μM IPTG to the medium. After 3 h, cells were harvested for 1 min at $13,000 \times g$. Pellets were resuspended in 800 μl 50 mM Tris/HCl buffer, pH 7.3. Bacterial cells were disrupted by sonication using a Vial Tweeter Instrument (Hielscher, Teltow, Germany) at 80% amplitude and cycle of 0.5 for 5×30 s with 30 s breaks on ice. Crude extracts were stored at −20 °C. p-Nitrophenyl (p-NP) ester-based activity assays were prepared as described previously [26]. Briefly, activity of crude extracts towards p-NP butyrate and p-NP palmitate (Sigma-Aldrich) was tested in 50 mM Tris/HCl buffer pH 7.3 at 25 °C. The hydrolysis reaction was started by the addition of p-NP ester to a final concentration of 50 μM. The reaction was observed for 20 min, continuously measuring the absorption at 405 nm in a V-650 UV/Vis spectrophotometer (Jasco, Tokyo, Japan). The change of absorption over time (dA/dt) was calculated using the "Enzymatic Reaction Rate" module of the spectrophotometer's software (Jasco). The activity of crude extract from *E. coli* AK50 (expressing Lipase A from *Bacillus subtilis*) was set to 100%.

Table 2 Strains used in this study

Escherichia coli strain	Genotype	Source
XL1-Blue	recA1 endA1 gyrA96 thi-1 hsdR17 supE44 relA1 lac [F′ proAB laclq ZΔM15 Tn10 (Tetr)]	Stratagene, La Jolla, CA
AK02	XL1-1Blue pCC	This work
AK50	XL1-Blue pCC_*lipA* (pCC containing *lipA* from *B. subtillis* between NdeI and EcoRI restriction sites)	This work
AK70	XL1-Blue pCC_*gos54*	This work
AK71	XL1-Blue pCC_*gos55*	This work
AK72	XL1-Blue pCC_*gos88*	This work
AK73	XL1-Blue pCC_*gos89*	This work

Additional files

Additional file 1: Distribution of HMM-profiles in 890 families created from more than one HMM.

Additional file 2: Families created based on HMMs and their associated HMMs.

Additional file 3: Families created based on HMMs and the corresponding proteins.

Additional file 4: Classes based on MCL clustering.

Additional file 5: Families based on MCL clustering and the corresponding proteins.

Additional file 6: 4129 novel Hidden Markov Models created in this study.

Additional file 7: Test set for performance testing of HMMs.

Additional file 8: Performance parameters of the 4129 novel HMMs and their taxonomic specificity.

Additional file 9: List of representatives for each family.

Additional file 10: List of all HMM- and MCL-based families and their corresponding proteins.

Abbreviations
DNA: Deoxyribonucleic acid; EMBL: the European molecular biology laboratory; GOS: Global Ocean Sampling; HMM: Hidden Markov Model; MAFFT: Multiple Alignment using Fast Fourier Transform; MCL: Markov Cluster; NCBI: National Center for Biotechnology Information, USA; Pfam: Protein families database; TIGRFAMs: The Institute for Genomic Research protein families database

Acknowledgements
We thank Björn Bourscheidt for the assistance in data acquisition and database setup. Nataliya Lupilova provided expert technical assistance in the wet lab. We thank Dr. Martin Eisenacher, Christopher Reher, and Michael Krafzik from the Medical Bioinformatics group of the Medizinisches Proteom-Center at the Ruhr-University Bochum for the allocation of computing and data services. We are grateful to Markus-Herrmann Koch, Dr. Michael Turewicz, and Dr. Axel Mosig for their helpful discussion of our project. We thank Dr. Julia E Bandow for carefully reading our manuscript and her invaluable suggestions.

Funding
Principal funding for this work was provided by the European Research Council under the European Union's Seventh Framework Programme (FP7/ 2007–2013)/ERC Grant agreement no. 281384–FuMe. Computational hardware was provided by the German network for bioinformatics infrastructure (https://www.deNBI.de) and funded by BMBF grant FKZ 031 A 534A. No funding body played any role in the design or conclusion of this study.

Authors' contributions
LIL and AK conceived the analyses and designed the experiments. AK performed the analyses and experiments. LIL and AK wrote the manuscript. Both authors read and approved the final manuscript.

Competing interests
The authors declare that they have no competing interests.

References
1. Yooseph S, Sutton G, Rusch DB, Halpern AL, Williamson SJ, Remington K, et al. The Sorcerer II Global Ocean Sampling expedition: Expanding the universe of protein families. PLoS Biol. 2007;5:432–66.
2. Koonin EV. Metagenomic sorcery and the expanding protein universe. Nat Biotechnol. 2007;25(5):540–2.
3. Mitchell A, Chang H-Y, Daugherty L, Fraser M, Hunter S, Lopez R, et al. The InterPro protein families database: the classification resource after 15 years. Nucleic Acids Res. 2015;43:D213–21.
4. Sharon I, Battchikova N, Aro E-M, Giglione C, Meinnel T, Glaser F, et al. Comparative metagenomics of microbial traits within oceanic viral communities. ISME J. 2011;5:1178–90.
5. Radivojac P, Clark WT, Oron TR, Schnoes AM, Wittkop T, Sokolov A, et al. A large-scale evaluation of computational protein function prediction. Nat Methods. 2013;10:221–7.
6. Punta M, Coggill PC, Eberhardt RY, Mistry J, Tate J, Boursnell C, et al. The Pfam protein families database. Nucleic Acids Res. 2012;40:D290–301.
7. Haft DH, Selengut JD, White O. The TIGRFAMs database of protein families. Nucleic Acids Res. 2003;31:371–3.
8. Eddy SR. Accelerated Profile HMM Searches. PLoS Comput Biol. 2011;7: e1002195–5.
9. Van Dongen S, Abreu-Goodger C. Using MCL to extract clusters from networks. Methods Mol Biol. 2012;804:281–95.
10. Katoh K, Misawa K, Kuma K-I, Miyata T. MAFFT: a novel method for rapid multiple sequence alignment based on fast Fourier transform. Nucleic Acids Res. 2002;30:3059–66.
11. Gouveia-Oliveira R, Sackett PW, Pedersen AG. MaxAlign: maximizing usable data in an alignment. BMC Bioinformatics. 2007;8:312.
12. Edgar RC. MUSCLE: a multiple sequence alignment method with reduced time and space complexity. BMC Bioinformatics. 2004;5:113.
13. Talavera G, Castresana J. Improvement of phylogenies after removing divergent and ambiguously aligned blocks from protein sequence alignments. Syst Biol. 2007;56:564–77.
14. Wu M, Chatterji S, Eisen JA. Accounting for alignment uncertainty in phylogenomics. PLoS One. 2012;7, e30288.
15. Zhang Y, Sun Y, Cole JR. A Sensitive and Accurate protein domain cLassification Tool (SALT) for short reads. Bioinformatics. 2013;29: 2103–11.
16. Tatusova T, Ciufo S, Fedorov B, O'Neill K, Tolstoy I. RefSeq microbial genomes database: new representation and annotation strategy. Nucleic Acids Res. 2014;42:D553–9.
17. Brister JR, Ako-Adjei D, Bao Y, Blinkova O. NCBI viral genomes resource. Nucleic Acids Res. 2015;43:D571–7.
18. Jaeger K-E, Eggert T. Lipases for biotechnology. Curr Opin Biotechnol. 2002;13:390–7.
19. Li W, Godzik A. Cd-hit: a fast program for clustering and comparing large sets of protein or nucleotide sequences. Bioinformatics. 2006;22:1658–9.
20. Anton BP, Chang Y-C, Brown P, Choi H-P, Faller LL, Guleria J, et al. The COMBREX project: design, methodology, and initial results. PLoS Biol. 2013; 11, e1001638.
21. Daiha K, De G, Angeli R, De Oliveira SD, Almeida RV. Are Lipases Still Important Biocatalysts? A Study of Scientific Publications and Patents for Technological Forecasting. PLoS One. 2015;10: e0131624.
22. Grote A, Hiller K, Scheer M, Münch R, Nörtemann B, Hempel DC, et al. JCat: a novel tool to adapt codon usage of a target gene to its potential expression host. Nucleic Acids Res. 2005;33:W526–31.
23. Sambrook J, Russell DW. Molecular Cloning. New York: Cold Spring Habor; 2001.
24. Bertani G. Studies on lysogenesis. I. The mode of phage liberation by lysogenic *Escherichia coli*. J Bacteriol. 1951;62:293–300.
25. Kouker G, Jaeger KE. Specific and sensitive plate assay for bacterial lipases. Appl Environ Microbiol. 1987;53:211–3.
26. Masuch T, Kusnezowa A, Nilewski S, Bautista JT, Kourist R, Leichert LI. A combined bioinformatics and functional metagenomics approach to discovering lipolytic biocatalysts. Front Microbiol. 2015;6:1110.
27. UniProt Consortium. UniProt: a hub for protein information. Nucleic Acids Res. 2015;43:D204–12.

3

Multi-scale structural community organisation of the human genome

Rasha E. Boulos[1,2], Nicolas Tremblay[1,3], Alain Arneodo[1,4], Pierre Borgnat[1] and Benjamin Audit[1]*

Abstract

Background: Structural interaction frequency matrices between all genome loci are now experimentally achievable thanks to high-throughput chromosome conformation capture technologies. This ensues a new methodological challenge for computational biology which consists in objectively extracting from these data the *structural motifs* characteristic of genome organisation.

Results: We deployed the fast multi-scale community mining algorithm based on spectral graph wavelets to characterise the networks of intra-chromosomal interactions in human cell lines. We observed that there exist structural domains of all sizes up to chromosome length and demonstrated that the set of structural communities forms a hierarchy of chromosome segments. Hence, at all scales, chromosome folding predominantly involves interactions between neighbouring sites rather than the formation of links between distant loci.

Conclusions: Multi-scale structural decomposition of human chromosomes provides an original framework to question structural organisation and its relationship to functional regulation across the scales. By construction the proposed methodology is independent of the precise assembly of the reference genome and is thus directly applicable to genomes whose assembly is not fully determined.

Keywords: Chromosome interaction network, Multi-scale community mining, Structural domain hierarchical organisation, Spectral graph wavelets, Human genome

Background

It is now well established that eukaryotic genome dynamics and 3D architecture have a fundamental role in the regulation of nuclear functions such as DNA replication and gene transcription [1–6]. At small scale ($\sim$ 200 bp), the crystal structure of the nucleosome core particle (the first level of eukaryotic DNA compaction formed by complexing $\sim$ 150 bp of DNA with 8 histone proteins) was determined 20 years ago [7]. At the scale of the nucleus, fluorescence imaging revealed the dominant structural organisation of the genome into *chromosome territories* reflecting a non-mixing compartmentalisation of the chromosomes [2]. However, until the emergence of Chromatin Conformation Capture (3C) technologies [8, 9], our knowledge of the structural organisation of DNA at the intermediary scales remained partial. High-throughput 3C protocol (Hi-C technique) has opened new perspectives in the study of these intermediary structures genome-wide in higher eukaryotes, closing the gap between the atomic and chromosomal resolutions [10–18]. Hi-C technique relies on high-throughput sequencing and allows to semi-quantitatively measure the co-localisation frequencies of all pairs of genomic loci (the spatial resolution of the most recent data [19, 20] is $\sim$ 1 − 10 kb for mammalian genomes of length $\sim$ 3 Gb). Inter-chromosome co-localisation frequencies are lower than intra-chromosome frequencies, following the nuclear organisation into chromosome territories [10]. Mean intra-chromosome frequencies decrease with the genomic distance as expected for a polymer [21]. Changes in the decreasing rate reflect the modifications of the global chromosome structure like the chromosome condensation observed during entry in metaphase [19]. Nevertheless Hi-C data also put into light a structural compartmentalisation of the genome at different scales that cannot be explained by

*Correspondence: benjamin.audit@ens-lyon.fr
[1]Univ Lyon, Ens de Lyon, Univ Claude Bernard Lyon 1, CNRS, Laboratoire de Physique, F-69342 Lyon, France
Full list of author information is available at the end of the article

simple homogeneous polymer models [22]. Principal component analysis of the correlation matrix between the co-localisation frequency profiles of each locus revealed the existence of two nuclear compartments, loci preferentially co-localising with other loci from the same compartment: compartment A is associated with gene rich and early replicating regions and compartment B with gene poor and late replicating regions [10]. Projected on the genome, this classification describes the chromosomes as the succession of A/B domains of length $\sim$ 10 Mb. Inspection of intra-chromosomal co-localisation frequency matrices reveals a finer structuring level characterised by diagonal blocks of length $\sim$ 0.1 − 1 Mb: co-localisation frequency is high between regions of the same block but weaker between regions belonging to different blocks [11] (Fig. 1). These blocks, named Topologically Associating Domains (TADs), underline a structural compartmentalisation of chromosomes whose link with genome functional organisation and dynamics is the subject of intense research activity [11, 15, 16, 19, 20, 23–29]. In order to carry out this research, methods allowing to objectively delineate structural domains from Hi-C data have been developed [11, 16, 26–34]. Most of these approaches look for structural domains that are intervals of the chromosomes. For example, chromosome structural partition was achieved using (i) 1D signals quantifying the balance between the

co-localisation frequencies of the locus of interest with upstream and downstream loci (directionality index) [11, 27], (ii) dynamic programming algorithms that also explicitly model structural domains as chromosome intervals [31, 32] and (iii) projecting on the genome the bisection obtained from a graph representation of the Hi-C data (see below) [28, 34]. As illustrated in Fig. 1, chromosome structural organisation can involve nested structures over a large range of scales [22, 29]. However only the method proposed in [31] explicitly includes the possibility to identify chromosome structural domains at diverse scales of observation and the method in [29] to hierarchically merge adjacent TADs into *metaTADs*.

Here we propose a novel method to analyse Hi-C data that allows a multi-scale identification of structural domains. Because it does not rely on the specific assembly of the reference genome, this method does not look for structural domains limited to chromosome intervals thereby relaxing our preconception about the nature of structural domains. Moreover, due to polymorphisms within a species or to chromosome rearrangements characteristic of cancer cells [35], the assembly of the reference genome does not necessarily corresponds to the true assembly for a cell line under investigation. In these situations, reduced sensitivity to genome assembly is likely to avoid erroneous structural domain predictions. A Hi-C co-localisation frequency matrix is positive and symmetric, it can thus be interpreted as the adjacency matrix of the genome interaction network where the nodes are the chromosome loci (typically non-overlapping windows) and the edges reflect the co-localisation frequency between these regions. This justifies the use of concepts and tools from graph theory to analyse Hi-C data [28, 30, 33, 34, 36–38]. This representation depends on genome assembly only up to the scale used to define the Hi-C matrix, the columns/rows of the Hi-C matrix can be permuted without affecting the output of graph algorithms. In graph theory, a set of nodes that share more connections between themselves than with the rest of the graph is called a *community* [39]. Hence we reformulate the question of structural domain mining as a search for community in the Hi-C interaction network. Note that Markov graph clustering was already experienced to delineate sub-segments within large A/B-like chromosomal domains obtained in a first step [30] and that extensions of graph stochastic block models were also applied to Hi-C matrices of human chromosome 4 and a segment of human chromosome 6 [33]. In order not to privilege any particular scale in the analysis, we performed the multi-scale partitioning of the full intra-chromosomal interaction networks into structural communities using a multi-scale community mining algorithm based on graph wavelets [40].

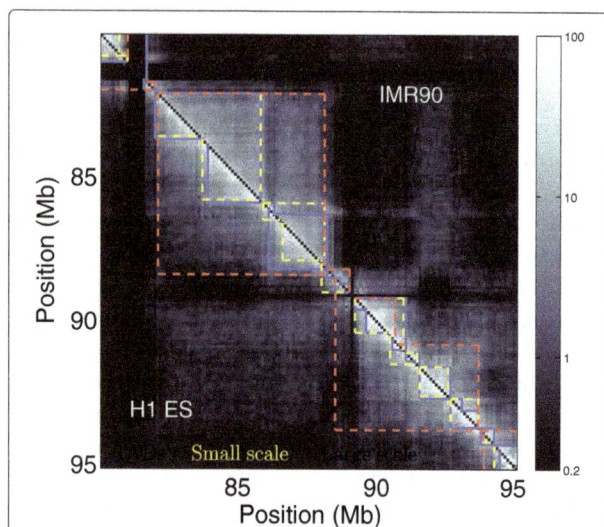

Fig. 1 Hi-C co-localisation maps reveal a multi-scale structural organisation. Hi-C co-localisation frequency matrices along a 15 Mb fragment of human chromosome 10 in H1 ES (resp. IMR90) under (resp. above) the *diagonal* with intensity of interactions colour coded according to colour map on the right. Blue lines represent TADs [11] in the two cell lines. Coloured dashed lines correspond to 2 partitions into communities obtained at small (*yellow*) and large (*red*) scales. Columns and rows in *black* correspond to masked regions (Methods and Additional file 1: Table S1)

Methods

Chromatin conformation capture data and topologically associating domains

Here we used Hi-C data obtained in different human cell lines:

- Embryonic stem cell line H1 (H1 ES) and foetal lung fibroblast cell line IMR90 Hi-C data for which TADs are available [11], allowing a direct comparison of our structural communities with what is considered as reference structural domains in the literature. Hi-C matrices at resolution 20 kb and 40 kb for two replicates in each cell lines as well as TADs predictions in these cell lines were downloaded from the GEO database under accession number GSE35156. These data are based on the hg18 assembly version of the human genome.

- Myelogenous leukemia cell line K562 and lymphoblastoid cell line GM06990 Hi-C data [10] for the analysis of the structural conservation between cell lines. Hi-C matrices at resolution 100 kb for the two cell lines were downloaded from the GEO database under accession number GSE18199. These matrices are based on the hg18 assembly version of the human genome.

- Cervical cancer cell line HeLaS3 Hi-C data [19] where the Hi-C experiments were performed on synchronised cells during mitosis and G1 allowing a study of the community structure during the cell cycle. The Hi-C reads alignment files to the human genome (hg19 assembly version) for the two stages of the cell cycle were downloaded from the ArrayExpress database under accession number E-MTAB-1948.

Hi-C intra-chromosomal co-localisation frequency matrices for non-overlapping 100 kb loci correspond to the downloaded matrices that were down-sampled to 100 kb when necessary or were constructed from the alignment files (Fig. 1). Unexpectedly low and unexpectedly high interacting loci that are likely to introduce noise were removed (Additional file 1: Table S1). The remaining 100 kb loci were concatenated resulting in new *masked positions*.

We compared the structural-communities described in this work to the TADs [11] that are considered as a reference for the structural description of Hi-C data. TADs were identified in H1 ES and IMR90 cell lines at both 20 and 40 kb resolutions [11]. Given our adopted resolution of 100 kb, we used the TADs dataset obtained at the 40 kb resolution, and we assigned each TAD border to the corresponding 100 kb pixel keeping only TADs larger than 200 kb (3 pixels). This led to a database of 2 993 (resp. 2 263) TADs in H1 ES (resp. IMR90), with 3 905 (resp. 3 096) distincts borders in H1 ES (resp. IMR90).

In this work one focus is to question the existence of a TAD-like structuration of the human genome in the intermediary scale range from the described TAD typical size up to the chromosome length. A second objective is to address the possible conservation of these structural motifs across cell lines. This led us to include the K562 and GM06990 datasets from the original Hi-C study [10]. These datasets are less resolutive than more recent ones in IMR90 and H1 ES cell lines [11] due to a limited sequencing depth and were analysed at best at 100 kb resolution by the original authors. This explains why we chose 100 kb as the resolution for all the analysis presented in our manuscript. However to check whether lower or higher resolution has significant impact on the results, the IMR90 dataset was also analysed at resolutions 40 and 200 kb.

Multi-scale community mining using graph wavelets

We used the multi-scale community mining algorithm based on spectral graph wavelets that we previously described and benchmarked against two other multi-scale community mining methods from the literature [40]. The purpose of detecting communities at different scales using graph wavelets instead of, say, cutting a hierarchical clustering at different levels, is to fit as close to the data as possible. Cutting a hierarchical clustering impose a hierarchical structure to the set of community obtained at the different scales (cutting levels). When using wavelets, we do not suppose beforehand that the data have a hierarchical structure: a community at a coarse scale does not necessarily have to contain communities found at a finer scale.

Our community mining algorithm [40] relies on the precise construction of graph wavelets in order to introduce the notion of scale [41] (Supplementary text: Graph wavelet transform and community mining and Figures S1 to S4 in Additional file 1). A graph wavelet centred on a node a is a function on the nodes of the graph whose values capture the *proximity* of each node to node a given a scale s of observation. As such, the set of graph wavelets at scale s characterise the local graph structure around each node over a "distance" controlled by the scale parameter s, as illustrated in Additional file 1: Figures S1 and S3. At a fixed scale, the similarity between the neighbourhood of 2 nodes (a and b) can be quantified as the correlation ($C^{(s)}(a,b)$, Additional file 1: Equation (S11)) between the wavelets centred on each of the two nodes at that scale. Computing the correlation distance $\mathcal{D}^{(s)}(a,b) = 1 - C^{(s)}(a,b)$ (Additional file 1: Equation (S12)) between all pairs of nodes results in a distance matrix capturing the similarity of node neighbourhood, which can in turn be used as the input of a hierarchical clustering algorithm in order to partition the nodes into communities for the scale of observation s (Additional file 1: Figure S4). To sum up, at each scale and for each intra-chromosomal interaction network, the community mining algorithm amounts to (i) compute the matrix of correlation distance $\mathcal{D}^{(s)}(a, b)$, (ii) apply average-linkage hierarchical clustering [42, 43], and (iii) finally cut the resulting dendrogram following the method prescribed in [40]. This results in a set of structural communities for a given scale and a given chromosome.

We used the fast implementation of this procedure [40]. On the one hand, the graph wavelet transform is computed using the fast algorithm proposed in [41] (Additional file 1: Equation (S17)). On the other hand, instead of computing the wavelets on the n nodes of the graph which requires n wavelet transforms of Dirac functions (using Additional file 1: Equation (S17) n times), the matrix of correlations between wavelets at scale s is approximated by the correlation between η ($\ll n$) wavelet transforms of random Gaussian functions on the graph (Supplementary text: Graph wavelet transform and community mining in Additional file 1). Importantly, the fast implementation of our multi-scale community mining protocol is applicable to large networks with $\gtrsim$ 10 000 nodes [40], allowing to consider its future application to intra-chromosomal interaction networks at high resolution ($\sim$ 10 kb) in mammals [20] but also to full genome interaction networks at the resolution used in the present work (100 kb). Note that graph spectral clustering can also be considered for these large interaction network settings thanks to recent algorithmic developments [44, 45].

Comparing sets of genomic domains

As discussed in Results and discussion, communities within intra-chromosomal interaction networks can be described in terms of genomic intervals i.e. sets of loci that form contiguous genomic domains and can thus be fully described by their two extreme positions, called domain borders. We adopted the three following points of view for the comparison of sets of genomic domains (chromosome intervals) of different origins. Note that because the sets of domains of interest here do not form partitions of the genome, we could not adopt the classical measures of similarity between partitions like Mutual Information and Adjusted Rand Index. Given two sets of domains $\mathcal{D}_1$ and $\mathcal{D}_2$ with two sets of associated borders $\mathcal{B}_1$ and $\mathcal{B}_2$ respectively, we used the following estimators:

- Mean best mutual coverage: We define the mutual coverage m_c between two domains $d_1 \in \mathcal{D}_1$ and $d_2 \in \mathcal{D}_2$ as their intersection length $L_{d_1 \cap d_2}$ divided by the maximum length of the two domain lengths L_{d_1} and L_{d_2}: $m_c(d_1, d_2) = L_{d_1 \cap d_2} / \max(L_{d_1}, L_{d_2})$.

The maximal value 1 of m_c is obtained when the two domains d_1 and d_2 are identical. Then, for each domain $d_1 \in \mathcal{D}_1$, we define its best mutual coverage with $\mathcal{D}_2$ domains ($bm_{c_{\mathcal{D}_2}}$) as its maximal mutual coverage with $\mathcal{D}_2$ domains: $bm_{c_{\mathcal{D}_2}}(d_1) = \max_{d_2 \in \mathcal{D}_2}(m_c(d_1, d_2))$. Sorting the $\mathcal{D}_1$ domains by size, we compute the mean best mutual coverage with $\mathcal{D}_2$ of groups of 50 $\mathcal{D}_1$ domains that we plot as a function of the mean length of the domains in the group. This results in an average mean best mutual coverage curve between domains in $\mathcal{D}_1$ and $\mathcal{D}_2$ as a function of $\mathcal{D}_1$ domain size.

- We say that a domain d has a match in $\mathcal{D}_2$ if $bm_{c_{\mathcal{D}_2}}(d) \geq 0.8$. $P_{\mathcal{D}_2}(\mathcal{D})$ is then defined as the proportion of domains $d \in \mathcal{D}$ that have a match in $\mathcal{D}_2$. Sorting the $\mathcal{D}_1$ domains by size, we consider them in groups $\mathcal{D}$ of 50 domains and plot $P_{\mathcal{D}_2}(\mathcal{D})$ as a function of the mean length of the domains in $\mathcal{D}$. This results in a matching proportion curve of domains in $\mathcal{D}_1$ and $\mathcal{D}_2$ as a function of $\mathcal{D}_1$ domain size.
- We say that a border b has a match in $\mathcal{B}_2$ when there is a border in $\mathcal{B}_2$ less than 100 kb away from b i.e $\pm$ 1 pixel away. $P_{\mathcal{B}_2}(\mathcal{B})$ is then defined as the proportion of borders $b \in \mathcal{B}$ that have a match in $\mathcal{B}_2$. Sorting the $\mathcal{B}_1$ borders according to their *associated lengths* (see below), we consider them in groups $\mathcal{B}$ of 100 borders and plot $P_{\mathcal{B}_2}(\mathcal{B})$ as a function of the average associated length of the borders in $\mathcal{B}$. This results in a matching proportion curve of borders in $\mathcal{B}_1$ and $\mathcal{B}_2$ as a function of $\mathcal{B}_1$ border associated length.

Domain length is an intuitive quantity to order a set of domains. In the same manner, we associated a length with each border of the genomic domains used in this work. TAD borders can be shared by at most 2 consecutive TADs, so we associated them with the length of the shortest TAD they border. At a fixed scale of analysis, a border of the novel interval-communities (see Section: Structural communities correspond to genome intervals) delimits two consecutive interval-communities, so (at that scale) we associated it with the minimum length of the two bordering communities. However these borders also present a strong pattern of conservation from one scale to another (Fig. 2; Additional file 1: Figure S5), so the largest of these lengths across the scales was retained as the final length associated with an interval-community border. In this way, border associated lengths allowed us to sort borders according to the importance (size) of the corresponding chromosome structures.

Results and discussion
Wavelet-based community detection in the DNA interaction network
As discussed above, Hi-C data can be represented as graphs where nodes represent DNA loci and the edges

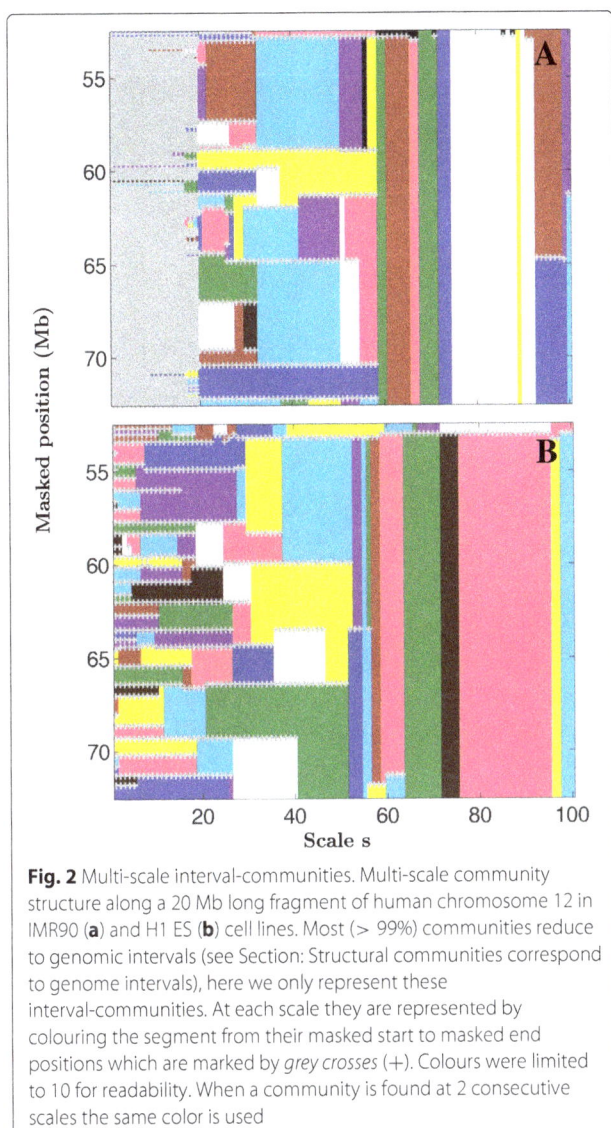

Fig. 2 Multi-scale interval-communities. Multi-scale community structure along a 20 Mb long fragment of human chromosome 12 in IMR90 (**a**) and H1 ES (**b**) cell lines. Most (> 99%) communities reduce to genomic intervals (see Section: Structural communities correspond to genome intervals), here we only represent these interval-communities. At each scale they are represented by colouring the segment from their masked start to masked end positions which are marked by *grey crosses* (+). Colours were limited to 10 for readability. When a community is found at 2 consecutive scales the same color is used

cell line was 5 h 40 mn using Matlab on a linux computing desktop with 8 Xeon CPU at 3.30 GHz. We first discuss the results obtained for human chromosome 12 in H1 ES and IMR90 cell lines as representative examples of the results obtained for all intra-chromosomal interaction networks. Chromosome 12 network initially contains 1 324 nodes. After the filtering procedure, 1 250 nodes are left in IMR90 and 1 249 in H1 ES (Methods). When applying the wavelet-based community detection method separately on the two interaction networks, we obtained 100 partitions of the masked genome for each cell line, one at each scale. Overall, we obtained 23 927 (resp. 4 266) communities for IMR90 (resp. H1 ES). As expected, the size of the resulting communities increases with the scale parameter (Fig. 3a). For H1 ES the increase of the mean community size with the scale is homogeneous suggesting that there is no characteristic size for the community structure. For IMR90 we observe a first range of scales where the communities reduce to singletons (mean size $\sim$ 1), followed by an abrupt transition to a community mean size $\sim$ 17 (Fig. 3a). The existence of singletons over a relatively large range of scales explains why the total number of communities in IMR90 is larger than in H1 ES. After removing the trivial communities (singletons), 3 342 (resp. 4 266) communities were kept in IMR90 (resp. H1 ES).

Structural communities correspond to genome intervals

The interaction frequencies outside the diagonal blocks characterising the structural compartimentalisation as described in [11] are not negligible (look for instance at the region around [82,89] Mb in IMR90 that highly interacts with the region around [92,93] Mb in Fig. 1). This suggests that structural communities may not necessarily reduce to intervals along the genome. Hence for each non trivial community (community of size > 1), we computed the proportion P_{int} of the largest set of successive 100 kb loci covered by the community over the size of the community: $P_{int} = 1$ when all the nodes of the community constitute an interval of the masked genome and $P_{int} = 1/N$ where N is the size of the community when the community do not contain any pair of consecutive loci of the genome. Considering $P_{int} \geq 0.95$ as a criterion for a community to constitute an interval along the genome, we observed for the 2 cell lines that more than 99% of the communities correspond to intervals of the genome. This property for the communities remains true for all the scales and whatever the size of the communities. This is consistent with the fact that at all scales, genomic neighbours tend to strongly co-localise resulting in higher frequency of interactions. These results demonstrate that the strongest motifs of structural organisation involve contiguous genomic segments. We will refer to the communities forming a genomic interval as *interval-communities*.

connect interacting loci, allowing us to reformulate the question of finding structural domains as a question of finding communities in the DNA interaction network. We used the fast implementation of the wavelet-based multi-scale community mining algorithm (Methods and Supplementary text: Graph wavelet transform and community mining in Additional file 1) with $\eta = 200$ random Gaussian functions to estimate the distance correlation matrix. For each Hi-C dataset, we considered the 22 autosomes' intra-chromosomal interaction networks constructed for non-overlapping 100 kb loci (Methods). We systematically applied the wavelet-based multi-scale community detection method to all the connected interaction networks scanning 100 scales logarithmically distributed in the range of available scales (Additional file 1: Equation (S13)) [40]. The average total running time per

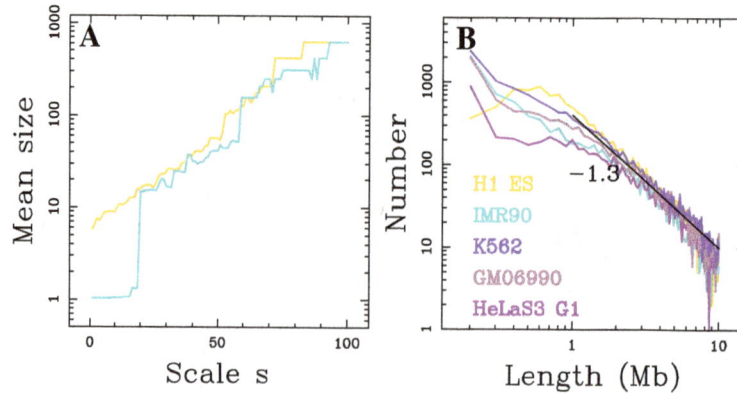

Fig. 3 Multi-scale communities in the DNA interaction network. **a** Mean structural community size (in 100 kb pixels) for chromosome 12 as a function of the scale index in IMR90 (*blue*) and H1 ES (*yellow*). **b** Histogram of interval-communities genomic length (*l*) calculated in 100 kb bins in a log-log representation for different cell lines: IMR90 (*blue*), H1 ES (*yellow*), GM06990 (*pink*), K562 (*purple*) and HeLa (G1) (*light purple*). The *black straight line* correspond to the power-law behaviour l^α with $\alpha = -1.3$

We only kept the communities that correspond to an interval ($P_{int} \geq 0.95$) reducing them to their main interval. This allowed us to adopt a simple representation of the structural-communities obtained across scales (Fig. 2). The differences observed between the resulting community size distributions in IMR90 and H1 ES (Fig. 3a) are visible in this representation. We clearly see a first range of scales ($s \leq 20$) where the interval-communities reduce to singletons in IMR90 (Fig. 2a) and not in H1 ES (Fig. 2b). Above this critical scale, non trivial interval-communities appear in IMR90. Note that the mean size of the interval-communities for this first meaningful partitioning in IMR90 is larger than the ones observed in H1 ES for its first meaningful partitioning (smallest scale). This results in a lack of small non trivial interval-communities in IMR90.

A hierarchical organisation of the genome

The representation in Fig. 2 reveals the hierarchical organisation of the communities. Across scales, small communities merge together to form bigger communities at larger scales. Hence the community borders present at the smallest scale progressively disappear at some larger scale allowing the emergence of bigger communities. Importantly, the conservation of borders from large scales to small scales is very high. For each pair of scales $s2 > s1$, we computed the proportion of borders at the larger scale $s2$ that are also present at the smaller scale $s1$. This proportion is close to 1 regardless of the scales (Additional file 1: Figure S5). The fact that the borders are conserved across scales means that there is no "new" structure that emerges and that only existent ones merge together, i.e. small structures are nested into bigger ones. This is consistent with the results of recent studies suggesting

that TADs hierarchically co-associate to form larger structures [16, 29, 46].

Another important property illustrated in Fig. 2 is the redundancy of the communities obtained across scales, underlining the robustness of the graph wavelet community mining protocol with respect to its stochasticity (usage of random vectors to estimate the graph wavelet correlation matrix; Methods). Hence, we kept only once each non trivial interval-communities (size ≥ 2 nodes and $P_{int} \geq 0.95$). We also filtered out the communities that more than double in size when reintegrating the masked regions of the genome, e.g. interval-communities spanning the centromers. This leads to 386 (resp. 537) non trivial interval-communities in IMR90 (resp. H1 ES) for the chromosome 12. When applied to the 6 Hi-C datasets considered (Methods), the methodology presented for human chromosome 12 in H1 ES and IMR90 resulted in few thousands interval-communities per dataset (Table 1), except for the mitosis HeLaS3 dataset (discussed below). Interestingly, the length distributions of the interval-communities for the IMR90, GM06990, K562 and HeLaS3 G1 datasets are very similar, but they display differences with the one obtained for H1 ES dataset for small interval-communities (Fig. 3b): there are more interval-communities involving only 2-3 nodes (200–300 kb) in the 4 differentiated cell lines datasets and a deficit in interval-communities of length ~ 500 kb to ~ 1.5 Mb relative to H1 ES. A possible interpretation of this excess of interval-communities of size ~ 1 Mb in H1 ES, compared to differentiated cell lines, is that cell differentiation is accompanied by the merging of the small structural communities in a *structural consolidation* scenario. For larger communities, the interval-community size distributions in these 5 Hi-C datasets are almost identical. Indeed, for

Table 1 Number of structural communities

Cell line	N	N (filtered)	Remaining communities	Distinct borders
H1 ES	12 343	65	12 278	5 751
IMR90	8 852	25	8 827	6 824
GM06990	10 279	60	10 219	6 967
K562	13 383	30	13 353	8 273
HeLaS3 G1	6 752	36	6 716	4 108
HeLaS3 M	1 059	4	1 055	885

For each cell line, N is the number of distinct non redundant and non trivial (size ≥ 2 i.e. 2 nodes) interval-communities. N(filtered) is the number of communities filtered out because (i) they do not correspond to a genomic interval or (ii) they double in size when going back to the original (not masked) positions. The last two columns correspond to the number of communities and distinct borders in the database

$l \gtrsim 2$ Mb, they display a power-law behaviour l^α with $\alpha \simeq -1.3$ (Fig. 3b). Note that if communities of length $\sim l$ would form a partition of the genome of length L, then the number of communities of this scale would be equal to L/l leading to $\alpha = -1$ ($\gtrsim -1.3$). This underlines the existence of domains at all scales up to the chromosome length without a characteristic size for genome structuring.

Are interval-communities structural domains?

To test the robustness of the wavelet-based community detection method with respect to the possible absence of a community structure over some range of scales, we compared the interval-communities obtained for the Hi-C datasets in synchronised HeLaS3 cells during G1 and M phase, respectively (Methods). The original study [19] showed that the highly compartmentalised organisation described before from non synchronous cells [10, 11, 13, 15, 16, 20, 26, 27] was restricted to interphase and that during a cell cycle, chromosomes transit from a decondensed and spatially organised state during interphase to a highly condensed and morphologically reproducible metaphase chromosome state. In the former phase, the Hi-C interaction maps display similar plaid patterns of regional enrichment or depletion of long range interactions (as the one shown in Fig. 1) while the maps in mitotic cells change and the plaid patterns disappear [19]. For HeLaS3 G1 (resp. mitosis) dataset, we obtained 6 716 (resp. 1055) non trivial communities and 4 108 (resp. 885) distinct borders (Table 1). For the mitosis HeLaS3 Hi-C dataset, we obtained 1 059 communities from which we filtered out 4 resulting in 885 distincts borders (Table 1). Consistently with non synchronous cells, G1 cells present a hierarchical structure into interval-communities that increase in size across scales (Additional file 1: Figures S6 and S7). Small scale singletons hierarchically group to form large interval-communities at larger scales. As discussed above, the length distribution of the G1 HeLaS3 interval-communities is similar to the

interval-communities size distribution obtained in the 3 other differentiated cell line datasets (Fig. 3b). In contrast, metaphase chromosomes do not present a hierarchical structural organisation. More specifically, chromosomes 16, 21 and 22 do not present any structure (each node constitutes a community on the full available range of scales, Additional file 1: Figure S6). In the 19 other autosomes, at small scales each node is a singleton and above a critical scale a sharp discontinuity of the community sizes distribution is observed: nodes are abruptly grouped in a small number (2–5) of communities (Additional file 1: Figure S7). For 12 out of these 19 chromosomes, when divided in two communities, these communities correspond to the two chromosomal arms, as illustrated for chromosome 17 in Additional file 1: Figure S7. These results demonstrate that the wavelet-based community detection method does not produce misleading intermediate scale communities when no structuration exists in that scale range.

To strengthen this point in a noisy situation, we simulated a structural interaction matrix between 2000 nodes (comparable to the largest human chomosomes at resolution 100 kb) organised in fully connected interval-communities with no specific organisation at scales larger than the community size: the matrix is built as a series of 40 pairs of domains of size 20 nodes and 30 nodes with internal domain interaction set to 60, with the two first (resp. second) sub-diagonals set to 80 (resp. 70) to assure connectivity and with an additive Poisson noise over all interaction pairs of mean value $\lambda = 50$ (Additional file 1: Figure S8 Left). When applying the graph wavelet community mining protocol, we recovered only trivial singleton communities at small and large scales. However in the intermediate scale range, we nicely recovered all the 20 and 30 nodes on a range of scales that depends on their size (Additional file 1: Figure S8 Right). This example shows that the method does not produce a fake hierarchical domain organisation by merging existing domains even in a noisy situation.

In order to verify that there are more interactions within interval-communities than between successive interval-communities, we compared the number of contacts between two 100 kb loci that are inside the same interval-community at equal distance from its center and the number of interactions between two loci at equal distance from one of the interval-community borders, as a function of the distance separating the pairs of loci. The ratio *vs* distance curves for different interval-community length categories show that on average there are more interactions within the communities than between communities, regardless of the cell line and the community length: the interaction ratio systematically increases to some maximal value at distances ~ 1–2 Mb, from a maximal value ~ 1.6 in GM06990 and K562, to ~ 2.2 in H1 ES and

~ 3 in IMR90. Over larger distances, the ratio remains rather constant in GM06990 and K562 and decreases to ~ 1.5 in H1 ES and IMR90 (Fig. 4). This property holds true even for communities larger than 10 Mb. As a comparison, we performed the same analysis for the original TAD datasets in H1 ES and IMR90 (Methods). Over the shared domain length range, the interaction ratio *vs* distance curves computed for the TAD datasets present very similar shapes as observed for interval-communities (Additional file 1: Figure S9), reaching maximal values ~ 3 in both H1 ES and IMR90. These results provide evidence that multi-scale interval-communities, very much like TADs, constitute units of 3D genome organisation bordered by structural barriers.

Are TADs interval-communities?

We next compared our communities to the TADs previously described in H1 ES and IMR90 [11], asking to which extent the TADs and TAD borders are recovered in our hierarchical database of interval-communities. The mean best mutual coverage *vs* TAD length curve (Methods) between TADs and interval-communities is slightly higher in H1 ES as compared to IMR90 for all TAD lengths (Fig. 5a), ranging from 62% (resp. 52%) at small length (300–500 kb) to 91% (resp. $\sim 89\%$) at larger length (~ 1–2 Mb) in H1 ES (resp. IMR90). This suggests a good recovery of the largest TADs by the interval-community classification. Given the 100 kb resolution used in this analysis, it is not surprising to observe lower mutual coverages at small lengths where 1 pixel error results in a dramatic lowering of mutual coverage. We also observed that the proportion of TADs that have a matching structural community (Methods) increases with the domain length (Fig. 5b). Only about 1/5 of the smallest TADs ($\lesssim 500$ kb) are recovered consistently with the fact that in this scale range a match has to be exact. For TADs longer than 1 Mb, the proportion of match is relatively high: in

IMR90 it increases from 40% for TADs ~ 1 Mb up to 70% for TADs ≥ 2 Mb and in H1 ES from 70% for TADs ~ 1 Mb up to 85% for TADs of ~ 2 Mb (Fig. 5b). Comparison of TAD borders to interval-community borders shows good concordance for the two datasets (Fig. 5c). For meaningful comparisons, we restricted the reference domain border set for each species to a subset of borders that at 100 kb resolution (± 1 pixel) collectively cover no more than 35% of the genome. Interval-community borders with the largest associated lengths (Methods) are selected first. Given the overlap between borders at that resolution, this process resulted in selecting a different number of distinct interval-community borders in each species: 3 468 in H1 ES, 2 834 in IMR90, 3 171 in GM06990 and 3 478 in K562. TAD borders are recovered from 50% up to $\sim 90\%$ in H1 ES and up to $\sim 80\%$ in IMR90, depending on the TAD border associated length, while the expected recovery rate by chance is 35% (Fig. 5c). These results quantify the high level of TAD recovery by interval-communities for domain length $\gtrsim 1$ Mb. Altogether, these results show that there is a significant agreement between TADs and the interval-communities. This provides evidence that interval-communities captures similar organisation principle of genome structure and, thus, extends this description up to chromosome size.

To test the robustness of our methodology with regards to the binning resolution, we reproduced the analysis of the IMR90 data at a finer (40 kb, total running time 8h48mn) and a coarser (200 kb, total running time 31 mn) resolution (Additional file 1: Figure S10). The lengths of interval-communities determined at these 2 resolutions nicely reproduce the distribution obtained at the 100 kb resolution (Additional file 1: Figure S10A). Intervals-communities of size $\gtrsim 1 - 2$ Mb strongly match between these intervals communities datasets (recovery proportion $\simeq 70 - 90\%$, Additional file 1: Figure S10B). For smaller community size, recovery proportion between

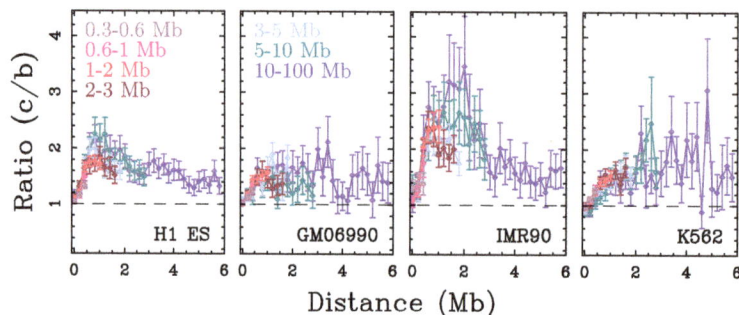

Fig. 4 Are interval-communities structural domains? Ratio (c/b) of the number of interactions between two 100 kb loci that are inside the same community at equal distance from its center (c) and the number of interactions between loci in different communities at equal distance from a community border (b), versus the distance between them. Different colours correspond to different community size categories: $0.3 \leq L < 0.6$ Mb (*light pink*), $0.6 \leq L < 1$ Mb (*pink*), $1 \leq L < 2$ Mb (*magenta*), $2 \leq L < 3$ Mb (*dark pink*), $3 \leq L < 5$ Mb (*light blue*), $5 \leq L < 10$ Mb (*blue*) and $10 \leq L < 100$ Mb (*purple*)

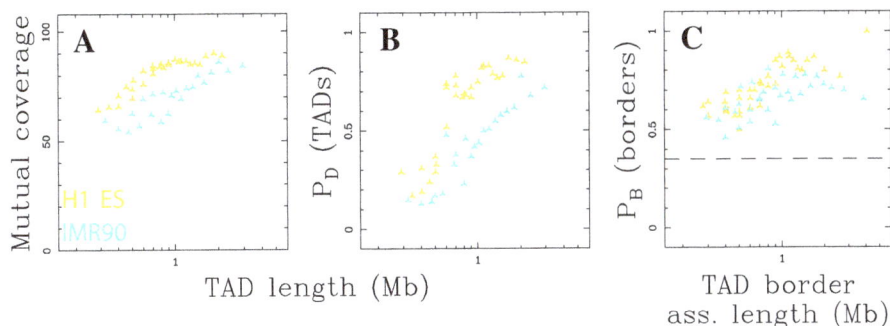

Fig. 5 TADs are interval-communities. **a** Mean best mutual coverage of TADs with interval-communities, and (**b**) proportion of TADs that have a match in the interval-community database, as functions of the average TAD size (Methods). **c** proportion of TAD boundaries that have a matching interval-community border as a function of the average TAD border associated length (minimum of the length of the two bordering domains, Methods); only the set of interval-community borders with largest associated length covering 35% of the genome were used (see text); the *horizontal dashed line* marks this expected border matching proportion of 35%. In (**a**, **b**, **c**), *yellow marks* the analysis in H1 ES and *blue* in IMR90

datasets decreases with community size in the same manner for the 3 resolution pairs. This can be understood when noting that the isolation strength of community borders significantly weakens when decreasing the genomic distance below ~ 1 Mb (Fig. 4). Finally, the proportion of TADs that have a match in the interval-community database is similar at 40 kb resolution than at 100 kb resolution (Additional file 1: Figure S11). This demonstrates that the results do not depend on the choice of the 100 kb resolution and further underlines that the lower structural domain recovery rate generally observed for small domain sizes ($\lesssim 1$ Mb) is likely related to the weaker isolation strength of structural domain borders over short distances ($\lesssim 1$ Mb).

Conservation of structural communities across cell lines

In the pioneering study [11], TADs were described to be conserved between cell lines. We observed that interval-communities in different cell lines present similar size distributions (Fig. 3b). This led us to investigate to which extent they are conserved across cell lines. To compare the communities obtained in different cell lines, we used each of the interval-community database obtained in H1 ES, GM06990, IMR90, K562, as a reference domain set and computed the proportion of matching interval-communities of the 3 other cell lines relative to this reference set (Methods). We observed that small interval-communities ($\lesssim 600$ kb) are not well conserved between different cell lines (Fig. 6). This might result from the fact that Hi-C data are average over cell populations and that some regions may present different structural organisations from cell to cell blurring the insulator property of structural domain borders at small scales. However, when considering interval-communities of larger sizes, higher conservation was observed (Fig. 6).

More than 60% of intervals-communities of length $L \gtrsim 0.6$ Mb in the differentiated cell lines correspond to an interval-community in H1 ES (Fig. 6a). H1 ES interval-community dataset thus contains a large proportion of the interval-communities observed in the differentiated cell lines above ~ 600 kb. When using one differentiated cell line interval-community database as reference, we observed a maximal recovery rate that is similar for the 3 other cell lines: 45% for sizes $\gtrsim 2$ Mb in IMR90, 65% for sizes $\gtrsim 1.5$ Mb in GM06990 and 70% for sizes $\gtrsim 1.5$ Mb in K562 (Fig. 6b, c and d). The observed differences likely reflect the excess of interval-communities in the size range 0.5-1.5 Mb observed in H1 ES relative to the differentiated cell lines (Fig. 3b). As a comparison, we performed the same analysis for the TADs that were claimed to be conserved between H1 ES and IMR90 cell lines [11] (Additional file 1: Figure S12). Like for interval-communities, the correspondance between TADs in the two cell lines decreases for domain sizes $\lesssim 600$ kb. For larger domain sizes, we observed that H1 ES TAD dataset contains more (maximal value $\sim 60\%$) of the IMR90 TADs than the IMR90 TAD dataset contains H1 ES TADs ($\sim 45\%$). These results corroborate the conservation of structural domains of length ~ 1–2 Mb between cell lines in the 45–70% range but also extend this conservation to the largest interval-communities up to length $\gtrsim 10$ Mb.

Conclusions

We introduced a fast multi-scale community mining algorithm based on spectral graph wavelets [40] to identify structural motifs from high-throughput chromatin conformation capture data (Hi-C) [10]. Hi-C data were represented as intra-chromosomal interaction networks and structural motifs were delineated as communities of these networks. The novelty of this approach relies on the

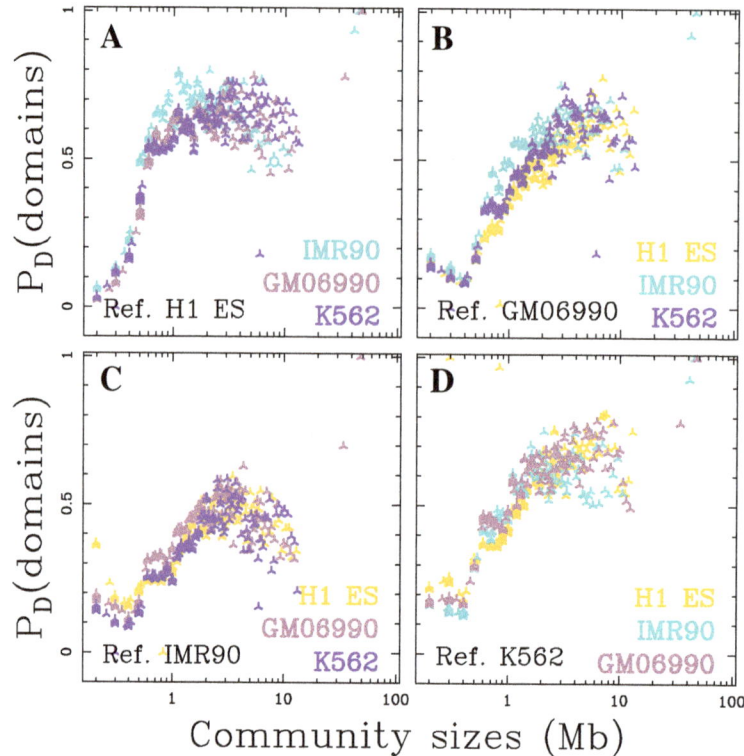

Fig. 6 Conservation of interval-communities between cell lines. Proportion of interval-communities in the query cell lines H1 ES (*yellow*), IMR90 (*blue*), GM06990 (*pink*) and K562 (*purple*) that have a matching interval-communities in the reference cell line indicated in each plot: H1 ES (**a**), GM06990 (**b**), IMR90 (**c**) and K562 (**d**). Proportion of interval-community matches is computed over groups of 50 query interval-communities ordered by length (Methods)

combination of a multi-scale procedure and a representation of the data that is independent of the exact assembly of the reference genome over length scales larger than the window size used to construct the interaction network. The proposed methodology has no a priori on the size and on the nature of the structural motifs. The application of this protocol to 6 Hi-C datasets led to a database of several thousands structural communities (Table 1). The database of interval-communities in mitotic HeLaS3 cells that were described not to present a TAD-like structural organisation [19], does not contain any intermediary scale structural communities, illustrating the robustness of the proposed methodology with regards to the absence of structural motifs. Consistently with the recent usage of Hi-C data for genome sequence assembly [47, 48], we observed that structural-communities in unsynchronised and G1 cells form hierarchies of chromosome intervals of length ranging from the resolution (100 kb) to the chromosome lengths ($\gtrsim$ 10 Mb) (Fig. 3). The prevalence of interval-communities underlines that chromosome folding is mainly driven by interactions between neighbouring loci, at all scales of observation. This constitutes a justification that TAD-like structural motifs indeed correspond to chromosome intervals. For

domains significantly larger than the resolution of the analysis ($\gtrsim$ 600 kb), a majority of the TADs [11] are recovered as interval-communities (Fig. 5) and, whatever the interval-community length, their borders present an insulator-like behaviour (Fig. 4) as expected for TAD-like structural motifs. Hence interval-communities capture similar structural organisation patterns as TADs but over the *full* chromosome range of scales.

This novel multi-scale structural decomposition of human chromosomes provides an original framework to question structural organisation and its relationship to functional regulation. It allowed us to reformulate the question of structural domain conservation between different cell lines across the scales: a high level of structural conservation between cell lines up to the largest scales becomes apparent. For example, $\sim$ 65% of the differentiated cell lines interval-communities larger than 600 kb were also found to be structural-communities in H1 ES cell line (Fig. 6a). It was previously noted that there likely exists some links between structural domains and replication domains [23, 25, 27, 49] including the so-called replication timing U-domains [24, 50]. U-domains are bordered by early replicating *master* replication initiation zones that present similar insulating properties as the ones observed for TADs and interval-communities borders (Fig. 4 and Additional file 1: Figure S9) [24]. In Human ES cells, master replication initiation zones are enriched in CTCF and pluripotent transcription factors NANOG and OCT4 that were recently shown to contribute to the overall folding of embryonic stem cells genome via specific long-range contacts [51, 52], and appear to be fundamental determinants of pluripotency maintenance [53, 54]. In particular they are at the heart of the so-called consolidation phenomenon [17, 23, 55, 56] corresponding to early to late transitions from embryonic stem cells to differentiated cells coinciding with the emergence of compact heterochromatin at the nuclear periphery [54]. ES cell line are characterised by smaller replication U-domains [24]. Here we observed in H1 ES cell line an excess of interval-communities in the range of scales from ∼ 500 kb to ∼ 1.5 Mb as compared to the differentiated cell lines (Fig. 3b). These domains not observed in differentiated cell lines might be subject to some structural consolidation scenario during cell differentiation, similar to the one described for replication timing domains. For example, the strutural community border present in H1 ES and absent in IMR90 at position ∼ 84 Mb in Fig. 1 correspond to a replication timing U-domain border specific of ES cell line. Further analysis of the structural consolidation scenario is likely to shed a new light on the role of structural organisation in the epigenetically regulated chromatin reorganisation that underlies the loss of pluripotency and lineage commitment [54]. It was shown that master origins of replication conserved between 6 cell lines are encoded in the DNA sequence via a local enrichment in nucleosome excluding energy barriers [57, 58]. This raises the question whether borders of the conserved structural community borders (Fig. 6) might be specified by a similar genetic mechanism.

A recent Hi-C experimental study at much higher (kb) resolution has provided some refined partitioning of the human genome by TADs of mean size ∼180 kb [20], much closer to the estimate ∼100-kb previously reported in *Drosophila* [16]. Interestingly, as in *Drosophila*, these refined TADs seem to have some specific epigenetic chromatin identity that can change dramatically their functional identity in different cell types [16, 20, 59]. Detecting interval-communities at higher resolution can provide better quantification of the chromatin state blocks as epigenetic communities. The wavelet-based community detection method provides us with a tool to investigate further the existence of some underlying rules for the association of structural/functional domains across scales. The robustness of the proposed protocol with respect to rearranged genomes is a key property to pursue this research.

Abbreviations
DNA: Deoxyribonucleic acid; Hi-C: High-throughput 3C; H1 ES: Embryonic stem cell line H1; TAD: Topologically associating domains; 3C: Chromatin conformation capture

Funding
This work was supported by Centre national de la Recherche Scientifique (CNRS), Ecole Normale Supérieure de Lyon (ENS de Lyon) and Agence National de la Recherche (ANR-14-CE27-0001 GRAPHSIP). BA acknowledges support from Science and Technology Commission of Shanghai Municipality (15520711500), Fondation pour la Recherche Médicale (DEI20151234404) and Agence National de la Recherche (ANR LightComb 2016-2018). The funding bodies had no role in the design of the study and collection, analysis, and interpretation of data and in writing the manuscript.

Authors' contributions
All authors designed the study. REB performed the analysis. NT and PB contributed to elaborate analysis tools. AA, PB and BA supervised the study. BA wrote the manuscript with contributions from all co-authors. All authors read and approved the final manuscript.

Competing interests
All authors declare that they have no competing interests.

Author details
[1] Univ Lyon, Ens de Lyon, Univ Claude Bernard Lyon 1, CNRS, Laboratoire de Physique, F-69342 Lyon, France. [2] Present address: Montpellier Cancer Institute (ICM), Montpellier Cancer Research Institute (IRCM) Inserm U1194, University of Montpellier, Montpellier, France. [3] Present address: CNRS, GIPSA-lab, Grenoble, France. [4] Present address: LOMA, Université de Bordeaux, CNRS, UMR 5798, 51 Cours de le Libération, 33405 Talence, France.

References
1. Cook PR. The organization of replication and transcription. Science. 1999;284(5421):1790–5.
2. Cremer T, Cremer C. Chromosome territories, nuclear architecture and gene regulation in mammalian cells. Nat Rev Genet. 2001;2:292–301.
3. Berezney R. Regulating the mammalian genome: the role of nuclear architecture. Adv Enzyme Regul. 2002;42:39–52.
4. Misteli T. Beyond the sequence: cellular organization of genome function. Cell. 2007;128(4):787–800. doi:10.1016/j.cell.2007.01.028.
5. Fraser P, Bickmore W. Nuclear organization of the genome and the potential for gene regulation. Nature. 2007;447(7143):413–7. doi:10.1038/nature05916.
6. Arneodo A, Vaillant C, Audit B, Argoul F, d'Aubenton-Carafa Y, Thermes C. Multi-scale coding of genomic information: From DNA sequence to genome structure and function. Phys Rep. 2011;498:45–188.
7. Luger K, Mäder AW, Richmond RK, Sargent DF, Richmond TJ. Crystal structure of the nucleosome core particle at 2.8 Å resolution. Nature. 1997;389(6648):251–60.
8. Dekker J, Rippe K, Dekker M, Kleckner N. Capturing chromosome conformation. Science. 2002;295:1306–11.
9. de Wit E, de Laat W. A decade of 3C technologies: insights into nuclear organization. Genes Dev. 2012;26(1):11–24. doi:10.1101/gad.179804.111.
10. Lieberman-Aiden E, van Berkum NL, Williams L, Imakaev M, Ragoczy T, Telling A, Amit I, Lajoie BR, Sabo PJ, Dorschner MO, Sandstrom R, Bernstein B, Bender MA, Groudine M, Gnirke A, Stamatoyannopoulos J, Mirny LA, Lander ES, Dekker J. Comprehensive mapping of long-range interactions reveals folding principles of the human genome. Science. 2009;326(5950):289–93. doi:10.1126/science.1181369.
11. Dixon JR, Selvaraj S, Yue F, Kim A, Li Y, Shen Y, Hu M, Liu JS, Ren B. Topological domains in mammalian genomes identified by analysis of chromatin interactions. Nature. 2012;485(7398):376–80. doi:10.1038/nature11082.
12. Hou C, Li L, Qin ZS, Corces VG. Gene density, transcription, and insulators contribute to the partition of the Drosophila genome into physical domains. Mol Cell. 2012;48(3):471–84. doi:10.1016/j.molcel.2012.08.031.

13. Kalhor R, Tjong H, Jayathilaka N, Alber F, Chen L. Genome architectures revealed by tethered chromosome conformation capture and population-based modeling. Nat Biotechnol. 2012;30(1):90–8. doi:10.1038/nbt.2057.

14. Moindrot B, Audit B, Klous P, Baker A, Thermes C, de Laat W, Bouvet P, Mongelard F, Arneodo A. 3D chromatin conformation correlates with replication timing and is conserved in resting cells. Nucleic Acids Res. 2012;40(19):9470–81. doi:10.1093/nar/gks736.

15. Nora EP, Lajoie BR, Schulz EG, Giorgetti L, Okamoto I, Servant N, Piolot T, van Berkum NL, Meisig J, Sedat J, Gribnau J, Barillot E, Bluthgen N, Dekker J, Heard E. Spatial partitioning of the regulatory landscape of the X-inactivation centre. Nature. 2012;485(7398):381–5. doi:10.1038/nature11049.

16. Sexton T, Yaffe E, Kenigsberg E, Bantignies F, Leblanc B, Hoichman M, Parrinello H, Tanay A, Cavalli G. Three-dimensional folding and functional organization principles of the Drosophila genome. Cell. 2012;148(3): 458–72. doi:10.1016/j.cell.2012.01.010.

17. Takebayashi SI, Dileep V, Ryba T, Dennis JH, Gilbert DM. Chromatin-interaction compartment switch at developmentally regulated chromosomal domains reveals an unusual principle of chromatin folding. Proc Natl Acad Sci USA. 2012;109(31):12574–9. doi:10.1073/pnas.1207185109.

18. Zhang Y, McCord RP, Ho YJ, Lajoie BR, Hildebrand DG, Simon AC, Becker MS, Alt FW, Dekker J. Spatial organization of the mouse genome and its role in recurrent chromosomal translocations. Cell. 2012;148(5): 908–21. doi:10.1016/j.cell.2012.02.002.

19. Naumova N, Imakaev M, Fudenberg G, Zhan Y, Lajoie BR, Mirny LA, Dekker J. Organization of the mitotic chromosome. Science. 2013;342(6161):948–53. doi:10.1126/science.1236083.

20. Rao SSP, Huntley MH, Durand NC, Stamenova EK, Bochkov ID, Robinson JT, Sanborn AL, Machol I, Omer AD, Lander ES, Aiden EL. A 3D map of the human genome at kilobase resolution reveals principles of chromatin looping. Cell. 2014;159(7):1665–80. doi:10.1016/j.cell.2014.11.021.

21. Fudenberg G, Mirny LA. Higher-order chromatin structure: bridging physics and biology. Curr Opin Genet Dev. 2012;22(2):115–24. doi:10.1016/j.gde.2012.01.006.

22. Gibcus JH, Dekker J. The hierarchy of the 3D genome. Mol Cell. 2013;49(5):773–82. doi:10.1016/j.molcel.2013.02.011.

23. Ryba T, Hiratani I, Lu J, Itoh M, Kulik M, Zhang J, Schulz TC, Robins AJ, Dalton S, Gilbert DM. Evolutionarily conserved replication timing profiles predict long-range chromatin interactions and distinguish closely related cell types. Genome Res. 2010;20(6):761–70. doi:10.1101/gr.099655.109.

24. Baker A, Audit B, Chen CL, Moindrot B, Leleu A, Guilbaud G, Rappailles A, Vaillant C, Goldar A, Mongelard F, d'Aubenton-Carafa Y, Hyrien O, Thermes C, Arneodo A. Replication fork polarity gradients revealed by megabase-sized U-shaped replication timing domains in human cell lines. PLoS Comput Biol. 2012;8(4):1002443. doi:10.1371/journal.pcbi.1002443.

25. Boulos RE, Julienne H, Baker A, Chen CL, Petryk N, Kahli M, d'Aubenton-Carafa Y, Goldar A, Jensen P, Hyrien O, Thermes C, Arneodo A, Audit B. From the chromatin interaction network to the organization of the human genome into replication N/U-domains. New J Phys. 2014;16:115014.

26. Le Dily F, Baù D, Pohl A, Vicent GP, Serra F, Soronellas D, Castellano G, Wright RHG, Ballare C, Filion G, Marti-Renom MA, Beato M. Distinct structural transitions of chromatin topological domains correlate with coordinated hormone-induced gene regulation. Genes Dev. 2014;28(19): 2151–62. doi:10.1101/gad.241422.114.

27. Pope BD, Ryba T, Dileep V, Yue F, Wu W, Denas O, Vera DL, Wang Y, Hansen RS, Canfield TK, Thurman RE, Cheng Y, Gulsoy G, Dennis JH, Snyder MP, Stamatoyannopoulos JA, Taylor J, Hardison RC, Kahveci T, Ren B, Gilbert DM. Topologically associating domains are stable units of replication-timing regulation. Nature. 2014;515(7527):402–5. doi:10.1038/nature13986.

28. Chen H, Chen J, Muir LA, Ronquist S, Meixner W, Ljungman M, Ried T, Smale S, Rajapakse I. Functional organization of the human 4D nucleome. Proc Natl Acad Sci USA. 2015;112(26):8002–7. doi:10.1073/pnas.1505822112.

29. Fraser J, Ferrai C, Chiariello AM, Schueler M, Rito T, Laudanno G, Barbieri M, Moore BL, Kraemer DCA, Aitken S, Xie SQ, Morris KJ, Itoh M, Kawaji H, Jaeger I, Hayashizaki Y, Carninci P, Forrest ARR, FANTOM Consortium, Semple CA, Dostie J, Pombo A, Nicodemi M. Hierarchical

30. folding and reorganization of chromosomes are linked to transcriptional changes in cellular differentiation. Mol Syst Biol. 2015;11(12):852.

30. Liu L, Zhang Y, Feng J, Zheng N, Yin J, Zhang Y. GeSICA: genome segmentation from intra-chromosomal associations. BMC Genomics. 2012;13:164. doi:10.1186/1471-2164-13-164.

31. Filippova D, Patro R, Duggal G, Kingsford C. Identification of alternative topological domains in chromatin. Algorithms Mol Biol. 2014;9:14. doi:10.1186/1748-7188-9-14.

32. Lévy-Leduc C, Delattre M, Mary-Huard T, Robin S. Two-dimensional segmentation for analyzing Hi-C data. Bioinformatics. 2014;30(17):386–92. doi:10.1093/bioinformatics/btu443.

33. Cabreros I, Abbe E, Tsirigos A. Detecting Community Structures in Hi-C Genomic Data. 2015. arXiv:1509.05121 [q-bio.GN]. http://arxiv.org/abs/1509.05121.

34. Chen J, 3rd Hero AO, Rajapakse I. Spectral identification of topological domains. Bioinformatics. 2016;32(14):2151–158. doi:10.1093/bioinformatics/btw221.

35. Negrini S, Gorgoulis VG, Halazonetis TD. Genomic instability–an evolving hallmark of cancer. Nat Rev Mol Cell Biol. 2010;11(3):220–8. doi:10.1038/nrm2858.

36. Botta M, Haider S, Leung IXY, Lio P, Mozziconacci J. Intra- and inter-chromosomal interactions correlate with CTCF binding genome wide. Mol Syst Biol. 2010;6:426. doi:10.1038/msb.2010.79.

37. Sandhu KS, Li G, Poh HM, Quek YLK, Sia YY, Peh SQ, Mulawadi FH, Lim J, Sikic M, Menghi F, Thalamuthu A, Sung WK, Ruan X, Fullwood MJ, Liu E, Csermely P, Ruan Y. Large-scale functional organization of long-range chromatin interaction networks. Cell Rep. 2012;2(5):1207–19. doi:10.1016/j.celrep.2012.09.022.

38. Boulos RE, Arneodo A, Jensen P, Audit B. Revealing long-range interconnected hubs in human chromatin interaction data using graph theory. Phys Rev Lett. 2013;111:118102.

39. Fortunato S. Community detection in graphs. Phys Rep. 2010;486:75–174.

40. Tremblay N, Borgnat P. Graph wavelets for multiscale community mining. IEEE Trans Signal Process. 2014;62(20):5227–39. doi:10.1109/TSP.2014.2345355.

41. Hammond DK, Vandergheynst P, Gribonval R. Wavelets on graphs via spectral graph theory. Appl Comput Harmon Anal. 2011;30:129–50.

42. King B. Step-wise clustering procedures. J Amer Statist Assoc. 1967;62(317):86–101.

43. Jain A, Murty M, Flynn P. Data clustering: A review. ACM Comput Surv (CSUR). 1999;31(3):264–323.

44. Tremblay N, Puy G, Borgnat P, Gribonval R, Vandergheynst P. Accelerated spectral clustering using graph filtering of random signals. In: Proceedings of the IEEE International Conference on Acoustics, Speech, and Signal Processing, ICASSP, 20-25 March 2016. Shangai: IEEE; 2016. doi:10.1109/ICASSP.2016.7472447.

45. Tremblay N, Puy G, Gribonval R, Vandergheynst P. Compressive spectral clustering. In: Proceedings of the Thirty-third International Conference on Machine Learning (ICML 2016), New York. JMLR Workshop and Conference Proceedings 2016;48:1002–11.

46. Sexton T, Cavalli G. The role of chromosome domains in shaping the functional genome. Cell. 2015;160(6):1049–59. doi:10.1016/j.cell.2015.02.040.

47. Kaplan N, Dekker J. High-throughput genome scaffolding from in vivo DNA interaction frequency. Nat Biotechnol. 2013;31(12):1143–47. doi:10.1038/nbt.2768.

48. Marie-Nelly H, Marbouty M, Cournac A, Flot JF, Liti G, Parodi DP, Syan S, Guillén N, Margeot A, Zimmer C, Koszul R. High-quality genome (re)assembly using chromosomal contact data. Nat Commun. 2014;5: 5695. doi:10.1038/ncomms6695.

49. Boulos RE, Drillon G, Argoul F, Arneodo A, Audit B. Structural organization of human replication timing domains. FEBS Lett. 2015;589(20 Pt A):2944–57. doi:10.1016/j.febslet.2015.04.015.

50. Audit B, Baker A, Chen CL, Rappailles A, Guilbaud G, Julienne H, Goldar A, d'Aubenton-Carafa Y, Hyrien O, Thermes C, Arneodo A. Multiscale analysis of genome-wide replication timing profiles using a wavelet-based signal-processing algorithm. Nat Protoc. 2013;8(1):98–110. doi:10.1038/nprot.2012.145.

51. Denholtz M, Bonora G, Chronis C, Splinter E, de Laat W, Ernst J, Pellegrini M, Plath K. Long-range chromatin contacts in embryonic stem cells reveal a role for pluripotency factors and polycomb proteins in

genome organization. Cell Stem Cell. 2013;13(5):602–16.
doi:10.1016/j.stem.2013.08.013.

52. de Wit E, Bouwman BAM, Zhu Y, Klous P, Splinter E, Verstegen MJAM,
Krijger PHL, Festuccia N, Nora EP, Welling M, Heard E, Geijsen N, Poot
RA, Chambers I, de Laat W. The pluripotent genome in three dimensions
is shaped around pluripotency factors. Nature. 2013;501(7466):227–31.
doi:10.1038/nature12420.

53. Julienne H, Zoufir A, Audit B, Arneodo A. Epigenetic regulation of the
human genome: coherence between promoter activity and large-scale
chromatin environment. Front Life Sci. 2013;7(1-2):44–62.
doi:10.1080/21553769.2013.832706.

54. Julienne H, Audit B, Arneodo A. Embryonic stem cell specific "master"
replication origins at the heart of the loss of pluripotency. PLoS Comput
Biol. 2015;11(2):1003969. doi:10.1371/journal.pcbi.1003969.

55. Hiratani I, Ryba T, Itoh M, Yokochi T, Schwaiger M, Chang CW, Lyou Y,
Townes TM, Schubeler D, Gilbert DM. Global reorganization of
replication domains during embryonic stem cell differentiation. PLoS Biol.
2008;6(10):245. doi:10.1371/journal.pbio.0060245.

56. Hiratani I, Ryba T, Itoh M, Rathjen J, Kulik M, Papp B, Fussner E,
Bazett-Jones DP, Plath K, Dalton S, Rathjen PD, Gilbert DM.
Genome-wide dynamics of replication timing revealed by in vitro models
of mouse embryogenesis. Genome Res. 2010;20(2):155–69.
doi:10.1101/gr.099796.109.

57. Drillon G, Audit B, Argoul F, Arneodo A. Ubiquitous human 'master'
origins of replication are encoded in the DNA sequence via a local
enrichment in nucleosome excluding energy barriers. J Phys Condens
Matter. 2015;27(6):064102. doi:10.1088/0953-8984/27/6/064102.

58. Drillon G, Audit B, Argoul F, Arneodo A. Evidence of selection for an
accessible nucleosomal array in human. BMC Genomics. 2016;17:526.

59. Ciabrelli F, Cavalli G. Chromatin-driven behavior of topologically
associating domains. J Mol Biol. 2015;427:248–58.
doi:10.1016/j.jmb.2014.09.013.

4

AnkPlex: algorithmic structure for refinement of near-native ankyrin-protein docking

Tanchanok Wisitponchai[1,2†], Watshara Shoombuatong[3†], Vannajan Sanghiran Lee[4,5], Kuntida Kitidee[2,6*] and Chatchai Tayapiwatana[1,2*]

Abstract

Background: Computational analysis of protein-protein interaction provided the crucial information to increase the binding affinity without a change in basic conformation. Several docking programs were used to predict the near-native poses of the protein-protein complex in 10 top-rankings. The universal criteria for discriminating the near-native pose are not available since there are several classes of recognition protein. Currently, the explicit criteria for identifying the near-native pose of ankyrin-protein complexes (APKs) have not been reported yet.

Results: In this study, we established an ensemble computational model for discriminating the near-native docking pose of APKs named "AnkPlex". A dataset of APKs was generated from seven X-ray APKs, which consisted of 3 internal domains, using the reliable docking tool ZDOCK. The dataset was composed of 669 and 44,334 near-native and non-near-native poses, respectively, and it was used to generate eleven informative features. Subsequently, a re-scoring rank was generated by AnkPlex using a combination of a decision tree algorithm and logistic regression. AnkPlex achieved superior efficiency with ≥ 1 near-native complexes in the 10 top-rankings for nine X-ray complexes compared to ZDOCK, which only obtained six X-ray complexes. In addition, feature analysis demonstrated that the van der Waals feature was the dominant near-native pose out of the potential ankyrin-protein docking poses.

Conclusion: The AnkPlex model achieved a success at predicting near-native docking poses and led to the discovery of informative characteristics that could further improve our understanding of the ankyrin-protein complex. Our computational study could be useful for predicting the near-native poses of binding proteins and desired targets, especially for ankyrin-protein complexes. The AnkPlex web server is freely accessible at http://ankplex.ams.cmu.ac.th.

Keywords: Ankyrin-protein complexes, Near-native docking pose, Machine learning methods, Decision tree, Logistic regression model, AnkPlex

* Correspondence: kitidee_010@hotmail.com; asimi002@hotmail.com
†Equal contributors
2Center of Biomolecular Therapy and Diagnostic, Faculty of Associated Medical Sciences, Chiang Mai University, Chiang Mai 50200, Thailand
1Division of Clinical Immunology, Department of Medical Technology, Faculty of Associated Medical Sciences, Chiang Mai University, Chiang Mai 50200, Thailand
Full list of author information is available at the end of the article

Background

Generally, antibodies have several applications in therapies and diagnostics due to the fact that they can be designed to have high affinity with a targeted protein [1, 2]. Due to the complications involved in generating specific antibodies and their large size, alternative scaffolds have been developed to overcome these limitations. One of those novel scaffolds is comprised of Designed Ankyrin Repeat Proteins (DARPins). These ankyrin-proteins have been used more frequently in medical applications [3–5] because of their stability and high affinity for protein targets [6–8]. Moreover, modification of the residues at the variable part of ankyrin allows for increased binding affinity towards the target protein without changes in the basic protein conformation [9]. The high affinity of ankyrin-proteins could be achieved due to random modifications at variable residues *in vitro* [9] and *in silico* prediction of the residues based on the structure of 3-dimensional (3D) complexes [4, 10]. The 3D protein complexes could be determined by X-ray crystallography or NMR spectroscopy, yet few 3D structures of ankyrin-protein complexes have been reported. Most of the structures were monomeric structures or genomics surveys. Therefore, a computational approach, called protein–protein docking, can be used to generate protein complex structures because there were no available reports on the protein complex.

Protein–protein docking is a well-known method for generating protein–protein complexes (poses) using computational methods. The challenging task of identifying the exact bound state of a pair of proteins must consider the following factors: (i) there are several potential ways that a pair of proteins can interact, (ii) the flexibility of the protein, and (iii) changes in the protein conformation after binding [11]. Currently, several software programs have been developed, such as Gramm-X, DOT, ClusPro, and ZDOCK, that provide a rational complex for a pair of proteins, [12]. The ZDOCK program includes initial-stage docking (ZDOCK algorithm) and refinement methods (RDOCK algorithm). The initial-stage docking is designed for searching all possible docking poses [13, 14]. In the refinement stage, the side chains of the docking poses from the ZDOCK algorithm are minimized [15]. The scoring functions (features) of the docking poses are energy terms, such as pairwise shape complementarity (PSC), desolvation (DE), electrostatics (ELEC), and van der Waals. This program has been demonstrated to be one of the most accurate prediction programs in the Critical Assessment of Predicted Interactions (CAPRI) [16].

ZDOCK has successfully predicted several near-native complexes (poses) of antibody-antigen, enzyme-inhibitor and other pairings *via* assessing the CAPRI criteria in the 10 top-rankings based on the features of ZDock, ZRank, or E_RDock. However, successful predictions do not occur for all cases [17–20]. Moreover, the near-native predictors are not selected from an easy ranking of those features, and manual inspections are often needed as well. Note that manual inspections include cluster, density, favourable contact, charge complementarity, buried hydrophobic residues, and overall agreement with the biological data in the literature. Importantly, all protein–protein cases do not agree with the manual inspections. Similar to other reports, the complex-type-dependent combinatorial scoring function was introduced and indicated that the weights of the scoring function were different between protease-inhibitor, antibody-antigen, and enzyme-inhibitor pairings [21]. Therefore, a complicated strategy has to be adopted for obtaining a near-native complex based on certain types of protein–protein complexes.

The near-native docking pose of Ankyrin-Her2 was successfully predicted using ZDOCK and an extra scoring function [10]. Recently, the universal criteria for obtaining the near-native complex of ankyrin-proteins have not been reported, and there was only a computational method that was applied to identify the repeat number of ankyrin-proteins [22]. According to different types of protein-protein complexes, the ankyrin-protein complex requires an individual strategy. Therefore, we aimed to search for explicit criteria to obtain a near-native pose using a set of features generated from one program to avoid using complicated methods or combining scores from several software programs.

In this study, we made a systematic attempt to develop a computational approach for achieving near-native predictors in 10 top-rankings of ankyrin-protein docking poses, which we named AnkPlex. Moreover, this method was generated for (i) analysing and characterizing ankyrin-protein complexes by using a set of informative features that have potential applications and (ii) establishing a user-friendly web server to obtain the desired results without the need to follow complicated mathematical equations generated by the research scientist. The docking poses of seven X-ray complexes of APKs, which had ankyrins with 3 internal domains, were generated using the reliable docking tool ZDOCK. The construction of the docking poses calculated by PSC alone and summation of PSC + DE + ELEC demonstrated there were different numbers of near-native docking poses. The steps for AnkPlex establishment included (i) balancing the near-native and non-near-native poses; (ii) processing the dataset through machine learning of a decision tree algorithm (DT) and a logistic regression (LG) with a combination of 11 features; (iii) selecting the efficient predictive models of DT and LG; and (iv) processing the dataset by combining models of DT and LG.

Method

Datasets

X-ray crystal structures of ankyrin-protein complexes (APKs) were collected from the *Protein Data Bank* (PDB) database for 41 APKs reported up to May 2014. Analyses of the 41 APKs were performed through data pre-processing using the following steps: (i) APKs containing 3-internal-domain were included; (ii) redundant APKs were excluded; (iii) APKs were filter based on the recognition areas [7]; and (iv) alpha, beta, and alpha–beta proteins were selected using the SCOP database [23]. Nine X-ray crystal structures of APKs (called Ank9) were obtained, as summarized in Additional file 1: Table S1. Subsequently, seven of the APKs were randomly selected as training complexes (Ank-TRN), including complex 1 (C_1), complex 2 (C_2), complex 3 (C_3), complex 4 (C_4), complex 5 (C_5), complex 6 (C_6), and complex 7 (C_7). At the same time, the rest of the APKs, including unknown 1 (U_1) and unknown 2 (U_2), were designated the test group (Ank-TEST). In order to avoid the distinct results from the different selections of training and test sets, other 35 possible datasets were constructed and were used to generate the predictive models for the identification of near-native poses.

The docking poses of Ank-TRN and Ank-TEST were regenerated by using the protein docking software ZDOCK [13, 14]. Two versions of the docking poses were generated, which were different in terms of energy calculations (especially PSC) and the combination of PSC, DE, and ELEC (PSC + DE + ELEC). Then, all the generated-docking poses were superimposed with the original X-ray crystal structures and were calculated for the root-mean-square deviation of the Cα atom (Cα-RMSD) value. The docking poses that presented Cα-RMSD values ≤10 Å were designated to be near-native poses or positive samples, whereas the docking poses that presented Cα-RMSD values >10 Å were defined as non-near-native poses or negative samples [24]. The numbers of near-native poses for the two versions of the docking poses were compared. In addition to screening near-native poses by the Cα-RMSD value, eight binding residues of the APKs on the second domain of ankyrin (Fig. 1) were used for filtering near-native poses based on the recognition areas (regKp).

Feature extraction

Based on observations of the generation of the features, ankyrin-protein docking poses were generated for the energy features using the ZDOCK protocol [13, 14] (a set of 5 features) and the RDOCK protocol [15] (a set of 6 features). Five features, including ZDock, ZRankElec, ZRnakSolv, ZRank, and ZRankVdw, were obtained from the protein-docking protocol (ZDOCK) using the CHARMm force field [25]. At the same time, six features, including E_vdw1, E_elec1, E_vdw2, E_elec2, E_sol, and E_RDock, were calculated from the docking refinement protocol (RDOCK) using the CHARMm polar H force field [25]. The energy equation used in RDOCK was the same as ZDOC. However, the ankyrin-protein

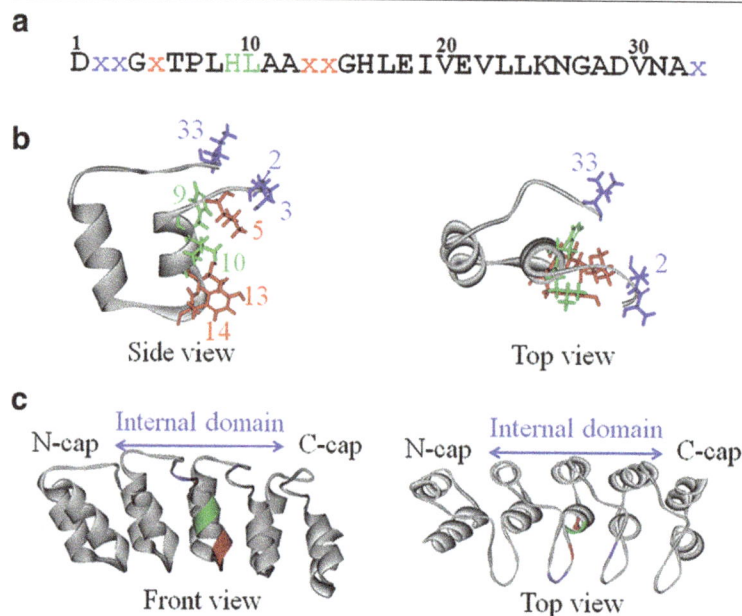

Fig. 1 The molecular architecture of ankyrin and its three recognition areas, as shown in ribbon style. **a** Amino acid sequence of an internal repeat [7] in which the recognition residues are shown in three colours. **b** The ribbon style of an internal repeat of ankyrin related to the above sequence. **c** The structure of the 3 internal domains of ankyrin flanked by the N-cap and C-cap. The recognition area consisted of six variable residues [7] (*red* and *blue* are positioned on the helix and turn, respectively) and two constant amino acids (*green*) on the second domain

docking poses were minimized before calculation. The details of the 11 features (A, B, C, D, E, F, G, H, I, J, K) are described below:

- ZDock (A) is the Pairwise Shape Complementarity (PSC) score and it was optionally augmented with the electrostatics (ELEC) and the desolvation energy (DE). In this study, the *ZDock* score was calculated using the following equation:

$$ZDock\ score = \alpha PSC + DE + \beta ELEC \tag{1}$$

where α and β have the default values of 0.01 and 0.06, respectively.

- ZRankElec (B) is the long-range electrostatic energy and the only fully charged side-chain, as represented in the following equation:

$$ZRankElec(i,j) = 332\frac{q_i q_j}{r^2_{ij}} \tag{2}$$

where q_i and q_j are the charges on ankyrin and the protein atoms, respectively. The r_{ij} in the equation stands for the distance between the atoms of ankyrin and the protein.

- ZRankSolv (C) is the desolvation term based on the Atomic Contact Energy (ACE).

$$ZRankSolv(i,j) = a_{ij} \tag{3}$$

where a_{ij} is the ACE score.

- ZRank (D) is a linear combination of ZRankVdw, ZRankElec, and ZRankElec.

$$ZRank\ score = ZRankElec + ZRankSolv + ZRankVdw \tag{4}$$

- ZRankVdw (E) is the van der Waals and short-range electrostatics energy with a distance between the atom pair being less than 5.0 Å. This calculation was based on the parameters of the CHARMm 19 polar hydrogen potential. The *ZRankVdw* score was calculated as follows:

$$ZRankVdw(i,j) = \varepsilon_{ij}\left[\left[\frac{\sigma_{ij}}{r_{ij}}\right]^{12} - 2\left[\frac{\sigma_{ij}}{r_{ij}}\right]^{6}\right] \tag{5}$$

where ε_{ij} and σ_{ij} are the depth and the width, respectively, of the coefficient for the CHARMm 19 polar H.

- E_vdw1 (F) and E_vdw2 (H) are the van der Waals energy, as presented in Equation (5), of the 1st and the 2nd minimized structure of the ankyrin-protein docking poses, respectively.
- E_elec1 (G) and E_elec2 (I) are the electrostatic energy, as presented in Equation (2), of the ankyrin-protein docking poses processed for the 1st and the 2nd minimization, respectively.
- E_sol (J) is the desolvation energy, as shown in Equation (4), of the 2nd minimization of the ankyrin-protein docking poses.
- E_RDock (K) is the summation of E_sol and (0.9 × E_elec2).

Construction of learning method

Several learning models were constructed including decision tree (DT), logistic regression (LG), artificial neural network (ANN), and support vector machine (SVM) using Ank-TRN (C_1-C_7). As shown in Additional file A1: Table S5, SVM yielded 100% near-native poses in the internal testing sets but could not obtain any near-native pose in the external testing sets. The DT and ANN provided the near-native poses from both internal and external testing sets. According to a dataset of C_5, the DT was superior in achieving the near-native poses of internal testing sets than the ANN. The LG provided a weighted summation that could rank the docking poses to achieve the near-native poses in the 10 top-rankings. As a consequence, the DT and the LG were selected to construct an ensemble model.

To identify the near-native docking poses of APKs, a learning method named AnkPlex was established by combining a decision tree (DT) and a logistic method (LG). The decision trees and the logistic regression methods were selected due to the fact that they provide a high number of predicted positive values (true near-native poses). The logistic regression especially provided a weighted summation that was finally ranked to search for near-native poses in 10 top-rankings. All 11 features and all datasets were used to build the DT and LG models. The Ank-TRN (7 APKs) and the Ank-TEST (2 APKs) were evaluated by AnkPlex using the following steps, as shown Fig. 2:

1. The number of near-native poses and non-near-native poses were balanced. ZDOCK using PSC + DE + ELEC and regKp provided 699 near-native poses and 44,334 non-near-native poses of Ank-TRN. The non-near-native poses were randomly clustered into 65 groups (≈44,334/669). Therefore, each training set was composed of the same near-native poses and different groups of non-near-native poses.

2. A predictive model using the DT and the LG models was established. All 11 features were combined and generated as feature subsets (i.e., A, B, C, D, E, F, G, H, I, J,

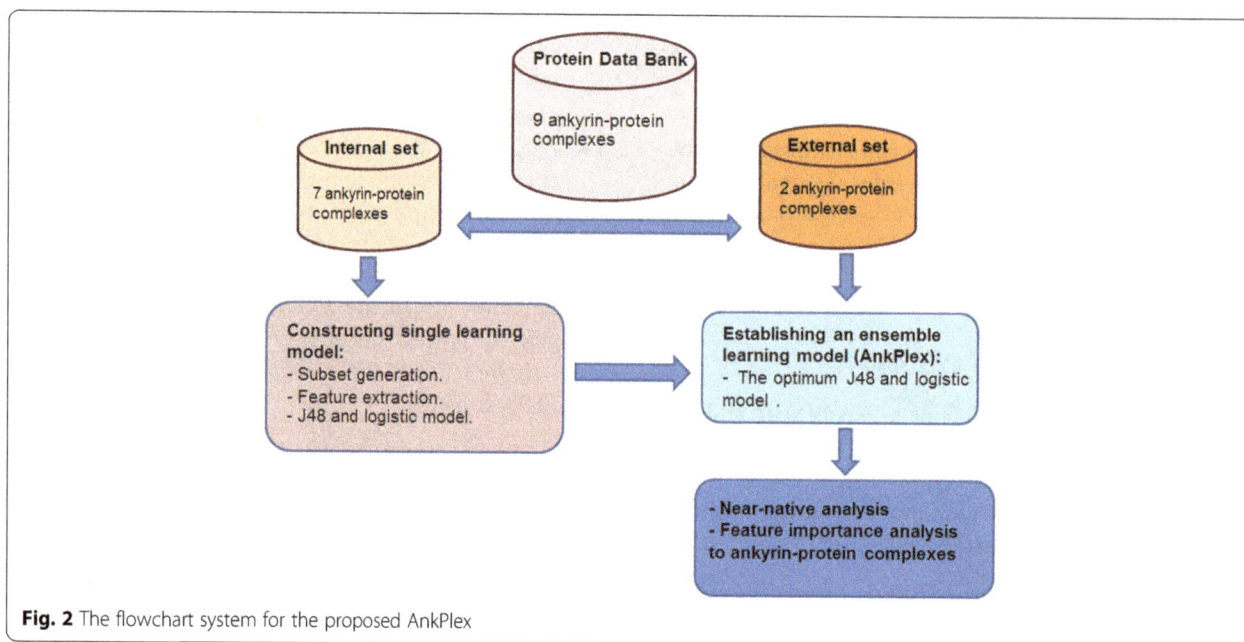

Fig. 2 The flowchart system for the proposed AnkPlex

K, AB, AC, AD, AE, AF, AG, AH, AI, AJ, AK, BC, BD, ..., ABCDEFG). The total number of feature subsets was calculated to be 4,095 by following this equation:

$$L = \sum_{r=1}^{11} \frac{11!}{r!(11-r)!}. \tag{6}$$

The DT model was established from 4,095 feature subsets and 65 training sets using the J48 algorithm [26, 27]. The parameters of the DT model were set with the confidence factor, the minimum number of objects, and the number of folds for reduced error pruning of 0.25, 2, and 3, respectively. Additionally, the LG model was constructed from the same feature subsets and training set with a ridge estimator [28] in which the maxis and the ridge were defined as −1 and 1.oE-8, respectively. Subsequently, the learning methods were generated by implementation of the DT and the LG models using the WEKA program [27].

3. An efficient predictive model of the DT and the LG models was selected. Ank-TRN consisted of 7 APKs and was submitted to the learning method for predicting the near-native poses. True positive rates (TPrate) greater than 50% were used as the cut-off value for an efficient learning method. The learning methods that demonstrated a TPrate greater than 50% were selected to further establish an ensemble learning model.

4. Ensemble methods were established. The ensemble learning method, named AnkPlex, was constructed by randomly integrating the DT-based learning models (OLM$_{DT}$) and the LG-based learning models (OLM$_{LG}$) from Step 3 for reducing the number of non-near-native docking poses. The main process of the proposed

method, AnkPlex, for increasing the number of TPs (reducing non-near-native poses) consisted of the following steps: (i) only predicted positive samples (PPV$_{DT}$) derived from OLM$_{DT}$ were select, (ii) a logistic score (LGS) on PPV$_{DT}$ using the LG model was calculated, and (iii) PPV$_{DT}$ was ranked according to LGS and the 10 top-ranking poses that demonstrated the highest LGS were selected. The near-native pose(s) or the true positive (TP) in the 10 top-ranking poses were our targets. The summation score of AnkPlex was defined in the equation given as Equation (7) on C_i, where $i = 1, 2, ..., 7$, and Y_i would be set as 1 in case TP was found in the 10 top-ranking poses. Otherwise, Y_i would be set as 0. Finally, the score of AnkPlex was the summation product, as defined in the following equation:

$$\# PP = \sum_{i=1}^{7} Y_i \tag{7}$$

where $\# PP$ belongs to Ank-TRN containing seven complexes ($C_1, C_2, ..., C_7$) and Y_i would be set as 1 when TP appears in the 10 top-ranking poses. Otherwise, Y_i would be set as 0. The number of #PP$_{TRN}$ indicated the sample of Ank-TRN in which the LGS score was among the 10 top-ranking near-native poses. The number of #PP$_{TEST}$ showed the LGS score of the Ank-TEST was among the 10 top-ranking poses.

Validation

The prediction performance of the AnkPlex method was evaluated by using 10-fold cross-validation (10-fold CV). The method validation parameters, including accuracy

(ACC), sensitivity (SEN), and precision (PRES), were calculated using the following equations:

$$\text{Accuracy} = \frac{TP + TN}{(TP + TN + FP + FN)} \times 100 \qquad (8)$$

$$\text{Sensitivity} = \frac{TP}{(TP + FN)} \times 100 \qquad (9)$$

$$\text{Precision} = \frac{TP}{(TP + FP)} \times 100 \qquad (10)$$

where TP, TN, FP, and FN are the numbers of true positive, true negative, false positive, and false negative results, respectively.

Results and discussion

Analysis of ankyrin-protein docking dataset

There were a few ankyrin-protein complexes (APKs) reported in the PDB database. Forty-one ankyrin complexes, with the number of internal domains ranging from 2–7, have been reported (up to May, 2014). The highest number, 19 complexes, of ankyrin-proteins contained 3 internal domains. Furthermore, the APKs that reacted with the target using recognition areas were selected. Focusing on target proteins, only proteins with common folding structures, i.e., alpha-, beta- and alpha-beta structures, were considered. Therefore, nine complexes, which included 1SVX, 4ATZ, 3Q9N, 1AWC, 2BKK, 2Y1L, 4DRX, 2P2C, and 4HNA, were used in this study. These nine complexes were randomly divided into two groups, i.e., 7 complexes as Ank-TRN and 2 complexes as Ank-TEST.

To optimize the ZDOCK calculation, X-ray crystal structures of APK-TRN, including seven APKs, were calculated with different feature calculations, including the PSC and the PSC + DE + ELEC. The total number of docking poses, including near-native and non-near-native poses, was 54,000 poses (54Kp). Subsequently, the numbers of near-native poses calculated by PSC and PSC + DE + ELEC were compared. As shown in Table 1, the average number of near-native poses calculated by PSC + DE + ELEC (116.57 ± 51.05) was twice as high as the number calculated using PSC (63.29 ± 41.43). To increase the predictive accuracy, binding sites on the second domain of Ank-TRN defined by Bintz et al. [7] were used for filtering near-native poses based on the recognition areas (regKp). The number of regKp calculated by PSC + DE + ELEC was observed to be slightly reduced (95.57 ± 52.58). According to the ZDOCK program suggestion, the near-native poses were identified in the top 2,000 poses (2Kp) ranked by the ZRank feature [13, 14]. The 2Kp were selected from the total docking poses of regKp compared to the near-native poses from regKp. The near-native poses of 2Kp (47.14 ± 31.49)

Table 1 Number of docking poses classified as near-native and non-near-native in Ank-TRN (C_1-C_7) and Ank-TEST (U_1 and U_2)

Complex	Near-native				Non-near-native	
	PSC[a]	PSC + DE + ELEC[b]			PSC + DE + ELEC[b]	
	54Kp	54Kp	regKp	2Kp	54Kp	regKp
C_1	81	157	136	83	53,843	6,626
C_2	4	72	53	19	53,921	4,865
C_3	125	194	179	54	53,806	7,055
C_4	83	131	104	74	53,869	4,450
C_5	31	101	59	8	53,891	8,565
C_6	84	42	28	18	53,958	6,177
C_7	35	119	110	74	53,881	6,596
Total	443	816	669	330	384,169	44,334
Mean	63.29	116.57	95.57	47.14	53,881	6,333.43
std.	41.43	51.05	52.58	31.49	49.74	1,377.39
U_1	ND[c]	83	83	57	ND[c]	6,833
U_2	ND[c]	183	32	13	ND[c]	7,183

[a]ZDOCK was calculated by PSC alone. [b]ZDOCK was calculated by combining PSC, DE, and ELEC. [c]The data were not used for analysis

substantially decreased two-fold compared to regKp. Thus, it can be concluded that 2Kp ranked by the ZRank feature was not suitable for screening near-native poses because of the exclusion of some near-native poses. Interestingly, screening by regKp resulted in a high number of near-native poses and an extremely reduced number of non-near-native poses. The results suggested that the ZDOCK calculation using PSC + DE + ELEC and the screening based on the recognition areas (regKp) were the optimal calculations because this procedure was capable of incorporating near-native poses and eliminating non-near-native poses. However, the number of non-near-native poses generated with regKp still remained high, which indicated that an alternative learning method is necessary for ruling out non-near-native poses.

Establishing learning methods

According to the ZDOCK calculations of Ank-TRN, 11 features of near-native poses and non-near-native poses were generated. Univariate statistical approaches were employed to perform exploratory data analysis using average and standard deviations for summarizing important patterns. As shown in Table 2, five features that were generated by the ZDOCK protocol demonstrated the significant differences between the near-native poses and the non-near-native poses with a *p-value* <0.001. As presented in Table 2, the five top-ranked features included E_RDock (−11.96 ± 9.40/1.72 ± 12.10), ZRankElec (8.22 ± 16.56/29.17 ± 22.41), ZRank (−54.03 ± 25.84/ −21.82 ± 31.33), E_elec2 (−18.21 ± 8.85/−8.18 ± 10.25), and E_sol (4.43 ± 6.26/9.07 ± 8.87). Almost all the

Table 2 Summary of statistical analysis of near-native and non-near-native poses of ankyrin-target complexes

Feature	Near-native	Non-near-native	p-value
ZDock	36.73 ± 6.32	33.41 ± 4.87	<0.001
ZRankElec	8.22 ± 16.56	29.17 ± 22.41	<0.001
ZRank	−54.03 ± 25.84	−21.82 ± 31.33	<0.001
ZRankSolv	3.66 ± 6.99	7.88 ± 10.10	<0.001
ZRankVdw	−65.91 ± 20.44	−58.87 ± 20.98	<0.001
E_vdw1	−56.21 ± 57.81	−47.46 ± 101.79	<0.001
E_elec1	-1.05 ± 2.18	-0.94 ± 2.37	0.18
E_vdw2	−70.01 ± 42.72	−74.33 ± 43.00	0.01
E_elec2	−18.21 ± 8.85	−8.18 ± 10.25	<0.001
E_sol	4.43 ± 6.26	9.07 ± 8.87	<0.001
E_RDock	−11.96 ± 9.40	1.72 ± 12.10	<0.001

Table 3 Comparison of performances of 10 top-ranking OLM_{DT} among various types of features and datasets in terms of 10-fold cross-validation

Rank	OLM_{DT}	PRES(%)	REC(%)	ACC(%)	$\#PP_{TRN}$
1	ABDEHIJK_g14	82.10 ± 8.49	70.58 ± 15.03	6.96 ± 4.82	6
2	ADEFHIK_g25	81.64 ± 9.08	72.05 ± 14.65	7.05 ± 5.06	6
3	ABDEGHIJK_g14	81.14 ± 8.83	73.66 ± 16.61	6.83 ± 4.68	6
4	BCEGHIJK_g13	81.04 ± 8.37	72.88 ± 14.37	6.03 ± 3.31	6
5	AEFGHIJK_g36	81.03 ± 8.71	73.85 ± 13.05	6.91 ± 5.17	6
6	ABFIJK_g16	80.84 ± 8.48	69.17 ± 14.68	6.33 ± 4.32	6
7	ABDEFGIJK_g25	80.84 ± 7.98	75.54 ± 8.88	6.41 ± 3.80	6
8	ABIJK_g16	80.82 ± 8.43	69.17 ± 14.68	6.31 ± 4.31	6
9	ABCGHJK_g13	80.80 ± 7.86	73.55 ± 12.44	6.30 ± 3.98	6
10	ABDEFGIK_g25	80.78 ± 8.09	75.67 ± 9.12	6.38 ± 3.73	6

11 feathers (A, B, C, D, E, F, G, H, I, J, K) are ZDock, ZRankElec, ZRank, ZRankSolv, ZRankVdw, E_vdw1, E_elec1, E_vdw2, E_elec2, E_sol, E_RDock

features that were calculated using the RDOCK protocol were significantly different, except E_elec1 ($p = 0.178$) and E_vdw2 ($p = 0.010$). Subsequently, 11 features calculated with the ZDOCK calculation were applied to establish the learning methods.

Eleven features of each of the near-native poses (669) and non-near-native poses (44,334) calculated from Ank-TRN based on the recognition areas (regKp) were used to establish the learning methods. Based on the unbalanced number of docking poses, training sets were generated by clustering the non-near-native poses and the near-native poses into 65 sets (44,334/669). Eleven features were calculated from each training set and were ordered to generate 4,095 feature sets. The DT-based learning models (OLM_{DT}) and the LG-based learning models (OLM_{LG}) were established using the 4,095 feature sets. The learning methods demonstrated the average of the true positive rate to be greater than 50% (TPrate $\geq$ 50%), and consisted of 4,762 OLM_{DT} and 2,688 OLM_{LG}. The learning models that represented TPrate $\geq$ 50% with the 10 top-ranking poses of %ACC are shown in Table 3 (10 top-rankings of OLM_{DT}) and Table 4 (10 top-rankings of OLM_{LG}). As a result, ABDE-HIJK_g14 of OLM_{DT} exhibited the highest %ACC with %TPrate $\geq$ 50%. This learning method consisted of sequential combination feature sets that included ZDock (A), ZRankElec (B), ZRank (D), ZRankVdw (E), E_vdw2 (H), E_elec2 (I), E_sol (J), and E_RDock (K) calculated from non-near-native dataset number 13. In addition, CDFGJ_g10 of OLM_{LG} also demonstrated the highest %ACC with %TPrate $\geq$ 50%. The percentage of precision (%PRES) for all the 10 top-ranking poses of OLM_{DT} and OLM_{LG} was low, which indicated that there was a high number of false positive results (FP). To diminish the number of FP, only the 10 top-ranking poses based on the ZRank score were selected to represent the true positive poses (TP). If the TP were found in 10 top-

ranking poses from each Ank-TRN, #PP was designated 1. Thus, the #PP-values of seven Ank-TRN ($\#PP_{TRN}$) were in the range of 0 to 7. As shown in Tables 3 and 4, the maximum values of the $\#PP_{TRN}$ of OLM_{DT} and OLM_{LG} were only 6. Therefore, the individual learning method of OLM_{DT} or OLM_{LG} was not capable of providing the maximum value for $\#PP_{TRN}$.

Ensemble learning method to generate AnkPlex

To enhance the prediction efficacy of the generated learning methods, 4,762 of the DT-based learning models (OLM_{DT}) and 2,688 of the LG-based learning models (OLM_{LG}) were randomly combined to generate an ensemble model. Interestingly, the combination of the ensemble model from ABEHIJ_g56 of OLM_{DT} and CDFGHJ_g30 of OLM_{LG} demonstrated superior prediction efficiency due to the fact that this ensemble model (ABEHIJ_g56- CDFGHJ_g30) achieved maximum values

Table 4 Comparison of performances of 10 top-ranking OLM_{LG} among various Types of features and datasets in terms of 10-fold cross-validation

Rank	OLM_{LG}	PRES(%)	REC(%)	ACC(%)	$\#PP_{TRN}$
1	CDFGJ_g10	74.76 ± 11.62	67.83 ± 14.80	4.91 ± 3.75	6
2	CDFJ_g10	74.75 ± 11.59	67.66 ± 15.08	4.90 ± 3.75	6
3	CDJ_g10	74.74 ± 11.55	67.83 ± 14.80	4.91 ± 3.76	6
4	CDGJ_g10	74.72 ± 11.55	67.83 ± 14.80	4.91 ± 3.76	6
5	BCDEFGJ_g10	74.64 ± 10.30	68.06 ± 17.77	4.47 ± 2.86	4
6	BCDEGHJ_g10	74.60 ± 10.41	68.27 ± 17.78	4.48 ± 2.86	4
7	BCDEFGHJ_g10	74.54 ± 10.17	68.45 ± 17.85	4.46 ± 2.85	4
8	BCDEFHJ_g10	74.51 ± 10.20	68.69 ± 17.58	4.47 ± 2.84	4
9	ACDGJ_g36	74.48 ± 11.91	68.96 ± 15.52	4.94 ± 3.80	5
10	BCDEGJ_g41	74.48 ± 10.78	67.93 ± 16.77	4.57 ± 3.21	4

11 feathers (A, B, C, D, E, F, G, H, I, J, K) are ZDock, ZRankElec, ZRank, ZRankSolv, ZRankVdw, E_vdw1, E_elec1, E_vdw2, E_elec2, E_sol, E_RDock

for #PP$_{TRN}$ and #PP$_{TEST}$ of 7 and 2, respectively. Therefore, the ensemble model, ABEHIJ_g56- CDFGHJ_g30, was designated to be an ensemble computational model for predicting the near-native docking pose of APKs or "AnkPlex" (Fig. 3). To compare the prediction efficiency of the ensemble model, AnkPlex with the single learning models, the total number of TP and the first TP of each Ank-TRN were used for the evaluation. As shown in Table 5, the single learning models of OLM$_{DT}$ (ABE-HIJ_g56) and OLM$_{LG}$ (CDFGHJ_g30) provided a #PP$_{TRN}$ value of 6. The first TP of C$_5$ predicted by ABE-HIJ_g56 and the C$_6$ predicted by CDFGHJ_g30 were found at pose numbers 14 and 19. This result indicated that a single learning model could not produce all the true positive poses. In the case of the Ank-TEST, OLM$_{DT}$ could not provide the value for the #PP$_{TEST}$, whereas the #PP$_{TEST}$ of OLM$_{LG}$ was comparable to AnkPlex. Consequently, it can be concluded that the ensemble model, AnkPlex, was capable of including a #PP$_{TRN}$ value of 7 and a #PP$_{TEST}$ value of 2, which suggested that the prediction efficacy of AnkPlex was superior to the single learning model. In addition, the predictive models generated from other 35 possible datasets demonstrated the average number of #PP$_{TRN}$ and

#PP$_{TEST}$ value of 6.78 ± 0.42 and 2 ± 0.00, respectively. This indicated that different selections of training and test sets had no effect in the generation of the learning models for predicting the near-native poses.

According to the ZDOCK program recommendations, near-native docking poses could be found in 2Kp, as indicated by a high ZDock score, low E_RDock, or low ZRank [15, 17–20]. Particularly, the ZRank score provided a #PP$_{TRN}$ value of 6, which was higher compared to the values for other features (Table 6). Thus, 2Kp ranked by the ZRank score was selected to identify #PP. As shown in Table 5, the #PP$_{TRN}$ and the #PP$_{TEST}$ of 2Kp could not reach the maximum value. In addition, the first TP of 2Kp was found to be a lower order number compared to AnkPlex. These results indicated that ZRANK was able to identify the most accurate near-native poses. Nevertheless, it would not be applied for all cases. Thus, the combined feature, AnkPlex, could be used to adjust this solution.

To apply the AnkPlex for investigating the ankyrin-protein complex, Ank^{GAG}1D4 was used to study this learning model. Ank^{GAG}1D4 is an artificial ankyrin that contains 3 internal domains and was designed as an

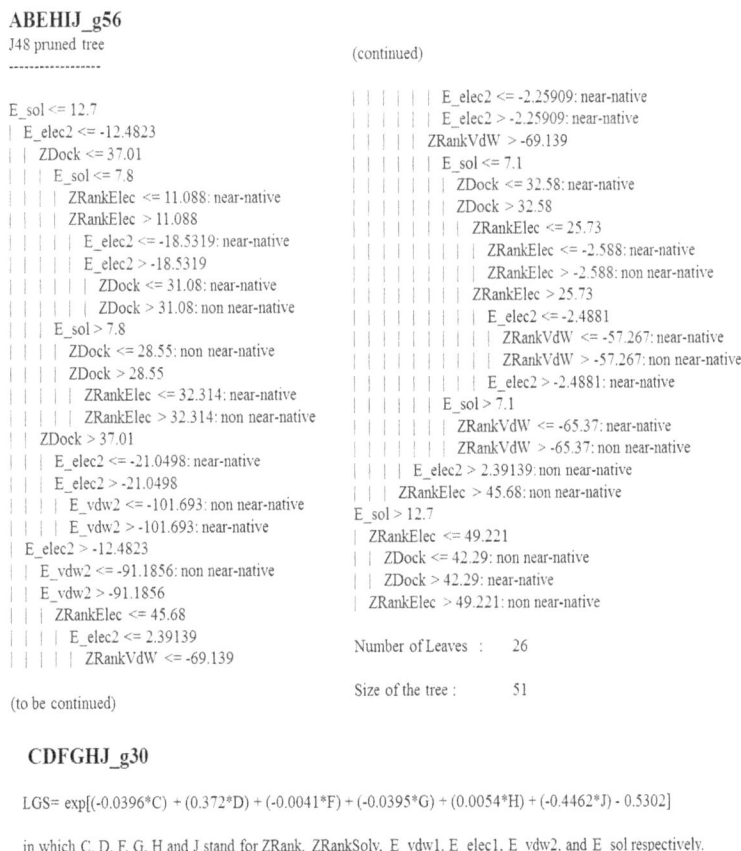

ABEHIJ_g56

J48 pruned tree

E_sol <= 12.7
| E_elec2 <= -12.4823
| | ZDock <= 37.01
| | | E_sol <= 7.8
| | | | ZRankElec <= 11.088: near-native
| | | | ZRankElec > 11.088
| | | | | E_elec2 <= -18.5319: near-native
| | | | | E_elec2 > -18.5319
| | | | | | ZDock <= 31.08: near-native
| | | | | | ZDock > 31.08: non near-native
| | | E_sol > 7.8
| | | | ZDock <= 28.55: non near-native
| | | | ZDock > 28.55
| | | | | ZRankElec <= 32.314: near-native
| | | | | ZRankElec > 32.314: non near-native
| | ZDock > 37.01
| | | E_elec2 <= -21.0498: near-native
| | | E_elec2 > -21.0498
| | | | E_vdw2 <= -101.693: non near-native
| | | | E_vdw2 > -101.693: near-native
| E_elec2 > -12.4823
| | E_vdw2 <= -91.1856: non near-native
| | E_vdw2 > -91.1856
| | | ZRankElec <= 45.68
| | | | E_elec2 <= 2.39139
| | | | | ZRankVdW <= -69.139

(to be continued)

(continued)

| | | | | | E_elec2 <= -2.25909: near-native
| | | | | | E_elec2 > -2.25909: near-native
| | | | | ZRankVdW > -69.139
| | | | | | E_sol <= 7.1
| | | | | | | ZDock <= 32.58: near-native
| | | | | | | ZDock > 32.58
| | | | | | | | ZRankElec <= 25.73
| | | | | | | | | ZRankElec <= -2.588: near-native
| | | | | | | | | ZRankElec > -2.588: non near-native
| | | | | | | | ZRankElec > 25.73
| | | | | | | | | E_elec2 <= -2.4881
| | | | | | | | | | ZRankVdW <= -57.267: near-native
| | | | | | | | | | ZRankVdW > -57.267: non near-native
| | | | | | | | | E_elec2 > -2.4881: near-native
| | | | | | E_sol > 7.1
| | | | | | | ZRankVdW <= -65.37: near-native
| | | | | | | ZRankVdW > -65.37: non near-native
| | | | E_elec2 > 2.39139: non near-native
| | | ZRankElec > 45.68: non near-native
E_sol > 12.7
| ZRankElec <= 49.221
| | ZDock <= 42.29: non near-native
| | ZDock > 42.29: near-native
| ZRankElec > 49.221: non near-native

Number of Leaves :   26

Size of the tree :   51

CDFGHJ_g30

LGS= exp[(-0.0396*C) + (0.372*D) + (-0.0041*F) + (-0.0395*G) + (0.0054*H) + (-0.4462*J) - 0.5302]

in which C, D, F, G, H and J stand for ZRank, ZRankSolv, E_vdw1, E_elec1, E_vdw2, and E_sol respectively.

Fig. 3 Characteristics of the optimal AnkPlex

Table 5 Comparison of performances of AnkPlex with single learning method and ZDOCK program[a]

Method	Number of TP docking poses (rank)								
	C_1	C_2	C_3	C_4	C_5	C_6	C_7	U_1	U_2
2Kp(ZDOCK)	0 (13)	1 (10)	6 (1)	7 (1)	1 (10)	1 (10)	8 (1)	0 (14)	0 (104)
OLM$_{DT}$ (ABEHIJ_g56)	1 (4)	1 (4)	8 (1)	7 (1)	0 (14)	2 (1)	7 (1)	0 (13)	0 (160)
OLM$_{LG}$ (CDFGHJ_g30)	2 (1)	1 (3)	8 (1)	10 (1)	1 (9)	0 (19)	7 (1)	1 (6)	1 (5)
AnkPlex	2 (1)	1 (3)	8 (1)	10 (1)	1 (9)	3 (2)	7 (1)	1 (6)	1 (5)

[a]The number of TP docking poses is the summation of the TP docking poses found in 10 top-ranking poses, where the maximum and the minimum are 10 and 0, respectively. The rank is denoted by the order in which the first TP docking poses are found. For example, on C_6, AnkPlex yields three TP docking poses on the top 10 ranking poses, and the orders of the three TP docking poses are 2, 8, and 9. Thus, the rank of AnkPlex on C_6 is 2

antiretroviral agent. Ank^{GAG}1D4 was able to bind to the N-terminal domain of the capsid protein (CA^{NTD}) of HIV-1 [3]. Recently, the X-ray structure Ank^{GAG}1D4 was already constructed. However, the complex structure of Ank^{GAG}1D4-CA^{NTD} was not detected [4]. Thus, we generated the docking poses of Ank^{GAG}1D4-CA^{NTD} and performed re-scoring with AnkPlex. The results revealed that three near-native structures of Ank^{GAG}1D4-CA^{NTD} were found in the 10 top-rankings. The recognition residues of Ank^{GAG}1D4-CA^{NTD} interactions were further investigated by observing interacting distances ≤ 5 Å. As a result, one docking pose showed that residue R18 was located on the recognition areas of CA^{NTD} and two docking poses demonstrated residues R132 and R143 played key roles in the interaction with Ank^{GAG}1D4 (data not shown). This result correlated with previous ELISA results. A point mutation of R18A on helix 1 and R132A and R143A on helix 7 of CA^{NTD} showed negative binding to Ank^{GAG}1D4. Thus, R18, R132 and R143 were the key residues of CA^{NTD} binding to Ank^{GAG}1D4 [4]. According to computational analysis of Ank^{GAG}1D4 using this learning model, AnkPlex could not only discriminate the near native docking poses of Ank^{GAG}1D4-CA^{NTD} complex but also demonstrated the correct orientation of the recognition area to CA^{NTD}.

Feature importance analysis

Identification of informative features among the 11 features was critical for designing a powerful learning model and for understanding and obtaining insights into the ankyrin-protein docking poses. Based on the six features (CDFGHJ) used in the calculation of the LGS in AnkPlex, the Pearson correlation coefficients (R values) were used to identify the correlation between LGS and the weights of the six features to obtain the near-native poses. As shown in Fig. 4 and Additional file 1: Table S3, the three top-ranked R values of the six features consisted of ZRank ($R = 0.60$), ZRankSolv ($R = -0.56$), and E_sol ($R = 0.54$), which indicated that these three features played an important role in the AnkPlex model for distinguishing near-native poses.

The ZRank score was ranked as the 1st informative feature according to the highest R values (0.60). The characteristics of the ZRank score between the near-native and the non-near-native poses were significantly different, with $p < 0.001$, as shown in Table 2. To confirm the important roles of the ZRank score in AnkPlex, the ensemble learning method based on AnkPlex was constructed without ZRank (C). As a result, as demonstrated in Additional file 1: Table S2, the AnkPlex lacking ZRank (OLM$_{DT}$(ABEHIJ_g56)–OLM$_{LG}$(DFGHJ_g30)) was able to obtain #PP$_{TRN} = 1$

Table 6 Number of near-native poses in 10 top-ranking poses obtained from ZDOCK program with 2Kp

Feature	Number of near-native poses in 10 top-ranking poses										
	C_1	C_2	C_3	C_4	C_5	C_6	C_7	#PP$_{TRN}$	U_1	U_2	#PP$_{TEST}$
ZDock	0	0	0	3	0	0	4	2	0	0	0
ZRankElec	2	0	0	0	0	0	0	1	0	0	0
ZRank	0	1	6	7	1	1	8	6	0	0	0
ZRankSolv	0	0	0	7	1	0	0	2	0	0	0
ZRankVdw	0	0	4	0	0	3	1	3	0	0	0
E_vdw1	0	0	0	0	0	0	0	0	0	0	0
E_elec1	0	0	0	0	0	7	0	1	0	0	0
E_vdw2	0	0	0	0	0	0	0	0	0	0	0
E_elec2	0	0	0	0	0	6	0	1	0	0	0
E_sol	0	0	0	9	1	0	0	2	0	0	0
E_RDock	8	0	0	4	0	0	5	3	0	0	0

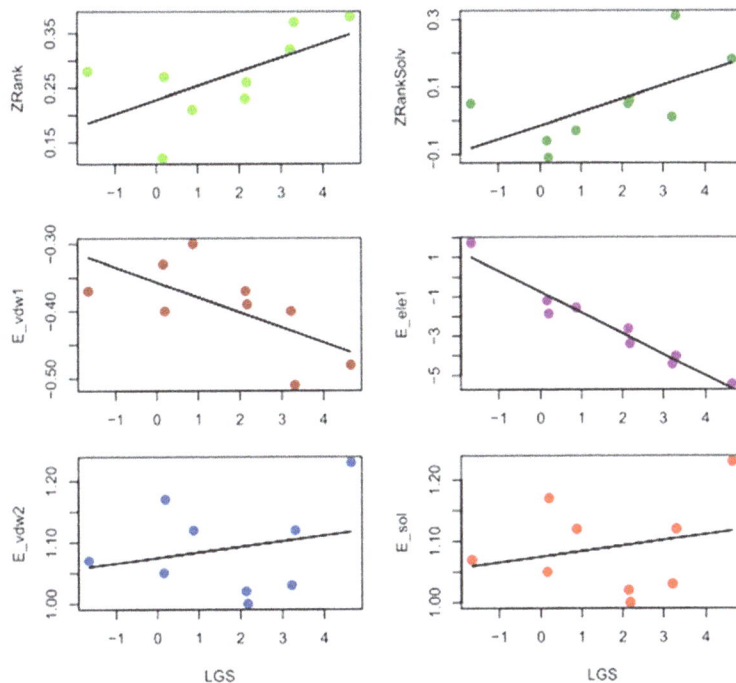

Fig. 4 The correlation coefficients between the dot products of the features and their weights of the best-ranking LGS for the near-native poses in the nine ankyrin-protein complexes

and $\#PP_{TEST} = 0$. However, AnkPlex (OLM_{DT}(ABE-HIJ_g56)–OLM_{LG}(DFGHJ_g30)) achieved success with $\#PP_{TRN} = 7$ and $\#PP_{TEST} = 2$. Therefore, ZRank was concluded to be an important feature of AnkPlex due to the fact that it could enhance the predictive performance of near-native poses.

Since ZRank is a linear combination of van der Waals (ZRankVDW), electrostatics (ZRankElec), and desolvation energy (ZRankSolv), one of them had to be identified as the most important. From Additional file 1: Table S3, it is evident that the ZRankVDW (van der Waals interaction) was more dominant than the ZRankElec and the ZRankSolv. Recently, ZRank was developed by correcting the weight of the energies and combining a pairwise interface potential in which the weight of van der Waals was higher than the original ZRank [29]. This result supports the theory that van der Waals is an important property for near-native docking poses of ankyrin-protein pairings.

ZRankSolv and E_sol were the desolvation energies estimated by the summation of the Atomic Contact Energy (ACE) in which the difference between the two features was in the force field calculation and side chain orientation. ZRankSolv and E_sol were the 2nd and the 3rd informative features with R values of –0.56 and 0.54, respectively. Moreover, these two features showed differences in their characteristics between near-native and non-near-native poses, with $p < 0.001$, as shown in

Table 2. Our experimental results (see Additional file 1: Table S2) demonstrated that AnkPlex lacked ZRankSolv (D), i.e., OLM_{DT}(ABEHIJ_g56)–OLM_{LG}(CFGHJ_g30) provided $\#PP_{TRN} = 6$ and $\#PP_{TEST} = 1$. Similar to E_sol, the performance of AnkPlex lacked E_sol, i.e., (OLM_{DT}(ABE-HIJ_g56)–OLM_{LG}(CDFGH_g30)) yielded $\#PP_{TRN} = 6$ and $\#PP_{TEST} = 1$. This result indicated that the absence of ZRankSolv and E_sol slightly reduced the predictive performance of AnkPlex. However, these two features were required for predicting near-native poses. ZRankSolv was a component of ZRank. This outcome emphasized that desolvation was important for obtaining near-native poses in the ankyrin-protein interaction. Additionally, it was also required for the accuracy of other protein–protein complexes [30–32].

The LGS score was a combination of the energy determined in the interaction area between ankyrin and proteins of ≤5 Å. In AnkPlex, the interaction area on ankyrin was located on variable and conserved residues of the L-shaped repeat belonging to the internal repeats, the N-terminal repeat and the C-terminal repeat [7]. The functional variable residues on ankyrin were required for the recognition of the target protein by using the available solvent-accessible surface [7, 33, 34]. To observe the variable area used for calculating the energy, analysis of the first TP of the near-native of Ank9 was carried out to count the variable and conserved residues in the interaction area. The result, which was presented in Additional file 1: Table S4, showed that there was $43.75 \pm 12.90\%$ of the interaction area on ankyrin belonging to the variable residues, and $58.50 \pm 11.36\%$ of this area represented the hydrophobic residues. This result indicated that the interaction energy was calculated on both the variable and the conserved residues. Therefore, computing the energy term at the interface of the variable residues could provide a score to distinguish between the near-native and the non-near-native docking poses. As a consequence, the calculation for evaluating the score based on the desired area could be applied in the docking algorithm.

According to the hydrophobicity on the interface in AnkPlex (see Additional file 1: Table S4), the interactions between ankyrin and proteins were comprised of $18.05 \pm 6.32\%$, hydrophobic–hydrophobic, $43.25 \pm 13.70\%$, hydrophobic–hydrophilic and $38.70 \pm 4.48\%$. hydrophilic–hydrophilic interactions. Moreover, the percentage of hydrophobic–hydrophobic interactions in the non-near-native pose was observed to be reduced by $12.69 \pm 7.31\%$, as shown in Additional file 1: Table S4. However, the percentage of hydrophobic–hydrophilic interactions in the near-native pose increased to $50.32 \pm 8.89\%$. This outcome indicated that the recognition site on ankyrin for the target protein was adopted to have hydrophobic and hydrophilic interactions, which promoted the solvent-accessible property [35]. Because LGS is modified from atom-based potential without considering the type of the hydrophobicity scale, the high LGS of the non-near-native docking pose could be calculated from the hydrophobic–hydrophilic interaction instead of the hydrophobic–hydrophobic interaction.

Conclusions

An ensemble method, named AnkPlex, was constructed for fast prediction of near-native states of ankyrin-protein complexes. The AnkPlex model was constructed based on a combination of features generated from the ZDOCK program without using manual inspections. AnkPlex successfully obtained the near-native poses of nine ankyrin-protein complexes in the 10 top-ranking poses. ZRank, which is a combination of electrostatic, desolvation, and van der Waals energy, was the most important feature in AnkPlex. In addition, van der Waals was the dominant feature for obtaining the near-native docking poses. To develop the method for predicting near-native poses of protein complexes, we have implemented easy access to the best models for the scientific community on a web server. AnkPlex (http://ankplex. ams.cmu.ac.th) is freely available online.

Additional files

Additional file 1: Table S1. Informative characteristics of nine ankyrin-protein complexes. **Table S2.** Comparison of performances of 10 top-ranking among ensemble learning models[a]. **Table S3.** Features values of the best-ranking LGS for the near-native poses in the nine ankyrin-protein complexes. **Table S4.** Types of interaction pair and hydrophobic residues on interface (≤5 Å) of the nine ankyrin-protein complexes. **Table S5.** The percentage of predictable true near-native poses of the internal and external testing sets on leaning methods of decision tree (DT), logistic regression (LG), artificial neural network (ANN), and support vector machine (SVM).

Additional file 2: Supplementary datasets used in this study.

Abbreviations
ACE: Atomic Contact Energy; APKs: ankyrin-protein complexes; Cα-RMSD: Carbon alpha root-mean-square deviation; DARPins: Designed Ankyrin Repeat Proteins; DE: desolvation; DT: decision tree; ELEC: electrostatics; LG: logistic model; LGS: logistic score; PPV: predicted positive value; PSC: shape complementarity

Acknowledgements
We would like to thank the editor and all anonymous reviewers for valuable suggestions and constructive comments. We acknowledged the Centre of Research in Computational Sciences and Informatics for Biology, Bioindustry, Environment, Agriculture and Healthcare (CRYSTAL) at the University of Malaya Research Centre for performing our study using ZDOCK program. We would like to thank Chiang Mai University Press for their editorial suggestions and springer nature author services for improving the English language of our work.

Funding
This work was supported by the Cluster and Program Management Office (CPMO), the National Science and Technology Development Agency (NSTDA), the Thailand Research Fund (TRF), the National Research Council of Thailand (NRCT), the Health Systems Research Institute (HSRI), the National Research University project under Thailand's Office of the Commission on Higher Education (NRU), the Centre of Biomolecular Therapy and Diagnostics (CBTD), the Mahidol University Talent Management Program.

Author's contributions
CT, KK and WS conceived the study. WS and TW participated in the design of the algorithms and experiments. TW, VSL and KK prepared the manuscript. All authors participated in manuscript preparation. All authors read and approved the final manuscript.

Competing interests
The authors declare that they have no competing interests.

Author details
[1]Division of Clinical Immunology, Department of Medical Technology, Faculty of Associated Medical Sciences, Chiang Mai University, Chiang Mai 50200, Thailand. [2]Center of Biomolecular Therapy and Diagnostic, Faculty of Associated Medical Sciences, Chiang Mai University, Chiang Mai 50200, Thailand. [3]Center of Data Mining and Biomedical Informatics, Faculty of Medical Technology, Mahidol University, Bangkok 10700, Thailand. [4]Thailand Center of Excellence in Physics, Commission on Higher Education, Bangkok 10400, Thailand. [5]Department of Chemistry, Faculty of Science, University of Malaya, Kuala Lumpur 50603, Malaysia. [6]Center for Research and Innovation, Faculty of Medical Technology, Mahidol University, Bangkok 10700, Thailand.

References

1. Binyamin L, Borghaei H, Weiner LM. Cancer therapy with engineered monoclonal antibodies. Update Cancer Ther. 2006;1(2):147–57.
2. Trikha M, Yan L, Nakada MT. Monoclonal antibodies as therapeutics in oncology. Curr Opin Biotechnol. 2002;13(6):609–14.
3. Nangola S, Urvoas A, Valerio-Lepiniec M, Khamaikawin W, Sakkhachornphop S, Hong SS, Boulanger P, Minard P, Tayapiwatana C. Antiviral activity of recombinant ankyrin targeted to the capsid domain of HIV-1 Gag polyprotein. Retrovirology. 2012;9:17.
4. Praditwongwan W, Chuankhayan P, Saoin S, Wisitponchai T, Lee VS, Nangola S, Hong SS, Minard P, Boulanger P, Chen CJ, et al. Crystal structure of an antiviral ankyrin targeting the HIV-1 capsid and molecular modeling of the ankyrin-capsid complex. J Comput Aided Mol Des. 2014;28(8):869–84.
5. Schweizer A, Rusert P, Berlinger L, Ruprecht CR, Mann A, Corthesy S, Turville SG, Aravantinou M, Fischer M, Robbiani M, et al. CD4-specific designed ankyrin repeat proteins are novel potent HIV entry inhibitors with unique characteristics. PLoS Pathog. 2008;4(7), e1000109.
6. Binz HK, Amstutz P, Kohl A, Stumpp MT, Briand C, Forrer P, Grutter MG, Pluckthun A. High-affinity binders selected from designed ankyrin repeat protein libraries. Nat Biotechnol. 2004;22(5):575–82.
7. Binz HK, Stumpp MT, Forrer P, Amstutz P, Pluckthun A. Designing repeat proteins: well-expressed, soluble and stable proteins from combinatorial libraries of consensus ankyrin repeat proteins. J Mol Biol. 2003;332(2):489–503.
8. Stumpp MT, Binz HK, Amstutz P. DARPins: a new generation of protein therapeutics. Drug Discov Today. 2008;13(15-16):695–701.
9. Kohl A, Binz HK, Forrer P, Stumpp MT, Pluckthun A, Grutter MG. Designed to be stable: crystal structure of a consensus ankyrin repeat protein. Proc Natl Acad Sci U S A. 2003;100(4):1700–5.
10. Epa VC, Dolezal O, Doughty L, Xiao X, Jost C, Plückthun A, Adams TE. Structural model for the interaction of a designed Ankyrin Repeat Protein with the human epidermal growth factor receptor 2. PLoS One. 2013;8(3), e59163.
11. Dobbins SE, Lesk VI, Sternberg MJE. Insights into protein flexibility: The relationship between normal modes and conformational change upon protein–protein docking. Proc Natl Acad Sci U S A. 2008;105(30):10390–5.
12. Tuncbag N, Kar G, Keskin O, Gursoy A, Nussinov R. A survey of available tools and web servers for analysis of protein-protein interactions and interfaces. Brief Bioinform. 2009;10(3):217–32.
13. Chen R, Li L, Weng Z. ZDOCK: an initial-stage protein-docking algorithm. Proteins. 2003;52(1):80–7.
14. Pierce B, Weng Z. ZRANK: reranking protein docking predictions with an optimized energy function. Proteins. 2007;67(4):1078–86.
15. Li L, Chen R, Weng Z. RDOCK: refinement of rigid-body protein docking predictions. Proteins. 2003;53(3):693–707.
16. Janin J, Henrick K, Moult J, Eyck LT, Sternberg MJ, Vajda S, Vakser I, Wodak SJ. CAPRI: a Critical Assessment of PRedicted Interactions. Proteins. 2003; 52(1):2–9.
17. Hwang H, Vreven T, Pierce BG, Hung JH, Weng Z. Performance of ZDOCK and ZRANK in CAPRI rounds 13-19. Proteins. 2010;78(15):3104–10.
18. Vreven T, Pierce BG, Hwang H, Weng Z. Performance of ZDOCK in CAPRI rounds 20-26. Proteins. 2013;81(12):2175–82.
19. Wiehe K, Pierce B, Mintseris J, Tong WW, Anderson R, Chen R, Weng Z. ZDOCK and RDOCK performance in CAPRI rounds 3, 4, and 5. Proteins. 2005;60(2):207–13.
20. Wiehe K, Pierce B, Tong WW, Hwang H, Mintseris J, Weng Z. The performance of ZDOCK and ZRANK in rounds 6-11 of CAPRI. Proteins. 2007;69(4): 719–25.
21. Li CH, Ma XH, Shen LZ, Chang S, Zu Chen W, Wang CX. Complex-type-dependent scoring functions in protein–protein docking. Biophys Chem. 2007;129(1):1–10.
22. Chakrabarty B, Parekh N. Identifying tandem Ankyrin repeats in protein structures. BMC bioinformatics. 2014;15(1):6599.
23. Gough J, Karplus K, Hughey R, Chothia C. Assignment of homology to genome sequences using a library of hidden Markov models that represent all proteins of known structure. J Mol Biol. 2001;313(4):903–19.
24. Mendez R, Leplae R, De Maria L, Wodak SJ. Assessment of blind predictions of protein–protein interactions: current status of docking methods. Proteins. 2003;52(1):51–67.
25. Momany FA, Rone R. Validation of the general purpose QUANTA® 3.2/ CHARMm® force field. J Comput Chem. 1992;13(7):888–900.
26. Quinlan JR. C4.5: programs for machine learning. 1993.
27. Mark H, Eibe F, Geoffrey H, Bernhard P. The WEKA Data Mining Software: An Update. SIGKDD Explor. 2009;11(1):10–8.
28. Cessie LS, van Houwelingen JC. Ridge Estimators in Logistic Regression. Appl Statist. 1992;41(1):191–201.
29. Pierce B, Weng Z. A combination of rescoring and refinement significantly improves protein docking performance. Proteins. 2008;72(1):270–9.
30. Camacho CJ, Kimura S, DeLisi C, Vajda S. Kinetics of desolvation-mediated protein–protein binding. Biophys J. 2000;78(3):1094–105.
31. Camacho CJ, Weng Z, Vajda S, DeLisi C. Free energy landscapes of encounter complexes in protein-protein association. BIOPHYS J. 1999;76(3): 1166–78.
32. Comeau SR, Gatchell DW, Vajda S, Camacho CJ. ClusPro: an automated docking and discrimination method for the prediction of protein complexes. Bioinformatics. 2004;20(1):45–50.
33. Magliery TJ, Regan L. Sequence variation in ligand binding sites in proteins. BMC Bioinforma. 2005;6(1):240.
34. Sedgwick SG, Smerdon SJ. The ankyrin repeat: a diversity of interactions on a common structural framework. Trends Biochem Sci. 1999;24(8):311–6.
35. Lins L, Thomas A, Brasseur R. Analysis of accessible surface of residues in proteins. Protein Sci. 2003;12(7):1406–17.

5

Spectral imaging toolbox: segmentation, hyperstack reconstruction, and batch processing of spectral images for the determination of cell and model membrane lipid order

Miles Aron[1] ⓘ, Richard Browning[1], Dario Carugo[1,2], Erdinc Sezgin[3], Jorge Bernardino de la Serna[3,4], Christian Eggeling[3] and Eleanor Stride[1]*

Abstract

Background: Spectral imaging with polarity-sensitive fluorescent probes enables the quantification of cell and model membrane physical properties, including local hydration, fluidity, and lateral lipid packing, usually characterized by the generalized polarization (GP) parameter. With the development of commercial microscopes equipped with spectral detectors, spectral imaging has become a convenient and powerful technique for measuring GP and other membrane properties. The existing tools for spectral image processing, however, are insufficient for processing the large data sets afforded by this technological advancement, and are unsuitable for processing images acquired with rapidly internalized fluorescent probes.

Results: Here we present a MATLAB spectral imaging toolbox with the aim of overcoming these limitations. In addition to common operations, such as the calculation of distributions of GP values, generation of pseudo-colored GP maps, and spectral analysis, a key highlight of this tool is reliable membrane segmentation for probes that are rapidly internalized. Furthermore, handling for hyperstacks, 3D reconstruction and batch processing facilitates analysis of data sets generated by time series, z-stack, and area scan microscope operations. Finally, the object size distribution is determined, which can provide insight into the mechanisms underlying changes in membrane properties and is desirable for e.g. studies involving model membranes and surfactant coated particles. Analysis is demonstrated for cell membranes, cell-derived vesicles, model membranes, and microbubbles with environmentally-sensitive probes Laurdan, carboxyl-modified Laurdan (C-Laurdan), Di-4-ANEPPDHQ, and Di-4-AN(F)EPPTEA (FE), for quantification of the local lateral density of lipids or lipid packing.

Conclusions: The Spectral Imaging Toolbox is a powerful tool for the segmentation and processing of large spectral imaging datasets with a reliable method for membrane segmentation and no ability in programming required. The Spectral Imaging Toolbox can be downloaded from https://uk.mathworks.com/matlabcentral/fileexchange/62617-spectral-imaging-toolbox.

Keywords: Spectral imaging, Lipid order, Lipid packing, Membrane viscosity, Membrane segmentation, Laurdan

* Correspondence: eleanor.stride@eng.ox.ac.uk
[1]Department of Engineering Science, Institute of Biomedical Engineering, University of Oxford, Oxford OX3 7DQ, UK
Full list of author information is available at the end of the article

Background

An increasing body of evidence suggests that the dynamic reorganization of lipids in cellular membranes can compartmentalize membrane proteins, influencing a cell's response to extracellular stimuli and its membrane permeability [1, 2]. It follows that drug-carrying agents, such as liposomes or gas microbubbles, with optimized lipid compositions can exploit these processes for enhanced drug-delivery via membrane fusion or membrane permeabilization [3–5]. To facilitate the characterization of such drug-delivery devices and to deepen our understanding of the fundamental biology of the cell membrane, a non-destructive method for evaluating intrinsic membrane physico-chemical properties is required. As an example, packing or molecular order of membrane lipids can be sensed by fluorescent polarity-sensitive probes such as Laurdan or Di-4-ANEPPDHQ, whose emission spectrum shifts in response to changes in the molecular order of the membrane environment, usually quantified by a parameter denoted Generalized Polarization (GP) [6–11]. With the advent of commercial microscopes equipped with spectral detectors, shifts in the fluorescence emission spectra, and thus the GP parameter, can now be determined with much higher spatial accuracy using spectral imaging [10]. Owing to the internalization of many polarity-sensitive fluorescent probes in living cells, however, membrane segmentation must be performed to accurately measure membrane lipid packing and to remove cytosolic contributions [7, 12]. Membrane segmentation is often performed using a secondary fluorophore which increases experimental cost and complexity.

To this end, we have developed the Spectral Imaging Toolbox, a toolbox for spectral analysis with reliable membrane segmentation without the need for a secondary imaging probe. In the Spectral Imaging Toolbox, we have included batch and hyperstack processing as well as 3D reconstruction of confocal z-stacks to facilitate processing of large datasets and experiments with multiple exposures. We demonstrate the utility of this tool with images of giant plasma membrane vesicles (GPMVs, cell-derived vesicles) labelled with either polarity-sensitive Laurdan or Di-4-ANEPPDHQ, images of live cancer cells and microbubbles labelled with carboxyl-modified Laurdan (C-Laurdan), and giant unilamellar vesicles (GUVs) labelled with Di-4-AN(F)EPPTEA (FE). In addition to the more commonly employed Laurdan and Di-4-ANEPPDHQ dyes, we chose FE and C-Laurdan for their superior photostability and emission spectrum range [8, 12].

Implementation

The Spectral Imaging Toolbox was designed for spectral analysis of high magnification images of single or sub-confluent cells, vesicles and microbubbles in MATLAB [13].

Inputs and outputs

In spectral imaging, a stack of images of a sample region is recorded with each image in the stack monitoring a different wavelength range, such that the information from the whole stack discloses the spectrum of emitted fluorescence for each image pixel [10]. The Spectral Imaging Toolbox is designed for batch processing and 3-4D stacks. Using the Spectral Imaging Toolbox, we were able to process and analyze a dataset containing over 1500 cells in a few hours [3]. To our knowledge, this is the largest study using the GP parameter of cell membranes as a metric for membrane lipid order, highlighting the utility of our toolbox. For an input directory of spectral image stacks, the Spectral Imaging Toolbox outputs pseudo-colored GP maps, fitted GP histograms, and plotted spectra at the whole image, whole object, and segmented membrane levels for each image in the folder, as well as a spreadsheet summarizing the results. Input images and metadata are automatically converted to the OME-TIFF data standard using the Bio-Formats Library (144 image formats currently supported) [14]. Options for automatic 3D reconstruction of confocal spectral z-stacks [15] and plotted size distributions of spherical vesicles are also available.

Graphical user interface (GUI)

A graphical user interface (GUI) guides the user through the analysis such that no programming skills are required. The GUI has a three panel design whereby the left panel displays instructions and menu items, the center panel allows for navigation through the images and user interaction (i.e., cropping and region of interest selection), and the right panel displays a gallery of images providing an overview of the results. The processing allows for user interaction at three steps. First, the user selects settings for which to run the Spectral Imaging Toolbox, such as whether to include membrane segmentation or a GP correction factor. Then following automatic object detection, the user has the option to segment each detected object further using one or more of several segmentation routines. Finally, the user can review the results and remove unwanted objects from the analysis as necessary.

Segmentation

Spectral image stacks are thresholded using an intensity threshold determined automatically by Otsu's method [16]. Objects of interest are then segmented and cropped using connected-component labelling of the binary thresholding mask [17]. The resultant cropped images are displayed for the user to discard off-target cropped images as necessary.

If a cropped image contains connected objects, such as fused GUVs or touching cell membranes, the user can readily separate them using a watershed-based segmentation approach. Prior to taking the watershed transform which identifies objects as catchment basins separated by watershed lines [18], a series of operations are conducted to improve performance. Namely, the distance transform of the complement of the binarized image is computed. The watershed transform of the negated distance transform is then taken. This process is demonstrated in Fig. 1d. By our method, the threshold level for suppressing shallow minima in this image is chosen such that the watershed transform labels n objects for segmentation, where n is input through the GUI. In other words, if the user specifies that a cropped image contains n cells, n cells are segmented. Owing to the sensitivity of this method to non-convex shaped cells and intracellular intensity variations, the user also has the option to segment manually using lasso-segmentation. Furthermore, lasso-segmentation can be used to conduct

spectral analysis on any user-defined region of interest, including intracellular vesicles.

Since the cropped images contain only a single object, the membrane segmentation is simple and reliable. The objects in the binarized cropped images are filled and the membranes detected using Sobel edge detection [19–21]. The membranes are then segmented using the edge-detected pixels following dilation with a horizontal line element [22]. The Spectral Imaging Toolbox also has a spherical object mode designed for microbubbles and spherical vesicles, where objects are segmented by finding circles using the circular Hough transform [23, 24].

Generalized polarization (GP)

As highlighted before, the GP parameter is introduced for quantification of the spectral shift in emission of a polarity-sensitive probe due to differences in lipid membrane order. GP is commonly calculated using fluorescence intensities collected at two emission wavelengths, λ_B and λ_R, occurring at the emission maxima of the probe in a liquid-ordered and liquid-disordered

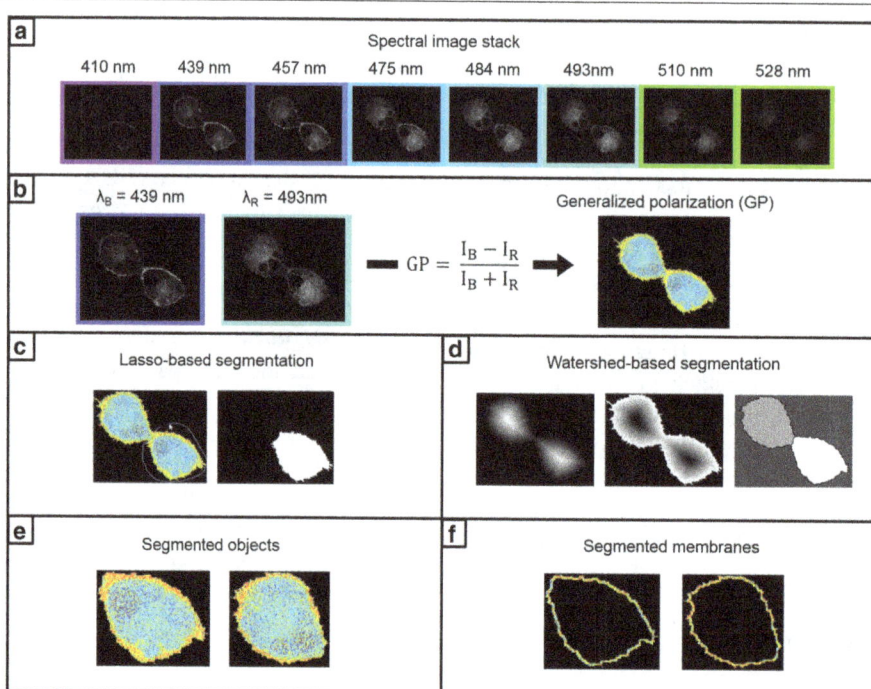

Fig. 1 The Spectral Imaging Toolbox. **a** An auto-thresholded spectral image stack containing images of C-Laurdan fluorescence emission from labelled A-549 cells collected at wavelengths ranging from 410 to 528 nm. **b** Generalized polarization (GP) is then calculated at each pixel using the intensities (I_B and I_R) from the images collected at λ_B and λ_R (*left*) using the equation (*center*). Pseudocolored GP maps can then be generated (*right*, color bar same as Fig. 2). Segmentation can then performed on the GP maps using lasso-based segmentation (**c**), where the user draws a region-of-interest (ROI) (*left*) used to generate a segmentation mask (*right*). Segmentation can alternately be performed using a watershed-based approach (**d**). From *left* to *right* in (**d**), the distance transform, the negated distance transform, and the labelled components following the watershed transform. Either segmentation routine will result in the segmented objects (**e**), from which a given number of border pixels are taken as the segmented membranes (**f**)

reference solution respectively [9, 10]. GP, which varies from -1 to 1, is calculated for each pixel in the spectral image from the following equation,

$$GP = \frac{I_B - I_R}{I_B + I_R},$$ (1)

where I_B and I_R correspond to the fluorescence intensity at λ_B and λ_R emission wavelengths, respectively. Consequently, low GP values indicate more disordered environments.

To clarify, only the intensities of the images at λ_B and λ_R are required for GP calculation, even with spectral image stacks consisting of images collected at many wavelengths (e.g. Fig. 1a). Thus, the spectral image stack is reduced to two images at λ_B and λ_R, and these two images are reduced to the single-valued GP parameter at each pixel (Fig. 1b).

The calculated GP values are then visualized using a pseudo-colored map with a look-up table scaled from -1 to 1 [11]. Finally, the distribution of GP values is fitted to either a one or two-peak Gaussian chosen by the lower root-mean squared error. The resultant GP histogram can be used to calculate changes in mean lipid order or, for a well-defined two-peak Gaussian, to indicate the presence of two phases [6]. To facilitate additional spectral analysis, spectra are generated from the mean intensities of images at each wavelength of the stack.

Generalized polarization (GP) correction factor

As GP is an intensity-based measurement, it is strongly influenced by microscope settings including detector gain and filter settings. When GP is calculated using intensities I_B and I_R from two channels detecting at wavelengths λ_B and λ_R, respectively, the relative intensities of the two channels must be calibrated to obtain absolute GP values. By acquiring an image of a reference solution with corresponding GP also measured with a fluorimeter (GP_{ref}), a correction factor, G, can be introduced,

$$G = \frac{I_{B,\,ref} \times (1 - GP_{ref})}{I_{R,\,ref} \times (1 + GP_{ref})},$$ (2)

where $I_{B,ref}$ and $I_{R,ref}$ correspond to the fluorescence intensity of the microscope image at λ_B and λ_R emission wavelengths, respectively [25]. GP is then calculated as follows,

$$GP = \frac{I_B - G \times I_R}{I_B + G \times I_R}.$$ (3)

In the Spectral Imaging Toolbox, GP_{ref} and a reference image can be specified in order to determine G for subsequent GP calculations.

Results

Here we present several examples of spectral imaging data processed with the Spectral Imaging Toolbox.

Spectral imaging by confocal microscopy

Spectral imaging was performed on a Zeiss LSM 780 confocal microscope equipped with a 32-channel gallium arsenide phosphide (GaAsP) detector array, as reported previously [10]. Laurdan, C-Laurdan, FE, and Di-4-ANEPPDHQ were excited at 405, 405, 488 and 488 nm respectively and the lambda detection ranges set between 410 nm and 695 nm, 415 nm and 691 nm, 500 nm and 650 nm, and 490 nm and 695 nm respectively. The resulting spectral image stacks were processed and analyzed using the Spectral Imaging Toolbox.

Sample preparation

A-549 cells, immortalized human alveolar adenocarcinomic epithelial cells, were grown in standard culture conditions with Dulbecco's modified eagle medium (DMEM) containing 10% fetal bovine serum (FBS) and 1% penicillin/streptomycin. Giant unilamellar vesicles (GUVs) made of dioleoyl phosphatidylcholine (DOPC), brain sphingomyelin (brain SM), and cholesterol from Avanti Polar Lipids were produced in a 2:2:1 molar ratio by electroformation by a modification of the protocol proposed by Angelova et al. [10, 26]. Phospholipid shelled microbubbles with a 9:1 molar ratio of 1,2-Distearoyl-sn-glycero-3-phosphocholine (DSPC, Avanti Polar Lipids, USA) and polyoxyethylene (40) stearate (PEG40S, Sigma Aldrich, UK) were produced using a batch sonication protocol previously reported [27]. Samples were labelled with either C-Laurdan (400 nM for A-549 cells and 100 nM for GUVs and microbubbles) or Di-4-AN(F)EPPTEA (FE) (100 nM for GUVs) in phosphate-buffered saline (PBS). Giant plasma membrane vesicles (GPMVs) were isolated from rat basophilic leukemia cells labelled with 100 nM Laurdan or 100 nM Di-4-ANEPPDHQ as described by Sezgin et al. [28]. Briefly, cells were exposed for 1 h at 37 °C to GPMV buffer (10 mM HEPES, 150 mM NaCl, 2 mM CaCl$_2$, pH 7.4) containing 25 mM paraformaldehyde and 2 mM dithiothreitol for inducing vesiculation. After vesiculation, the GPMV-rich supernatant was collected by pipetting and resuspended in GPMV buffer for imaging. For all samples, spectral imaging was performed with samples on 170 μm thick glass coverslips.

Segmentation of cells, GUVs, and microbubbles

Image segmentation and spectral analysis using the Spectral Imaging Toolbox are demonstrated in Fig. 2. Pseudo-colored Generalized Polarization (GP) maps, fluorescence spectra generated from the mean intensities of images at each wavelength of the spectral image stack, and histograms of GP values fitted with either a single

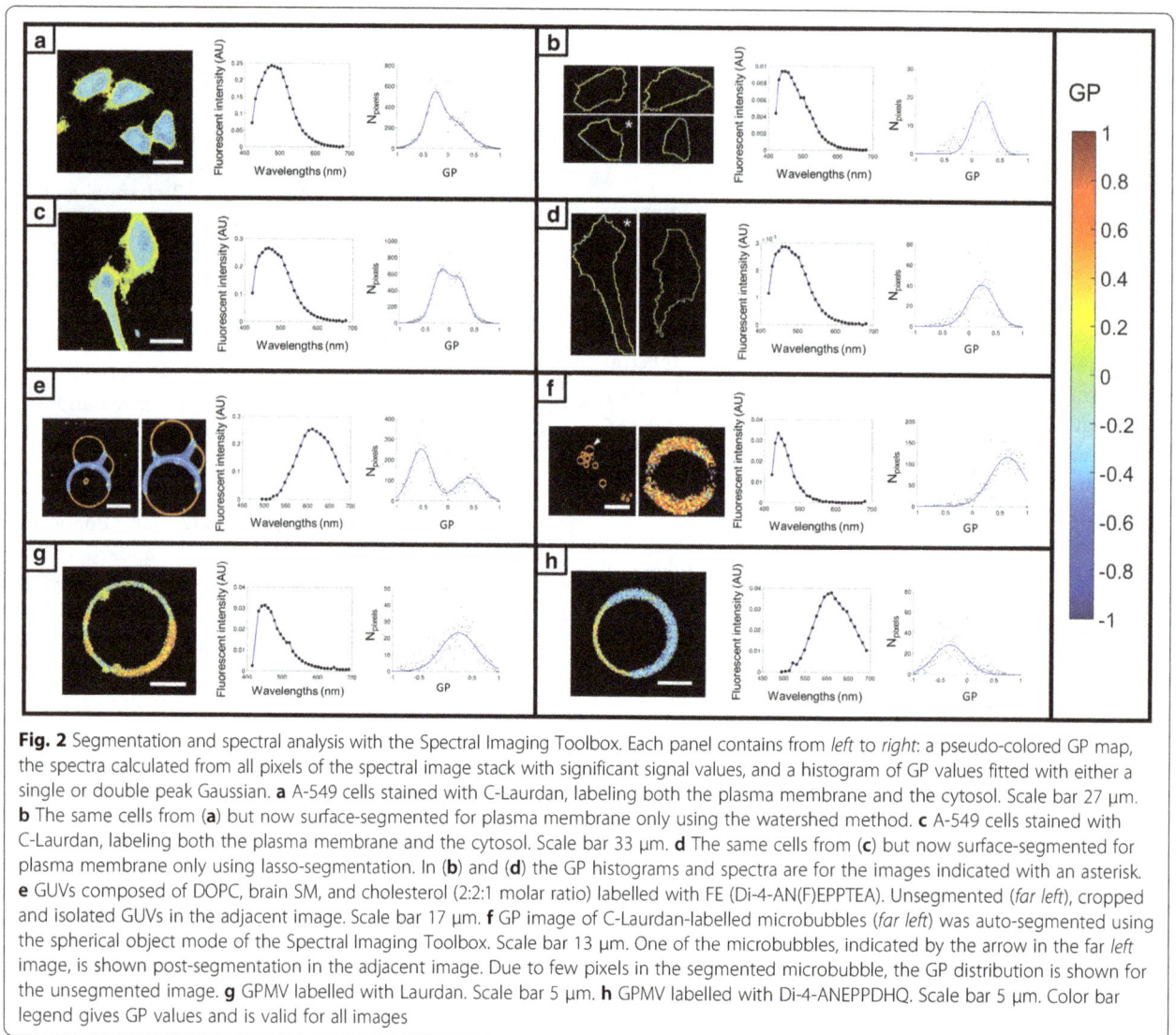

Fig. 2 Segmentation and spectral analysis with the Spectral Imaging Toolbox. Each panel contains from *left* to *right*: a pseudo-colored GP map, the spectra calculated from all pixels of the spectral image stack with significant signal values, and a histogram of GP values fitted with either a single or double peak Gaussian. **a** A-549 cells stained with C-Laurdan, labeling both the plasma membrane and the cytosol. Scale bar 27 µm. **b** The same cells from (**a**) but now surface-segmented for plasma membrane only using the watershed method. **c** A-549 cells stained with C-Laurdan, labeling both the plasma membrane and the cytosol. Scale bar 33 µm. **d** The same cells from (**c**) but now surface-segmented for plasma membrane only using lasso-segmentation. In (**b**) and (**d**) the GP histograms and spectra are for the images indicated with an asterisk. **e** GUVs composed of DOPC, brain SM, and cholesterol (2:2:1 molar ratio) labelled with FE (Di-4-AN(F)EPPTEA). Unsegmented (*far left*), cropped and isolated GUVs in the adjacent image. Scale bar 17 µm. **f** GP image of C-Laurdan-labelled microbubbles (*far left*) was auto-segmented using the spherical object mode of the Spectral Imaging Toolbox. Scale bar 13 µm. One of the microbubbles, indicated by the arrow in the far *left* image, is shown post-segmentation in the adjacent image. Due to few pixels in the segmented microbubble, the GP distribution is shown for the unsegmented image. **g** GPMV labelled with Laurdan. Scale bar 5 µm. **h** GPMV labelled with Di-4-ANEPPDHQ. Scale bar 5 µm. Color bar legend gives GP values and is valid for all images

or double peak Gaussian are provided for each example. Spectral analysis is demonstrated with images of cells stained with C-Laurdan (Fig. 2a and c) and segmented by either the watershed method (Fig. 2b) or manually by lasso-segmentation (see Fig. 2d). The value of membrane segmentation in spectral analysis is highlighted by comparing the spectra and GP distributions of the segmented cells in Fig. 2b and d with the pre-segmentation results in Fig. 2a and c. The segmented spectra are blue-shifted and the GP increased reflecting the higher lipid order of cell membranes compared to the intracellular milieu. This is also indicated by the double-peak Gaussian GP distributions in the pre-segmentation results in Fig. 2a and c. Spectral analysis is also demonstrated with images of GUVs composed of a mixture of DOPC, brain SM, and cholesterol (2:2:1 molar ratio) labelled with FE (Fig. 2e). The presence of DOPC, brain SM, and cholesterol clearly give rise to phase separation as indicated by

the distinct peaks in the GP histogram and in the pseudo-colored GP map. The Spectral Imaging Toolbox was used to auto-crop the fused GUVs and remove background objects for spectral analysis. In this case, membrane segmentation was not necessary because the interior of the GUVs was not fluorescent like in the examples with cells. The lower lipid order region on the GP map (blue pixels), however, is thicker due to this region having higher fluorescence intensity. Membrane segmentation could be used to take an equal-thickness sampling of pixels around the GUV, for consistency of analysis across a population of multiple GUVs. Vesicles derived from cell membranes (GPMVs) and labelled with Laurdan (Fig. 2g) or Di-4-ANEPPDHQ (Fig. 2h) also exhibit phase separation as indicated by their respective GP maps. The GP histograms from the GUVs and GPMVs illustrate a key difference between these two constructs. The phases present in GPMVs are much closer in lipid order than those present in GUVs. Spectral analysis with the spherical object segmentation mode of the Spectral Imaging Toolbox is demonstrated in Fig. 2f with an image of C-Laurdan-labelled microbubbles. The automated segmentation of a microbubble from a cluster of microbubbles is demonstrated.

3D reconstruction of pseudo-colored GP maps
A 3D reconstruction of pseudo-colored GP values calculated from a spectral z-stack of FE-labelled GUVs is demonstrated in Fig. 3. A single slice of the stack can be seen in Fig. 2e. The two phase-separated GUVs in the foreground are connected at their lower end through more lipid-ordered domains (GP < 0).

Microbubble size distribution
Ten spectral image stacks of DSPC-PEG 9:1 molar ratio microbubbles labelled with C-Laurdan were analysed with the spherical object segmentation mode of the Spectral Imaging Toolbox. The resultant pseudo-colored GP maps, size distribution of segmented microbubbles ($n = 71$), and distribution of mean GP values for the segmented microbubbles ($n = 71$) are displayed in Fig. 4.

Discussion
Novel aspects
The Spectral Imaging Toolbox is the first free and open-source software to accurately measure cell membrane lipid packing without cytosolic contributions using a single dye. Furthermore, by implementing batch and hyperstack processing as well as 3D reconstruction of confocal z-stacks, it addresses a growing need to process large spectral imaging datasets and data from experiments with multiple exposures. It is also the only

spectral imaging software to our knowledge to leverage different processing routines for vesicles, for adherent cells, and for regions of interest (i.e., sub-cellular) respectively. Finally, while the algorithms used are not individually novel, their implementation for spectral imaging is not available elsewhere to our knowledge.

Comparison with existing software
Without using membrane segmentation, it is common to decompose the GP histogram into two Gaussian components whereby the lower GP component corresponds primarily to the intracellular regions and the higher GP component to the cell membrane [29]. While this technique is valuable for localizing high and low lipid order regions, it is not appropriate for determining plasma membrane lipid order. Low lipid order domains in the membrane and high lipid order vesicles inside the cell, for instance, could not be attributed to their respective sub-cellular components without some form of segmentation. Thus, more advanced software is required for accurately determining membrane lipid order.

Existing tools of note for processing spectral imaging data with the GP parameter include the ImageJ plugins of Sezgin et al. and Owen et al., and SimFCS developed by Professor Enrico Gratton [10, 30, 31]. These tools all provide adequate means of calculating GP, generating GP visualizations, and histograms for a single spectral image.

The plugin of Owen et al. provides batch processing and enables membrane segmentation with the requirement of a secondary image acquisition and fluorescent membrane label. The Spectral Imaging Toolbox does not require an additional membrane label or image acquisition step to achieve membrane segmentation.

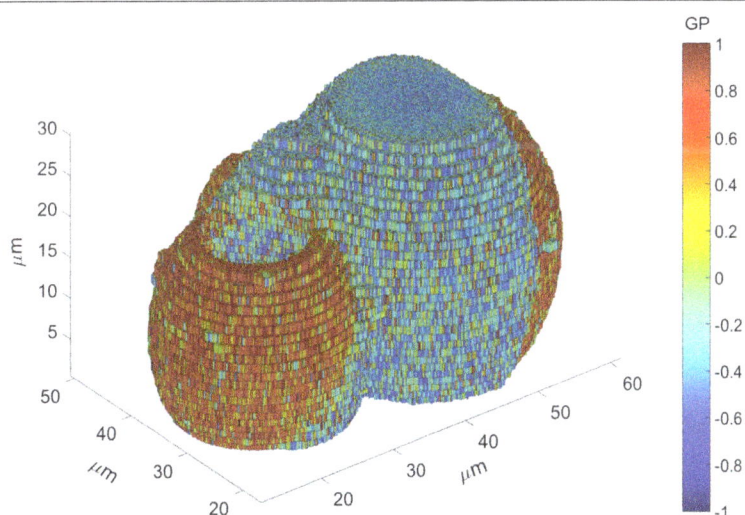

Fig. 3 3D reconstructed GP image calculated from a spectral image stack of FE-labelled GUVs using the Spectral Imaging Toolbox. Axes give spatial dimensions along all three dimensions and color bar legend indicates GP values

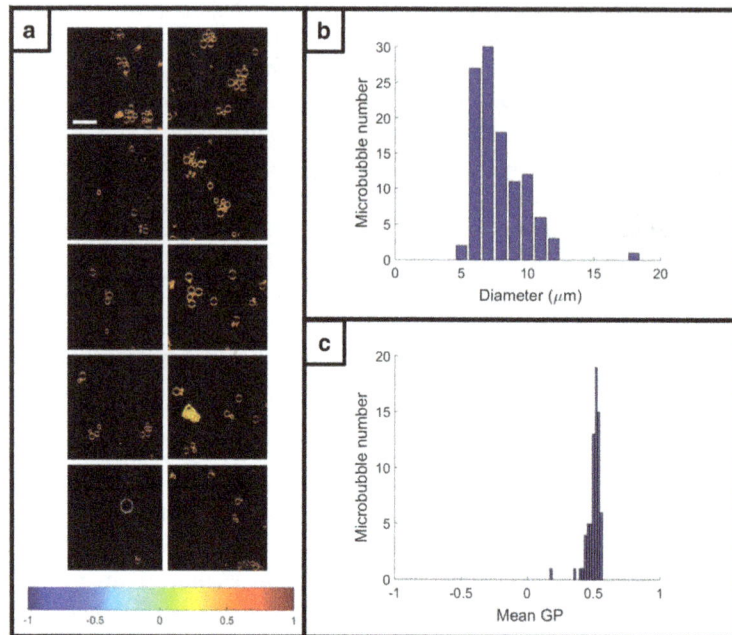

Fig. 4 Spectral analysis and size distribution of microbubbles. **a** Pseudo-colored GP images from 10 spectral image stacks of microbubbles labelled with C-Laurdan. Microbubbles were auto-segmented and analyzed using the spherical object mode of the Spectral Imaging Toolbox. Scale bar 30 μm. Color bar legend gives GP values. **b** Size distribution (diameter) of the segmented microbubbles ($n = 71$). **c** Distribution of mean GP values for the segmented microbubbles ($n = 71$)

Sezgin et al. allow for fitting the spectra of each pixel with either a Gaussian or gamma-variate function to interpolate the intensities, I_B and I_R, for reducing noise in the GP calculation. We found that the gamma-variate fit is most appropriate for spectral imaging data but was too computationally expensive for batch and hyperstack processing. The Spectral Imaging Toolbox instead allows for optionally smoothing the intensity images using a median filter prior to GP calculation, much like SimFCS.

The power of SimFCS is its ability to process many types of advanced imaging data with one software suite. SimFCS does not, however, support batch processing, ROI segmentation, membrane segmentation, or z-stack GP analysis and visualization - core features of the Spectral Imaging Toolbox.

Regarding availability, ImageJ is free [32], as is SimFCS 2 from Globals Software (although the laboratory license for the updated version, SimFCS 4, is $2000). Most research institutions have MATLAB licenses and without a site license, students can purchase MATLAB with the necessary add-ons for only $60.

Another benefit of our software is the ease of customization. SimFCS is not designed for user modification of the source code, and ImageJ provides only a limited macro language and plugin facility. Conversely, the Spectral Imaging Toolbox can be readily extended using MATLAB vector operations well-suited to rapid and complex image processing and analysis. The open-source

code will be maintained on the MATLAB Central File Exchange at the URL provided where updates and feature requests can be publicly discussed.

Conclusion

The Spectral Imaging Toolbox provides an easy-to-use means of analyzing large spectral imaging datasets. It requires no programming experience, outputs publication-quality figures, enables reliable membrane segmentation without the requirement of a counter stain, and incorporates batch and hyperstack processing. It is our intention to continue to develop this free and open-source toolbox with input from the community to further facilitate ambitious research with spectral imaging.

Acknowledgements
We would like to extend our gratitude to Dr. Shamit Shrivastava and Valerio Pereno for helpful discussions, James Fisk and David Salisbury for device fabrication, and Falk Schneider for assistance with GUV preparation.

Funding
This work has been supported by the Engineering and Physical Sciences Research Council (EPSRC, grant number EP/I021795/1) who have provided funding for the research materials and overall project of which this work is a part. Miles Aron gratefully acknowledges the support of the Institute of Engineering and Technology for funding contributions towards his PhD studentship. JBdIS acknowledges support from a Marie Curie Career Integration Grant. CE, JBdIS and ES acknowledge microscope support by the Wolfson imaging Centre and financial support by the Wolfson Foundation, the Medical Research Council (MRC, grant number MC_UU_12010/unit pro-grammes G0902418 and MC_UU_12025), MRC/BBSRC/EPSRC (grant number MR/K01577X/

1), and Wellcome Trust (grant ref 104924/14/Z/14). None of the funding bodies have played any part in the design of the study, in the collection, analysis, and interpretation of the data, or in the writing the manuscript.

Authors' contributions

MA and RB wrote and implemented the software. MA drafted the manuscript. MA and DC performed the measurements with cells. DC, ES, and JBdlS performed the measurements with GUVs. RB performed the experiments with microbubbles. CE and ES supervised and participated in the design of the project. All authors participated in revising the manuscript. All authors read and approved the final manuscript.

Competing interests

The authors declare that they have no competing interests.

Author details

[1]Department of Engineering Science, Institute of Biomedical Engineering, University of Oxford, Oxford OX3 7DQ, UK. [2]Faculty of Engineering and The Environment, University of Southampton, Southampton SO17 1BJ, UK. [3]MRC Human Immunology Unit, Weatherall Institute of Molecular Medicine, University of Oxford, Headley Way, Oxford OX3 9DS, UK. [4]Research Complex at Harwell, Central Laser Facility, Rutherford Appleton Laboratory, Science and Technology Facilities Council, Harwell-Oxford OX11 0FA, UK.

References

1. Lingwood D, Simons K. Lipid rafts as a membrane-organizing principle. Science. 2010;327:46–50.
2. Simons K, Gerl MJ. Revitalizing membrane rafts: new tools and insights. Nat Rev Mol Cell Biol. 2010;11:688–99.
3. Carugo D, et al. Modulation of the molecular arrangement in artificial and biological membranes by phospholipid-shelled microbubbles. Biomaterials. 2016;113:105.
4. Hosny NA, et al. Mapping microbubble viscosity using fluorescence lifetime imaging of molecular rotors. Proc Natl Acad Sci U S A. 2013;110:9225–30.
5. Lentacker I, et al. Understanding ultrasound induced sonoporation: Definitions and underlying mechanisms. Adv Drug Deliv Rev. 2014;72:49–64.
6. De La Serna Bernardino J, et al. Compositional and structural characterization of monolayers and bilayers composed of native pulmonary surfactant from wild type mice. Biochim Biophys Acta. 2013;1828:2450–9.
7. Dodes Traian MM, et al. Imaging lipid lateral organization in membranes with C-laurdan in a confocal microscope. J Lipid Res. 2012;53:609–16.
8. Kwiatek JM, et al. Characterization of a new series of fluorescent probes for imaging membrane order. PLoS One. 2013;8:1–7.
9. Parasassi T, et al. Two-photon fluorescence microscopy of laurdan generalized polarization domains in model and natural membranes. Biophys J. 1997;72:2413–29.
10. Sezgin E, et al. Spectral imaging to measure heterogeneity in membrane lipid packing. ChemPhysChem. 2015;16:1387–94.
11. Yu W, et al. Fluorescence generalized polarization of cell membranes: a two-photon scanning microscopy approach. Biophys J. 1996;70:626–36.
12. Sezgin E, et al. Measuring lipid packing of model and cellular membranes with environment sensitive probes. Langmuir. 2014;30:8160–6.
13. MATLAB. version 8.5.0 (R2015a) The Mathworks Inc., Natick, Massachusettes. 2015.
14. Linkert M, et al. Metadata matters: access to image data in the real world. J Cell Biol. 2010;189:777–82.
15. Aitkenhead A. Plot a 3D array using patch. MATLAB Central File Exchange. 2010. https://www.mathworks.com/matlabcentral/fileexchange/28497-plot-a-3d-array-using-patch. Retrieved March 25, 2016.
16. Otsu N. A threshold selection method from gray-level histograms. IEEE Trans Syst Man Cybern. 1979;9:62–6.
17. Haralock RM, Shapiro LG. Computer and robot vision Addison-Wesley Longman Publishing Co., Inc. 1991.
18. Meyer F. Topographic distance and watershed lines. Signal Process. 1994;38: 113–25.
19. Lim JS. Two-dimensional signal and image processing, vol. 710. Englewood Cliffs: Prentice Hall; 1990. p. 1.
20. Parker JR. Algorithms for image processing and computer vision John Wiley & Sons. 2010.
21. Soille P. Morphological image analysis: principles and applications Springer Science & Business Media. 2013.
22. van den Boomgaard R, van Balen R. Methods for fast morphological image transforms using bitmapped binary images. CVGIP Graph Model Image Process. 1992;54:252–8.
23. Atherton TJ, Kerbyson DJ. Size invariant circle detection. Image Vis Comput. 1999;17:795–803.
24. Yuen H, et al. Comparative study of Hough Transform methods for circle finding. Image Vis Comput. 1990;8:71–7.
25. Brewer J, et al. Multiphoton excitation fluorescence microscopy in planar membrane systems. Biochim Biophys Acta. 2010;1798:1301–8.
26. Angelova MI, Dimitrov DS. Liposome electro formation. Faraday Discuss Chem Soc. 1986;81:303–11.
27. Carugo D, et al. Biologically and acoustically compatible chamber for studying ultrasound-mediated delivery of therapeutic compounds. Ultrasound Med Biol. 2015;41:1927–37.
28. Sezgin E, et al. Elucidating membrane structure and protein behavior using giant plasma membrane vesicles. Nat Protoc. 2012;7:1042.
29. Golfetto O, et al. Laurdan fluorescence lifetime discriminates cholesterol content from changes in fluidity in living cell membranes. Biophys J. 2013;104:1238–47.
30. Owen DM, et al. Quantitative imaging of membrane lipid order in cells and organisms. Nat Protoc. 2012;7:24–35.
31. Sanchez S a, et al. Laurdan generalized polarization fluctuations measures membrane packing micro-heterogeneity in vivo. Proc Natl Acad Sci. 2012; 109:7314–9.
32. Schindelin J, et al. The ImageJ ecosystem: an open platform for biomedical image analysis. Mol Reprod Dev. 2014;82:518–29.

6

Generalizing cell segmentation and quantification

Zhenzhou Wang[*] ⓘ and Haixing Li

Abstract

Background: In recent years, the microscopy technology for imaging cells has developed greatly and rapidly. The accompanying requirements for automatic segmentation and quantification of the imaged cells are becoming more and more. After studied widely in both scientific research and industrial applications for many decades, cell segmentation has achieved great progress, especially in segmenting some specific types of cells, e.g. muscle cells. However, it lacks a framework to address the cell segmentation problems generally. On the contrary, different segmentation methods were proposed to address the different types of cells, which makes the research work divergent. In addition, most of the popular segmentation and quantification tools usually require a great part of manual work.

Results: To make the cell segmentation work more convergent, we propose a framework that is able to segment different kinds of cells automatically and robustly in this paper. This framework evolves the previously proposed method in segmenting the muscle cells and generalizes it to be suitable for segmenting and quantifying a variety of cell images by adding more union cases. Compared to the previous methods, the segmentation and quantification accuracy of the proposed framework is also improved by three novel procedures: (1) a simplified calibration method is proposed and added for the threshold selection process; (2) a noise blob filter is proposed to get rid of the noise blobs. (3) a boundary smoothing filter is proposed to reduce the false seeds produced by the iterative erosion. As it turned out, the quantification accuracy of the proposed framework increases from 93.4 to 96.8% compared to the previous method. In addition, the accuracy of the proposed framework is also better in quantifying the muscle cells than two available state-of-the-art methods.

Conclusions: The proposed framework is able to automatically segment and quantify more types of cells than state-of-the-art methods.

Keywords: Boundary filtering, Noise blob filtering, Threshold selection, Calibration, Iterative erosion

Background

Imaging of cells in biology are becoming more and more popular with the fast development of microscopy and nanotechnology [1–7]. In different applications, different ways had been utilized to separate the imaged cells and they usually took the researchers great effort. As a powerful tool, the image processing technology is becoming more and more important for the segmentation, quantification and analysis of microscopy data [8, 9]. In different applications, the forms, the dimensions of the cells and their gray-level distributions vary significantly, which makes the segmentation task challenging. In many applications, the cells are frequently neighboring or overlapping on each other, which makes the quantification difficult. In this paper, we propose a generalized framework for robust segmentation and quantification of different types of cells imaged in different biological applications.

In the past decades, image processing technology has been utilized widely in segmenting and quantifying different types of cells. The absence of a generalized framework for different types of cell images makes the

* Correspondence: zzwangsia@yahoo.com
State Key Laboratory for Robotics, Shenyang Institute of Automation, Chinese Academy of Sciences, Shenyang, China

research work application specific instead of convergent to a common solution. Different methods were proposed and claimed to be superior in segmenting a class of cells. These methods include watershed method [10–12], region growing based method [13], morphological method [14, 15], clustering based method [16], contour based method [17], multilayer segmentation based method [18], pattern modeling based method [19], supervised learning method [20], morphological watershed based method [21], inference based method [22] and methods that combine the threshold selection and morphology techniques [23–25]. However, the performance and applicability of most of these methods are very limited because they are diverging rather than convergent to a generalized solution to address so many types of cells. To overcome this drawback, the author has proposed a new approach to segment and quantify different types of cells or nanoparticles based on the general property of the cell images: global intensity distribution and local gradient [24], which is more versatile than the referenced state of the art methods. The approach proposed in [24] evolves the method proposed in [25] and makes it to be able to segment and quantify more types of cells or nanoparticles. One fundamental improvement of [24] compared to [25] is that the threshold selection method used in [25] was improved to be able to segment more types of cells or nanoparticles robustly. However, the details of how to apply the proposed threshold selection method with practical cell images are not addressed adequately in [24]. In this paper, we design the practical algorithm to apply the threshold selection method proposed in [24] to segment the practical cell images. In addition, we calibrate more parameters than [24] to guarantee the robust segmentation.

A more important goal of this paper is to propose a generalized framework to segment and quantify different types of cells imaged in different systems with higher accuracy compared to the past work [10–25]. To this end, we tested more cell images in addition to the muscle cell images in [25]. Some imaged cell images have artifacts or the segmentation results contain too much noise. Consequently, the segmented cells contain shape noises which will increase the number of the eroded seeds by the iterative erosion method proposed in [25], which will affect the final quantification accuracy. To eliminate these shape noises, we propose a Fourier Transformation based shape filter and it could decrease the wrong quantification effectively. In addition to the shape filter, we also propose a blob filter that could remove the line shape noise blobs effectively. For the muscle cell images [25], two cases are defined in the union method

based on the image characteristics. For the generalized framework to segment more types of cell images, three cases are defined in the union method in this paper. To verify the advantages of the proposed generalized framework over the past research work [10–25], we give both the qualitative results and the quantitative results.

Methods
The generalized framework
The proposed framework for segmentation and quantification of the cells is illustrated in Fig. 1. In the framework, the content in the ellipse vary depending on the input image to be processed while the content in the rectangle are the proposed algorithms and they remain the same for different types of cells. The input image denotes the original cell image. The gradient image is obtained after edge enhancement. Both the input image and the gradient image are segmented by the threshold selection method automatically to get the binarized image and the constraint edge image, respectively. The segmentation result is obtained by unifying the binarized image and the negative constraint edge image. The noise blob removing filter is used to eliminate the thread-like or small noise blobs. The boundary smoothing filter is used to remove the noise contained in the extracted

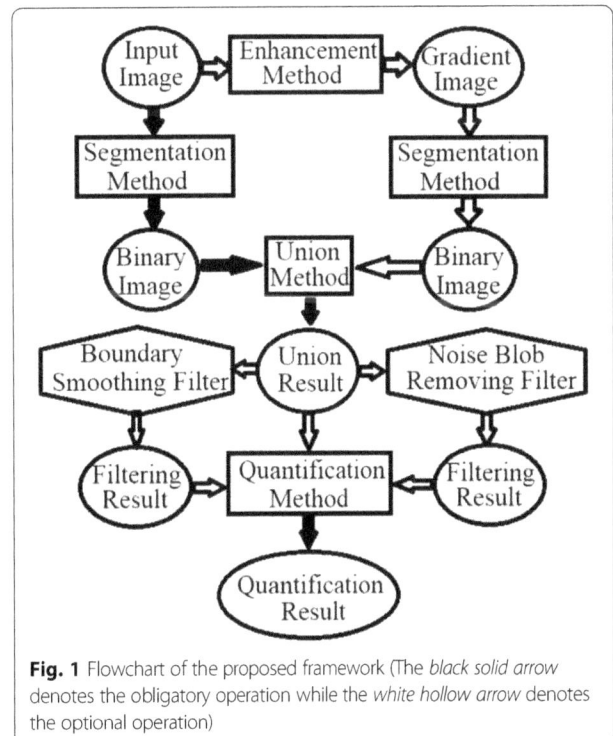

Fig. 1 Flowchart of the proposed framework (The *black solid arrow* denotes the obligatory operation while the *white hollow arrow* denotes the optional operation)

boundary. The quantification method is used to obtain the quantification result after identifying each cell individually. Except the segmentation method, the union method and the quantification method, all the other methods are optional based on the characteristics of the cells. For each type of cell, the methods are selected and then applied one by one in the framework and they need to be prepared and calibrated carefully before the framework could segment and quantify the cells automatically.

The enhancement method

The gradient image is generated by the enhancement method. In the current framework, we generate the gradient image using the Sobel operators. Firstly, the Sobel operator is applied to the cell image along the row direction to get the horizontal gradient components, I_x. Secondly, the Sobel operator is applied to the cell image along the column direction to get the vertical gradient components, I_y. Thirdly, the gradient image is formed by the following equation.

$$I_g(h,j) = \sqrt{I_x^2(h,j) + I_y^2(h,j)} \qquad (1)$$

where $h = 1, ..., H$; is the index of the pixel along the column direction and $j = 1, ..., J$; is the index of the pixel along the row direction. $H \times J$ denotes the dimension of the cell image.

The segmentation method

The segmentation method should be flexible and robust enough for a vast variety of cell images. We tested all the available state of the art segmentation methods [26–33] to segment the cell images and the generated gradient images. Unfortunately, we did not find any state of the art segmentation method that could yield adequate accuracy consistently for so many types of cell images. A more versatile, flexible and generalized image segmentation method has been proposed in [25] to produce acceptable segmentation results consistently for many types of muscle cell images. However, the histogram modalities of the muscle cell images are similar and so are the modalities of their gradient images, which makes the image segmentation less challenging compared to segmenting more divergent types of cell images. Fortunately, the flexibility of the previously proposed threshold selection method makes it adjustable for different types of images by varying the its parameters. Hence, we introduce the process of calibration in this paper to find the optimal parameters of the threshold selection method for each specific type of cell image. The

previously proposed threshold selection method could be summarized as follows.

The threshold is calculated from the slope difference distribution of the normalized histogram. The histogram is assumed as Gaussian-mixture distributions in this research work. We define the slope difference distribution of the image as the variation rate of the normalized histogram and it could be computed by the following steps.

Step 1, Assuming the image is non-negative, the cell image is modified by rearranging its gray-scale values in the interval [0, 255] with the following equation.

$$I'(h,j) = \frac{255 \times I(h,j)}{\max(I)}; \quad h = 1, ..., H; j$$
$$= 1, ..., J \qquad (2)$$

where $H \times J$ is the resolution of the cell image, I. h is the index of the pixel along the vertical direction of the cell image and j is the index of the pixel along the horizontal direction of the cell image. Here, 255 is used for convenience because most gray images have the maximum value of 255. 255 could be changed to other values based on the application requirements.

Step 2, the histogram distribution $P(x)$ of the modified cell image, I' is normalized by the following equations:

$$P(x = m) = \frac{N_m}{N_l}; m = 0, ..., 255; \qquad (3)$$

$$l = \operatorname*{argmax}_{l \in [0,255]} N_{\bar{l}} \qquad (4)$$

where N_l denotes the maximum frequency that occurs at l in the interval [0, 255]. N_m denotes the frequency of the pixel value m.

Step 3, after the histogram distribution is normalized, it is then filtered in the frequency domain. Firstly, the normalized histogram distribution, $P(x)$ is transformed into the frequency domain with the Discrete Fourier Transformation (DFT):

$$F(k) = \sum_{x=0}^{255} P(x) e^{-i\frac{2\pi kx}{255}}; k = 0, ..., 255 \qquad (5)$$

Then, we select the low frequency parts from 1 to L and eliminate the rest of high frequency parts with the following equation.

$$F'(k) = \begin{cases} F(k); k = 0, 1, ..., W \\ F(k); k = 255 - W, ..., 254, 255 \\ 0; k = W + 1, ..., 255 - W - 1 \end{cases} \qquad (6)$$

where W the bandwidth of the low pass DFT filter and it is going to be determined by the calibration process.

After the above equation is performed to filter histogram distribution in the frequency domain, we transform the smoothed histogram distribution back into spatial domain by the following equation.

$$P'(x) = \left| \frac{1}{255} \sum_{k=0}^{255} F'(k) e^{i\frac{2\pi x k}{255}} \right|; x = 0, ..., 255 \quad (7)$$

where $P'(x)$ is the filtered and smoothed histogram.

Step 4, for each point, i on $P'(x)$, there are two slopes, $a_1(i)$ and $a_2(i)$. They are on the left side and the right side of the point, i respectively. They could be computed by a fitted line model with N adjacent points at each side and the parameter N will also be determined by the calibration process. The line model is formulated as:

$$y_i = ax_i + b \quad (8)$$

$$[a, b]^T = (B^T B)^{-1} B^T Y \quad (9)$$

$$B = \begin{bmatrix} x_1 & 1 \\ x_2 & 1 \\ \vdots & \vdots \\ x_N & 1 \end{bmatrix} \quad (10)$$

$$Y = [y_1, y_2, ..., y_N]^T \quad (11)$$

When the N fitting points are on the left side of the point i, the slope a equals $a_1(i)$. When the N fitting points are on the right side of the point i, the slope a equals $a_2(i)$. Both slopes are computed by Eq. 9.

Accordingly, the slope difference of the point i is computed by the following equation.

$$s(i) = a_2(i) - a_1(i); i = N + 1, ..., 255 - N \quad (12)$$

The continuous version as $s(i)$ is defined as the slope difference distribution. Setting its derivative to zero, we could get the N_v valleys V_i; $i = 1, ..., N_v$ with greatest local variations and N_p peaks P_i; $i = 1, 2, ..., N_p$ with greatest local variations of the slope difference distribution. Not all peaks or valleys are caused by the histogram variations because the smoothing process by the low-pass DFT filter might produce small harmonics when significant parts of the original histogram remain the same or close to the horizontal axis. Consequently, these harmonics produce pseudo peaks and valleys. Fortunately, the pseudo peaks or valleys are much smaller compared to the real peaks or valleys. The real peaks or valleys could be distinguished from the pseudo ones easily based on their magnitudes. On the other hand, the produced harmonics avoid the possible ill-conditions of the matrix inverse

operation in Eq. 9. The matrix inverse operation will become ill-conditioned when the N fitting points are from a horizontal line. The horizontal parts in the histogram are replaced with harmonics after DFT filtering. We demonstrate the slope difference distribution with three synthesized images in Fig. 2. The first synthesized image is an image with two objects as shown in Fig. 2a. The grayscale of the background equals 50, the grayscale of the dark object equals 120 and the grayscale of the bright object equals 220. Its slope distribution is demonstrated in Fig. 2d. The original histogram distribution consists of three isolated peaks. After DFT filtering, the histogram distribution become continuous with small harmonics that produce many small pseudo peaks and valleys. There are three real peaks and six real valleys and their magnitudes are much greater than those of the pseudo ones. The peaks and valleys are denoted by the blue crosses and red circles respectively in Fig. 2d. The second image is synthesized by adding Gaussian noise to the first synthesized image and it is shown in Fig. 2b. Its slope difference distribution is shown in Fig. 2e. As can be seen, its original histogram is continuous with less parts on the horizontal axis. As a result, less harmonics and less pseudo peaks and valleys are generated. The third image is synthesized by blurring the second synthesized image with an iterative moving average filter and it is shown in Fig. 2c. Its slope difference distribution is shown in Fig. 2f. As can be seen, its original histogram is also continuous. However, many parts are close to the horizontal axis. As a result, many pseudo peaks and pseudo valleys occur. From all these results, it is seen that the real peaks or valleys could be easily distinguished from the pseudo peaks or valleys. For most practical images, their histograms are usually continuous without significant parts close to the horizontal axis or remain the same, thus no pseudo peaks or valleys will occur. For the image with known number of pixel classes K_c, the rule to select the peaks is as follows. Firstly, all the peaks are sorted in the magnitude descending order. Secondly, the first K_c peaks are then selected as the real peaks.

The slope difference distribution has three fundamental properties that help to design the threshold selection process.

Property 1: in situations where the histogram distribution of background and the histogram distribution of the cells are both Gaussian distributed, the valley positions between the background and the object on the slope difference distribution change monotonically with the number of the fitted points N in the line model while the peak positions are almost the same when the parameter, N is changed gradually. In

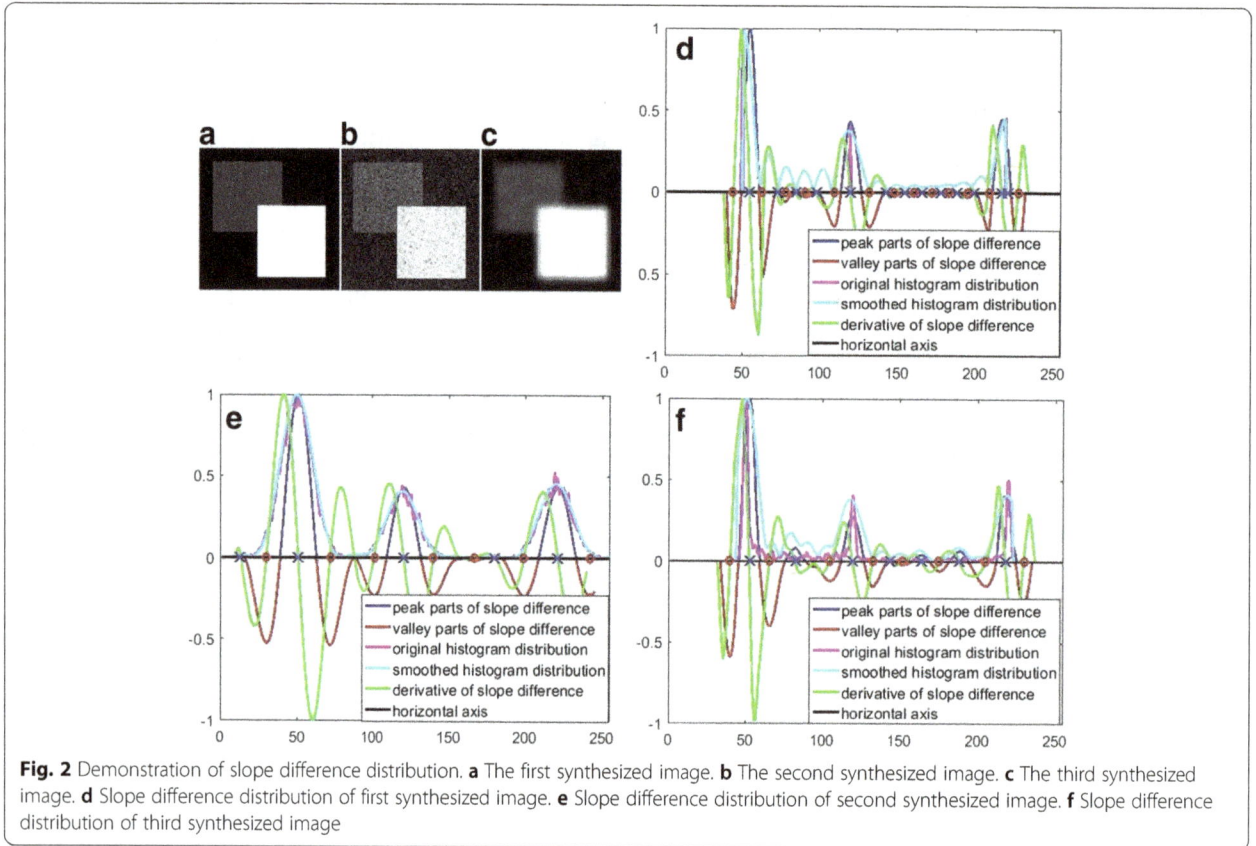

Fig. 2 Demonstration of slope difference distribution. **a** The first synthesized image. **b** The second synthesized image. **c** The third synthesized image. **d** Slope difference distribution of first synthesized image. **e** Slope difference distribution of second synthesized image. **f** Slope difference distribution of third synthesized image

the experiments, we found that this property holds only when the histogram is filtered by the designed filter with the bandwidth parameter W calibrated and chosen properly for each specific type of image. When we used other filters, for instance, the finite impulse response (FIR) filter and the infinite impulse response (IIR) filter, both the peaks and the valleys of the slope difference distribution change irregularly. Hence, we conclude that the Fourier Transformation based filtering is capable of removing the high frequency noises effectively while maintaining the shape of the histogram well. However, the FIR filter and IIR filter lack this capability and will change the shape of smoothed histogram undesirably. Consequently, they cause the peaks of the slope difference distribution to change randomly.

Property 2: the peaks of the slope difference distribution correspond to the cluster centers of the objects or the background while the valleys correspond to the thresholds that could separate the objects and the background.

Property 3: the fitting number N of line model determines the number of the peaks of the slope difference distribution. A large N value could suppress small peaks and unify adjacent peaks into one peak.

The proposed threshold selection method is flexible and has some changeable manual inputs that could be adjusted to meet different segmentation requirements. The first manual input defines how many pixel classes the image contains. The default value of it is 2, which indicates that there are one object class and one background class. The second manual input defines what classes to segment. When the user wants to separate the background class and all the objects classes, it is defined as ***Case 1***. When the user wants to separate the first object class and the second object class along the pixel increase direction, it is defined as ***Case 2***. ***Case 3*** is defined as the separation between the second object class and the third object class. In the same way, other cases are defined. ***Case 1*** is default case. The third input is how many points the line model uses to fit the line and the fourth input is the bandwidth of the low pass filter. To determine the third and fourth inputs for each type of cell images before segmentation, we calibrate the threshold selection method based on the popular F-measure. For a specific type of cell images, the calibration process is summarized as follows.

For a specific type of cell images, we select several typical images and obtain the ground truth manual segmentation results for these images.

Then, we vary the value of the third input, the parameter N in Eq. (11) from 3 to 60 and the fourth input, the parameter W in Eq. (6) from 2 to 50. We compute the F-measure, F_m of the automatic segmentation result by the threshold selection method and the manual segmentation result for each pair parameters (N, W) by the following equations.

$$F_m = \frac{2 \times P \times R}{P + R} \tag{13}$$

$$P = \frac{S_{SD} \cap S_m}{S_m} \tag{14}$$

$$R = \frac{S_{SD} \cap S_m}{S_{SD}} \tag{15}$$

where S_m is the ground truth manual segmentation and S_{SD} is the automatic segmentation result by the threshold selection method. We choose the pair of parameters (N, W) that yields the largest F_m.

During segmentation of a great of variety of cell images, it might be inconvenient to obtain the benchmark manual segmentation from the cytologist for each type of cell image. Here, we propose a rational calibration method in the absence of benchmark manual segmentation result based on **Property 3** of the slope difference distribution.

Step 1: we determine how many pixel classes the image contains rationally. Here, we give an example of cell image with three pixel classes: the black cell, the gray clutters and the brighter background as shown in Fig. 3a. There are small abrupt parts with pixel values close to 255 in the original histogram distribution, which affects the normalization of the histogram and makes most parts of the histogram below 0.5. After DFT filtering, this bad effect is removed and the normalized histogram becomes much better.

Step 2: we use the default value, $N = 15$ and $W = 10$ to calculate the thresholds visually as shown in Fig. 3b. It is seen that there are 7 peaks instead of 3

peaks existing in the calculated slope difference distribution.

Step 3: we increase the value of N until there are only 3 peaks in the calculated slope difference distribution as shown in Fig. 3c.

Step 4: we select the threshold according to rules described above from the calculated slope difference distribution with three peaks.

Please note that the proposed rational calibration method is used only when the benchmark manual segmentation is not available. When the benchmark data is available, the calibration method based on the F-measure is used because it is more robust than the proposed rational method.

The union method

We calculate the threshold, T_0 for the modified input cell image, I' with the efficiently calibrated threshold selection method. Then, the modified cell image is binarized by the following equation.

$$S_I = \begin{cases} 1; & I' \geq T_0 \\ 0; & I' < T_0 \end{cases} \tag{16}$$

We calculate the threshold, T_1 for the gradient image, I_g with the efficiently calibrated threshold selection method. Then, the gradient image is binarized as follows.

$$S_g = \begin{cases} 1; & I_g \geq T_1 \\ 0; & I_g < T_1 \end{cases} \tag{17}$$

After calculating the two segmentations, S_I and S_g, we compute their union segmentation S_u in three cases. For one specific type of cell image, the user need to decide which case it belongs to. For the cell images with a lot of overlapping/neighboring boundaries and their segmented boundaries are not closed for each

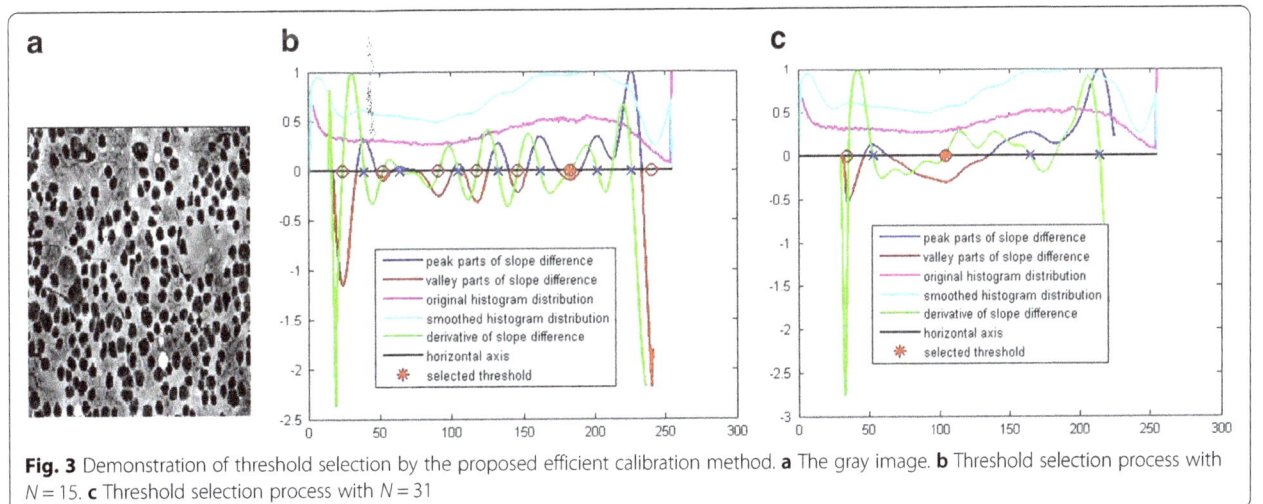

Fig. 3 Demonstration of threshold selection by the proposed efficient calibration method. **a** The gray image. **b** Threshold selection process with $N = 15$. **c** Threshold selection process with $N = 31$

cell, we define that they belong to *Case 1* and most cell images belong to this case. For this case, the segmentation method is formulated as follows to utilize the segmentations of the gradient image and original image, S_g and S_I.

$$S_u = \begin{cases} 1; & if\ (S_I = 1)\ and\ (S_g = 0) \\ 0; & else \end{cases} \quad (18)$$

For the cell images with many overlapping or neighboring boundary and the segmented boundary for each cell is closed, we define that they belong to *Case 2*. For instance, many muscle cell images belong to this case. For this case, the segmentation method is formulated as follows to utilize the segmentation of the gradient image, S_g alone.

$$S_u = \begin{cases} 1; & if\ S_g = 0 \\ 0; & else \end{cases} \quad (19)$$

For the cell images with little overlapping or neighboring boundary, we define that they belong to *Case 3*. In this case, we formulate the segmentation method as follows to make use of the segmentation of the original input image, S_I only.

$$S_u = \begin{cases} 1; & if\ S_I = 1 \\ 0; & else \end{cases} \quad (20)$$

The noise blob removing filter

In many situations, there are a lot of noise blobs in the union segmentation S_u, which might affect the accuracy of the automatic quantification process. One big difference between the noise blob and the cell blob is that the noise blob is usually more tenuous than the cell blob

as shown in Fig. 4a, where the noise blobs are threadlike while the cell blobs are relatively massive. Hence, we propose the following filter to remove this kind of noise blobs.

Step 1: Erode the union segmentation, S_u morphologically by the following equations.

$$S_u' = S_u \ominus B = \left\{ z | (B)_z \subseteq S_u \right\} \quad (21)$$

$$(B)_z = \{ c | c = p + z,\ p \in B \} \quad (22)$$

$$S_u = S_u' \quad (23)$$

where B is the 4-connected structure element with the disk shape and its radius is 1. p is the point in the structuring element B and z is the translation vector.

Step 2: Repeat **Step 1** N_l times. The value of N_l is determined by the user and its default value is 3.

Step 3: Dilate the union segmentation, S_u morphologically by the following equations.

$$S_u' = S_u \oplus B = \left\{ z | (B^s)_z \cap C_j^i \neq \varnothing \right\} \quad (24)$$

$$S_u = S_u' \quad (25)$$

where B^s denotes the symmetric or supplement of B.

Step 4: Repeat **Step 3** N_l times.

The functionality of the above filter is to remove the threadlike or small blobs by a repeating morphological erosion process at first. Then, a morphological dilation process with the same repeating times is used to restore the eroded cell blobs. Figure 4b shows the result of applying the above filter to the union segmentation shown in Fig. 3a. As can be seen, the tenuous noise blobs are removed effectively while the cell blobs are maintained well.

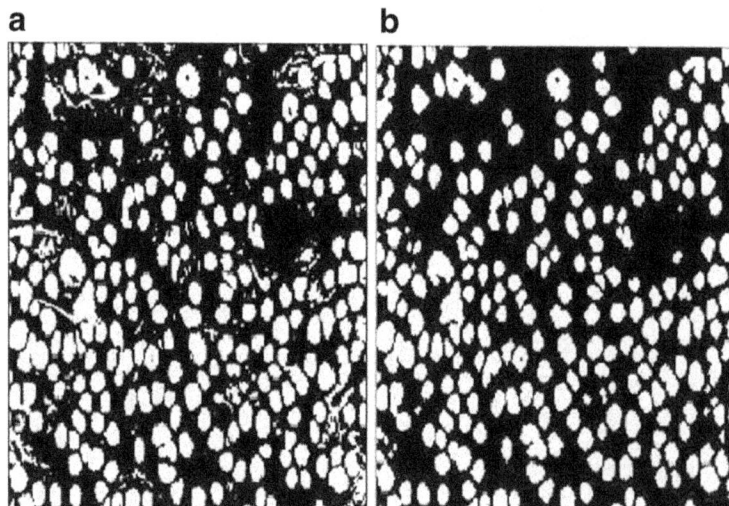

Fig. 4 Demonstration of removing the noise blobs by the proposed filter. **a** The result of union segmentation. **b** The filtering result by the proposed noise blob removing filter

The boundary smoothing filter

There are some imaged cell images with poor quality or with imaging artifacts. As a result, the segmentation results produce a lot of boundary noise that is defined as the elements e.g. the boundary roughness or holes inside the segmented cells that make the cells irregular. The irregular cells will produce more seeds during the iterative erosion process proposed in [24, 25], which will increase the quantified number. To eliminate the shape noises, we propose a boundary smoothing filter as follows.

Step 1: Exact the boundaries $\{x_i^j, y_i^j\}; i = 1, 2, ..., N_j$ of all the binary blobs and the holes inside the blobs in the union segmentation S_u. j denotes the index of the binary blobs and the holes inside the blobs. i denotes the index of the point in the j th extracted boundary for the j th binary blob or the hole. N_j denotes the total number of the points in the j th extracted boundary.

Step 2: For the j th boundary, if $N_j > T_{sn}$, the boundary is valid and will be kept. Otherwise, the boundary is invalid and will be removed. T_{sn} is the shape noise threshold and it could be computed based on the average size of all the segmented blobs in the image. For a specific type of cell images, the sizes of the cells and the sizes of the noise blobs usually change in different ranges. Offline analysis could find a more accurate size threshold, T_{sn} to separate cells and the noise blobs robustly.

Step 3: For all the valid boundaries, filter them by the Fourier filter defined by Eqs. (5-7). The input is changed from the normalized histogram to x coordinates and y coordinates of the valid boundaries respectively.

Step 4: Using the filtered boundaries to compute binary blobs again and form the filtered blob image, I_{fb}.

The quantification method

In most cases, there are cells separate from others and there are also cells connected with each other in the filtered blob image, I_{fb}. To identify the cell individually, the same iterative morphological erosion method proposed in [24, 25] is used here.

Step 1: Initialize the seeds of all the cells to be the filtered blob image, I_{fb}.

$$I_b^1 = I_{fb} \tag{26}$$

Step 2: Erode the seeds I_b^i morphologically with the structure element $B = \{(0, 0)\}$ as follows.

$$I_b' = I_b^i \ominus B = \{z|(B)_z \subseteq I_b^i\} \tag{27}$$

$$(B)_z = \{c|c = p + z, \ p \in B\} \tag{28}$$

where p is the point in the structuring element B and z is the translation vector.

Step 3: Then calculate the union of the separated cells that are determined according to their areas. Use them as the updated seeds.

$$I_c^{i+1} = \bigcup C(\tilde{j}); \tilde{j} = \arg_j area(C(j)) < S_0 \tag{29}$$

$$I_b^{i+1} = I_b' - I_c^{i+1} \tag{30}$$

S_0 that is defined as the area threshold to distinguish the area of the cell and the area of noise blob, is computed as the mean area of all the cells after a number of erosions on the segmented cells.

Step 4: Use the above steps to erode the segmented cells until the area of each cell is smaller than S_0. At last, the seeds are updated as:

$$I_s = \bigcup_{i=1}^{L} I_c^i \tag{31}$$

where L denotes the total number of the isolated cells. After all the cells are identified, the coordinate (x_c^k, y_c^k) of the k th cell's center is computed as:

$$x_c^k = \frac{1}{M} \sum_{j=1}^{M} x_j^k \tag{32}$$

$$y_c^k = \frac{1}{M} \sum_{j=1}^{M} y_j^k \tag{33}$$

where M is the total number of pixels in the segmented cell and j is the pixel index of the segmented cell.

The algorithm of the generalized framework

The generalized framework is summarized in Algorithm 1.

The calibration process based on the F-measure is summarized in Algorithm 2.

Results

In this section, we verify the robustness and the generality of the proposed framework with both the muscle cell images used in [24, 25] and other types of cell images.

One big difference between the proposed framework in this paper and the methods proposed in [24, 25] is the inclusion of the boundary smoothing filter. Here, we use two examples of muscle cells to demonstrate the advantages of the proposed framework in this paper over the methods proposed in [24, 25]. Two typical muscle images that have been used to testify the proposed method in [25] are used to show the superiority of the proposed framework in Figs. 5 and 6 respectively. Figure 5a shows the gradient image enhanced from the gray image by Eq. (1). Figure 5b shows the threshold selection process for the gradient image. The smoothed histogram is plotted in cyan. The original histogram is plotted in mauve. The peak part of the slope difference distribution is plotted in blue and the valley part of the slope difference is plotted in red. The derivative of the slope difference is plotted in green and its interception points with the horizontal axis are denoted as blue crosses when they correspond to the peaks of the slope difference. They are denoted as the red circles when they correspond to the valleys of the slope difference. The selected threshold is denoted as the red asterisk. After calibration, the optimal W value is chosen as 10 and the optimum N value is chosen as 17. Figure 5c shows the segmented edges with the selected threshold. Figure 5d shows the gray image of the muscle cell image and Fig. 4e shows the threshold selection process for it. Figure 5f shows the segmented edges from the gray image. Figure 5g shows the filtered boundary overlaying on the segmentation result by the case 1 union method. Figure 5h shows the cell quantification result overlaying on the original cell image.

Figure 6a-h show the segmentation and quantification results of another testified muscle cell image in [25]. To compare the quantification accuracy of the generalized framework in this research work and the method previously proposed in [25] more conveniently, we show the quantification results by [25] in Fig. 7. As can be seen, two missing cells in the quantification result of Fig. 7a are quantified correctly in Fig. 5h. In addition, the extra one false quantification in Fig. 7a is avoided in Fig. 5h. Similarly, the quantification result in Fig. 6h are better than that in Fig. 7b.

To demonstrate the advantage of the generalized framework over state of the art methods, we show the results of the two muscle cell images by the SMASH method [34] and the CELLSEGM method [35] in Fig. 8. As can be seen, the generalized framework yields

---

Algorithm 1 The generalized framework

Input: Image, I with grayscale values re-arranged in the interval [0, 255]; The number of fitting points of the line model, N; The bandwidth of the DFT filter, W; The number of pixel classes, K_c; The case in which the user wants the image to be segmented, C_s; The case in which the user selects the union method, C_u; The repeat times, N_l in the noise blob removing filter; The shape noise threshold, T_{sn} in the boundary smoothing filter; The area threshold, S_0 in the quantification method.

Generalized Framework:

    The enhancement method:

    1) Compute the gradient image by the enhancement method.

    The segmentation method:

    2) Compute the normalized histogram of the input image and smooth the normalized histogram with the DFT filter.

    3) Compute the slope difference distribution.

    4) Sort the peaks of the slope difference distribution in the magnitude decrease direction.

    5) Choose the first K_c peaks as valid peaks and delete other smaller peaks.

    6) Select two peaks according to the inputted case, C_s and find the valley with maximum absolute value between them.

    7) The threshold is selected at the position corresponding to the found valley, V_i.

    The union method:

    8) Select the optimal union case, C_u and obtain the corresponding segmentation result, S_u for the cell image.

    The noise blob removing filter:

    9) Apply the noise blob removing filter to S_u with the selected repeat times, N_l.

    The boundary smoothing filter:

    10) Apply the boundary smoothing filter to S_u with the selected shape noise threshold, T_{sn}.

    The quantification method:

    11) Apply the quantification method to S_u with the selected area threshold, S_0.

Output: The quantification result.

Algorithm 2 The calibration process

Input: A specific type of image, I with grayscale values re-arranged in the interval [0, 255]; The benchmark quantification result, I_q for the input image; The number of fitting points of the line model, N; The bandwidth of the DFT filter, W; The number of pixel classes, K_c; The case in which the user wants the image to be segmented, C_s; The case in which the user selects the union method, C_u; The repeat times, N_l in the noise blob removing filter; The shape noise threshold, T_{sn} in the boundary smoothing filter; The area threshold, S_0 in the quantification method.

Calibration:

for N = 5,6, ..., 59, 60; for W = 5,6, ..., 59, 30; for K_c = 1,2, 3; for C_s = 1,2; for C_u = 1,2,3; for N_l = 1,2,...,6; for T_{sn} = T_{min},..., T_{max}/2; for S_0 = S_{min},..., S_{max}/2.

The enhancement method:
1) Compute the gradient image by the enhancement method.

The segmentation method:
2) Compute the normalized histogram of the input image and smooth the normalized histogram with the DFT filter.
3) Compute the slope difference distribution.
4) Sort the peaks of the slope difference distribution in the magnitude decrease direction.
5) Choose the first K_c peaks as valid peaks and delete other smaller peaks.
6) Select two peaks according to the inputted case, C_s and find the valley with maximum absolute value between them.
7) The threshold is selected at the position corresponding to the found valley, V_i.

The union method:
8) Select the optimal union case, C_u and obtain the corresponding segmentation result, S_u for the cell image.

The noise blob removing filter:
9) Apply the noise blob removing filter to S_u with the selected repeat times, N_l.

The boundary smoothing filter:
10) Apply the boundary smoothing filter to S_u with the selected shape noise threshold, T_{sn}.

The quantification method:
11) Apply the quantification method to S_u with the selected area threshold, S_0. Get the quantification result S_q.

Accuracy evaluation:
12) Compute the F-measure between I_q and S_q.

end; end; end; end; end; end; end; end.

Outputs: Select N, W, K_c, C_s, C_u, N_l, T_{sn} and S_0 that yield the largest F-measure for this specific type of cell images.

---

significantly better results than state of the art methods [34, 35]. More comparisons are given with different types of cell images in Figs. 9 and 10. In Fig. 9, the muscle cell boundaries are much more unclear than those in Fig. 7. The generalized framework still achieves good result while state of the art methods performed significantly worse. We show the quantitative comparison with ten muscle cell images in Table 1. As can be seen, the proposed generalized framework achieves better accuracy than the two state of the art methods [34, 35] in segmenting muscle cell images. More importantly, the proposed framework is capable of segmenting other different types of cells besides the muscle cell images while the other two state of the art

methods [34, 35] might not be capable. In Fig. 10, we show the results of a different type of cell image by these three methods. It is seen that only the generalized framework yielded meaningful result while SMASH and CELLSEGM failed.

The effectiveness of the proposed boundary smoothing filter has been verified by the qualitative results shown in Figs. 5 and 6. Similarly, we show the effectiveness of the proposed noise blob removing filter in Figs. 11 and 12. Figure 11a shows the boundary extracted directly from the union segmentation without noise blob filtering and Fig. 11b shows the extracted boundary from the union segmentation after noise blob filtering. Figure 11c shows the final quantification results based on the

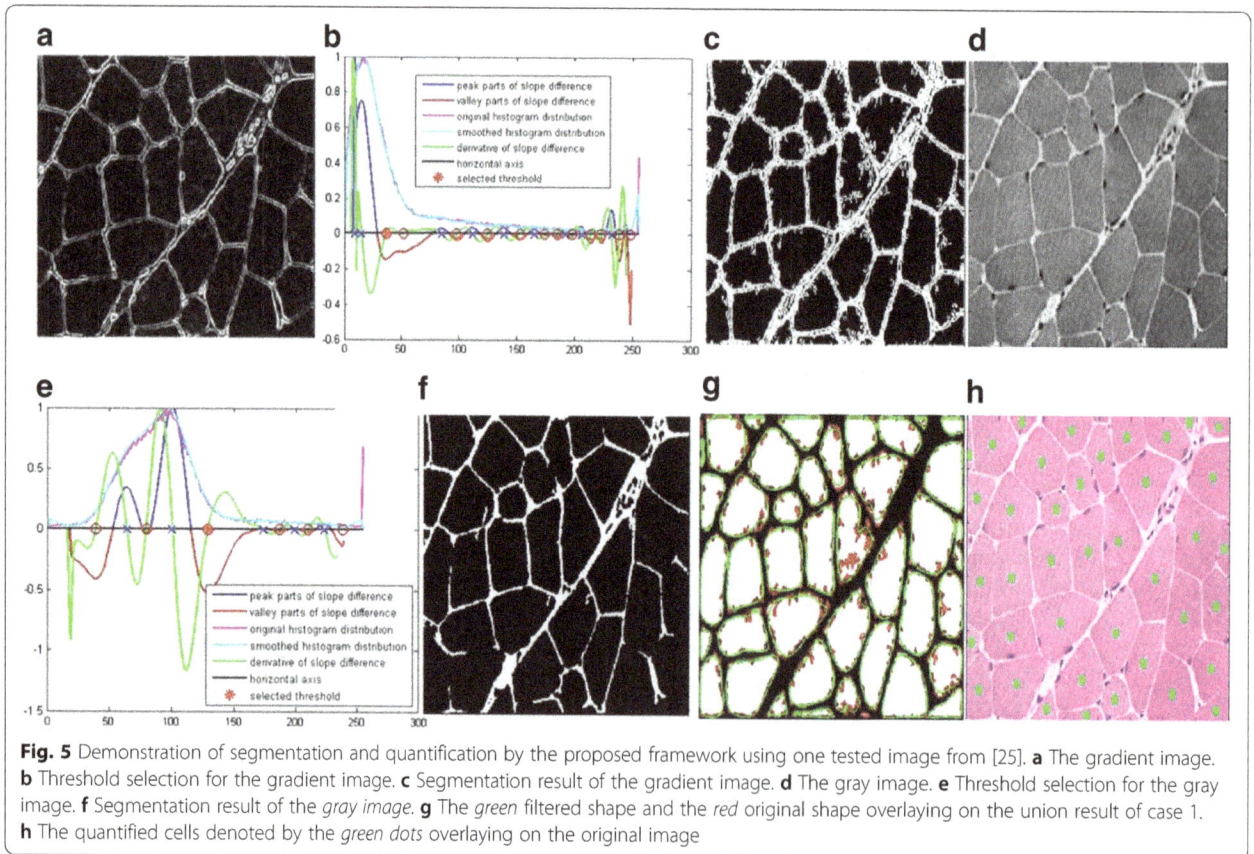

Fig. 5 Demonstration of segmentation and quantification by the proposed framework using one tested image from [25]. **a** The gradient image. **b** Threshold selection for the gradient image. **c** Segmentation result of the gradient image. **d** The gray image. **e** Threshold selection for the gray image. **f** Segmentation result of the *gray image*. **g** The *green* filtered shape and the *red* original shape overlaying on the union result of case 1. **h** The quantified cells denoted by the *green dots* overlaying on the original image

extracted boundary in Fig. 11a and d shows the final quantification results based on the extracted boundary in Fig. 11b. As can be seen, the quantification accuracy based on the filtered result by the noise blob filter is significantly higher that of the result without noise blob filtering. Figure 12 shows another example of muscle cell image. The extracted boundary without and with the noise blob filter affects the final accuracy of the quantification result obviously. There is one missing quantification in Fig. 12c, which is caused by the noise blobs.

For the quantitative result, the same cell image dataset used in [24, 25] is used for validation of the generalized framework proposed in this paper. The measure for accuracy evaluation is the same as [24, 25]. The true positive (TP) is defined as that there is one and only one identified cell inside each "ground-truth" boundary; The false positive (FP) is defined as that there is more than one identified cell inside each "ground-truth" boundary. The false negative (FN) is defined as that there is none identified cell inside each "ground-truth" boundary. The comparison is shown in Table 2. As can be seen, the robustness of the generalized framework is superior to the proposed method in [24, 25].

We use the 20 synthetic fluorescent cell images from the open access Broad Bioimage Benchmark Collection (BBBC) [36] for the general comparison with state-of-the-art methods. Among the referenced literatures, only [18] reports quantitative results based on the BBBC dataset. Hence, we compare the proposed method with [18] using the quantitative results in Table 3. The correct quantification rate which is denoted as TP in Table 2 is 93.5%, which is better than that of [18], 91.8%. Overall, the robustness and generality of the proposed framework is validated. We share the codes for testing the quantitative results with these 20 synthetic images in the section of *Data availability*. Since the generalized framework evolves and enhances the previous approaches [24, 25] and it inherits all their merits, more performance evaluation of the generalized framework could also be referred from the past work [24, 25].

Discussion

The microscopy imaging technology has been developed rapidly in recent years. Accordingly, image processing techniques for automatic cell segmentation and robust quantification are becoming more and more necessary. According to our investigation, we concluded that threshold selection is the most appropriate method in this application due to its good efficiency, good resistance to noise and easy implementation. State of the art

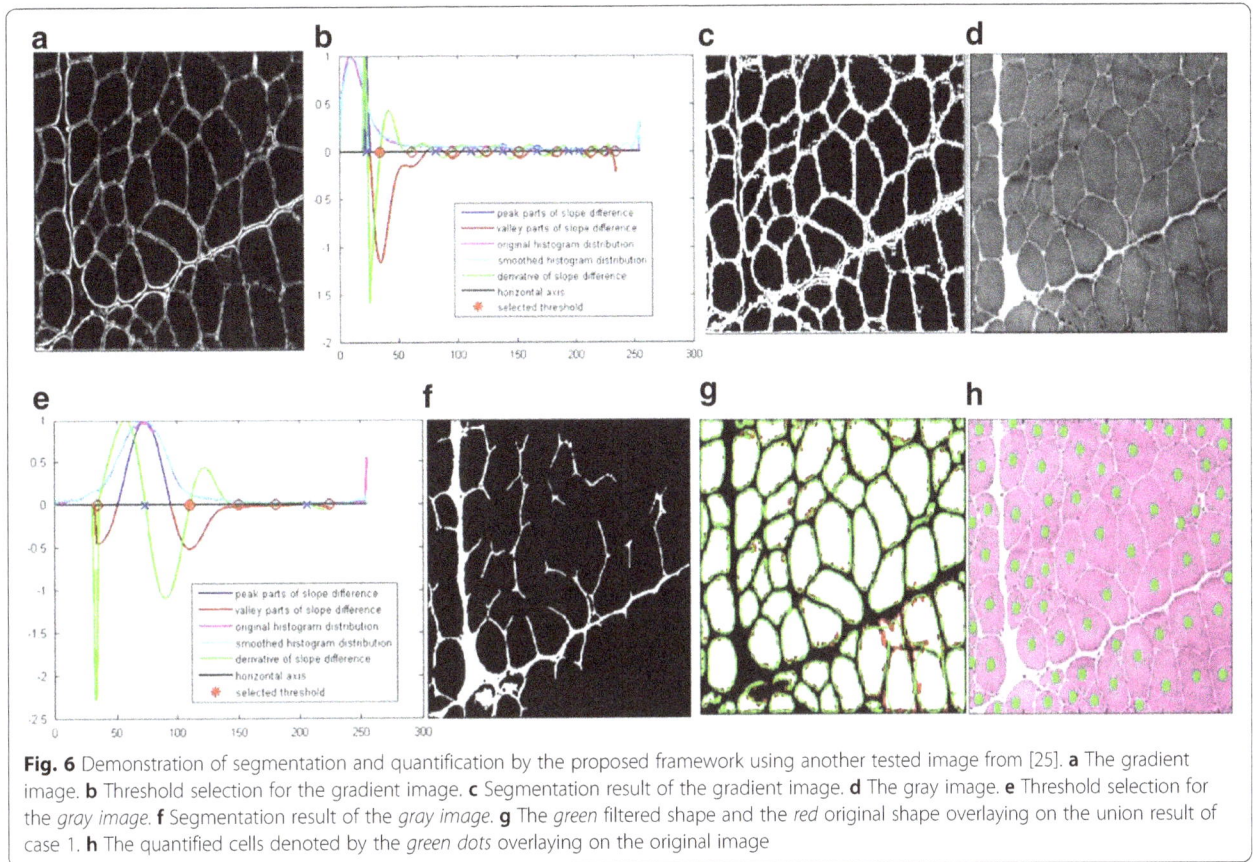

Fig. 6 Demonstration of segmentation and quantification by the proposed framework using another tested image from [25]. **a** The gradient image. **b** Threshold selection for the gradient image. **c** Segmentation result of the gradient image. **d** The gray image. **e** Threshold selection for the *gray image*. **f** Segmentation result of the *gray image*. **g** The *green* filtered shape and the *red* original shape overlaying on the union result of case 1. **h** The quantified cells denoted by the *green dots* overlaying on the original image

threshold selection methods [29–33] failed to select the threshold robustly for the gradient image as stated and proved in [24, 25]. As a result, the threshold selection method was evolved and utilized in [25] to segment the muscle cell images and its advantage over state of the art thresholding methods was also verified in [25]. Later, the threshold selection is improved further in [24] by adding the calibration procedure to the selection process. As a result, the threshold selection becomes flexible and could segment different types of cells

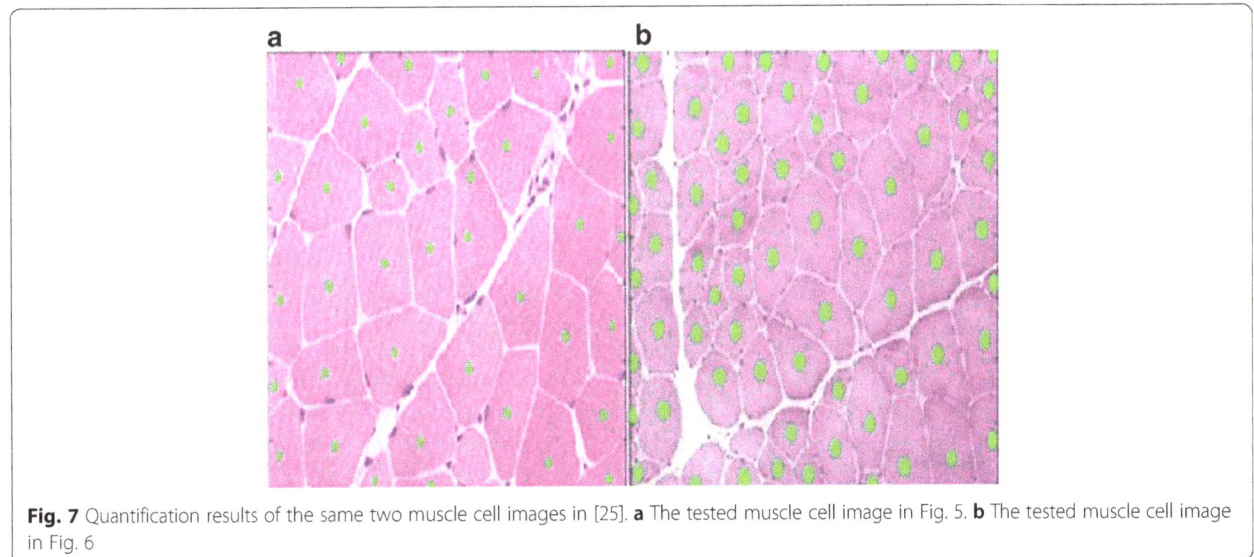

Fig. 7 Quantification results of the same two muscle cell images in [25]. **a** The tested muscle cell image in Fig. 5. **b** The tested muscle cell image in Fig. 6

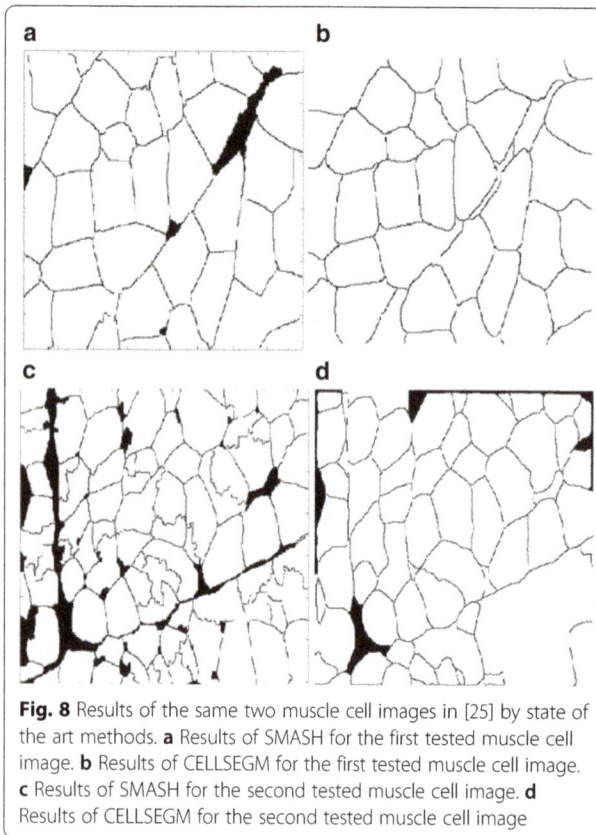

Fig. 8 Results of the same two muscle cell images in [25] by state of the art methods. **a** Results of SMASH for the first tested muscle cell image. **b** Results of CELLSEGM for the first tested muscle cell image. **c** Results of SMASH for the second tested muscle cell image. **d** Results of CELLSEGM for the second tested muscle cell image

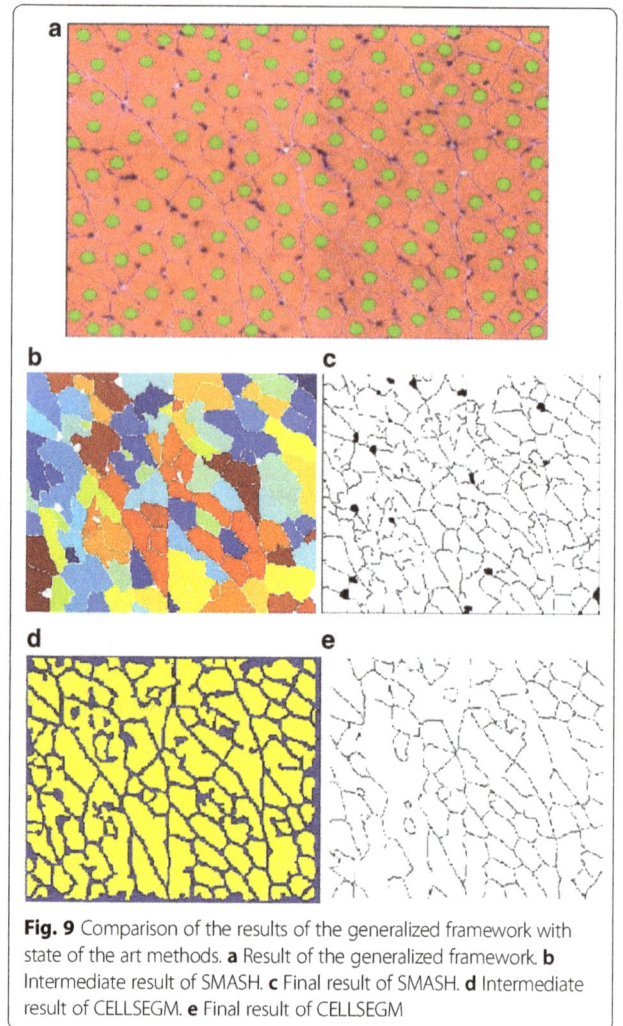

Fig. 9 Comparison of the results of the generalized framework with state of the art methods. **a** Result of the generalized framework. **b** Intermediate result of SMASH. **c** Final result of SMASH. **d** Intermediate result of CELLSEGM. **e** Final result of CELLSEGM

robustly. In this paper, we propose a simpler and more practical calibration method to determine the parameters for the threshold selection method based on the third property of the slope difference distribution.

Only with the thresholding method to guarantee the accurate enough and complete enough segmentation, we could proceed to high level applications, e.g. boundary extraction or quantification. There are two challenging aspects for automatic and reliable quantification of cells by the proposed iterative

erosion method in [24, 25]: (1), there are some noise blobs that might be identified as the cell seed by the iterative erosion method. (2), the extracted boundaries of the cell blobs are usually irregular with noise.

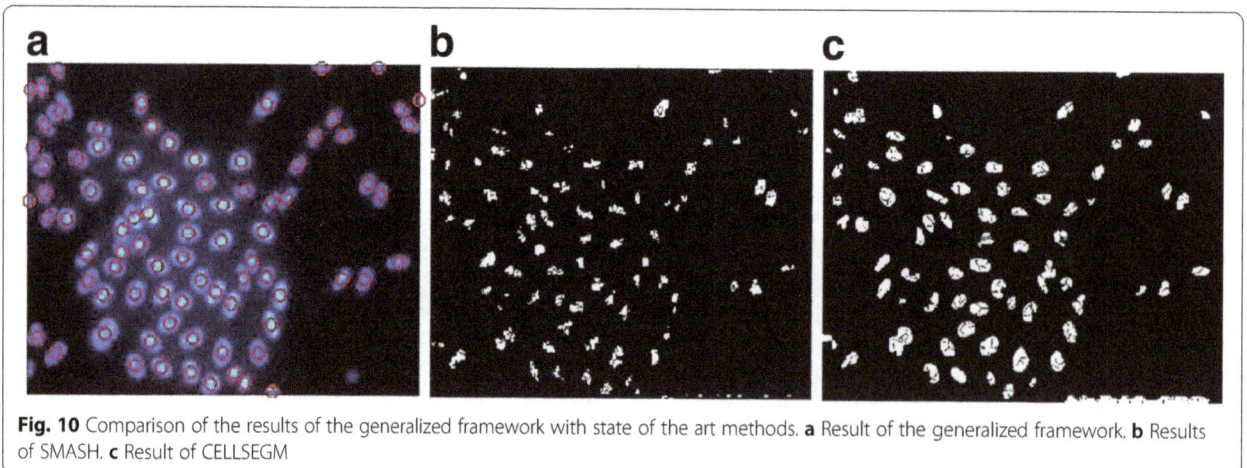

Fig. 10 Comparison of the results of the generalized framework with state of the art methods. **a** Result of the generalized framework. **b** Results of SMASH. **c** Result of CELLSEGM

Table 1 Quantitative comparison of the proposed approach with state of the methods [34, 35]

Methods	TP	FP	FN
SMASH [34]	84.27%	11.89%	16.08%
CELLSEGM [35]	82.69%	2.10%	17.31%
Proposed	95.28%	1.92%	4.72%

(3), there might be holes inside the cell blobs. All these three aspects will increase the number of the false seeds produced by the iterative erosion method proposed in [24, 25]. To solve these problems, we propose a noise blob removing filter to get rid of the threadlike or small noise blobs. We propose a boundary smoothing filter to smooth the extracted boundary of the cell blobs and also eliminate small holes inside the cell blobs.

To verified the proposed methods in this paper, both qualitative and quantitative experiments are conducted. As it turned out, the proposed framework is more versatile than other state of the art methods due to the fact that it utilizes the characteristics of the adjacent cells and the general property of the cell images: the global intensity distribution and the local gradient. The frequently occurring overlapping characteristics of the adjacent cells could be dealt effectively by the iterative erosion method. The intensity image and the gradient image could be segmented effectively by the proposed segmentation

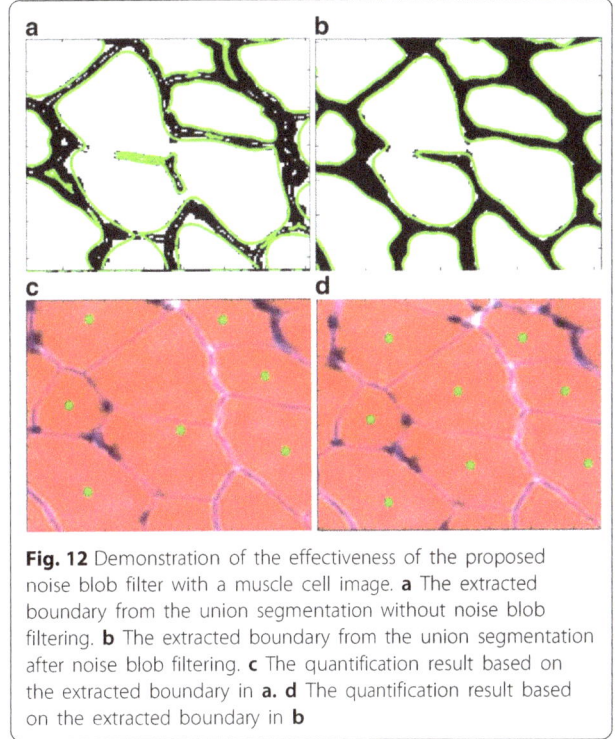

Fig. 12 Demonstration of the effectiveness of the proposed noise blob filter with a muscle cell image. **a** The extracted boundary from the union segmentation without noise blob filtering. **b** The extracted boundary from the union segmentation after noise blob filtering. **c** The quantification result based on the extracted boundary in **a**. **d** The quantification result based on the extracted boundary in **b**

method. The segmentation method is able to segment different kinds of images and their formed gradient images more accurately because of the introduced calibration process.

Conclusion

In this paper, we propose a generalized framework for automatically segmenting and quantifying different types of cells. To simplify the calibration process for the threshold selection, we proposed a practical calibration method. To improve the quantification accuracy over the past research, we proposed a noise blob filtering method and a boundary smoothing filtering method in this paper. Experimental results verified their effectiveness. As a generalized tool for automatic segmentation and quantification of different kinds of cells, it possible for the proposed framework to benefit a lot of automated microscopy applications in the future.

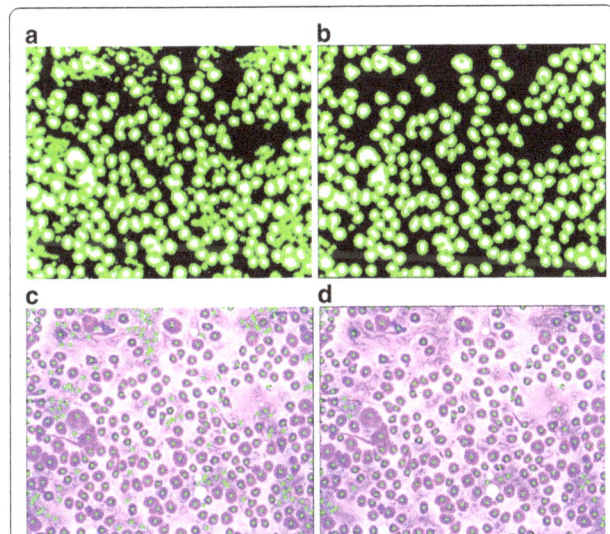

Fig. 11 Demonstration of the effectiveness of the proposed noise blob filter. **a** The extracted boundary from the union segmentation without noise blob filtering. **b** The extracted boundary from the union segmentation after noise blob filtering. **c** The quantification result based on the extracted boundary in **a**. **d** The quantification result based on the extracted boundary in **b**

Table 2 Quantitative comparison of the quantification accuracy with [24, 25]

Methods	TP	FP	FN
[24, 25]	93.4%	0.18%	6.6%
Proposed	96.8%	0.12%	3.2%

Table 3 Quantitative comparison of the quantification accuracy with State of the art methods

Methods	TP	FP	FN
[18]	91.8%	NA	8.2%
Proposed	93.5%	NA	6.5%

Abbreviations
DFT: Discrete Fourier transformation; FIR: Finite impulse response; IIR: Infinite impulse response

Acknowledgements
The authors would like to thank the anonymous reviewers for the helpful comments.

Funding
This work has been supported by the Chinese Academy of Sciences with the grant number: Y5A1270101.

Authors' contributions
ZW designed and implemented the generalized framework, wrote and proofread the paper. HL run the SMASH and CELLSEGM software to get their optimal results and compare them with those of the proposed framework both quantitatively and qualitatively. Both authors read and approved the final manuscript.

Competing interests
The authors declare that they have no competing interests.

References
1. Baker M. Cellular imaging: taking a long, hard look. Nature. 2010;466(26): 1137–40.
2. Landecker H. Seeing things: from microcinematography to live cell imaging. Nat Methods. 2009;6(10):707–9.
3. Miyazaki J, Tsurui H, Kawasumi K, Kobayashi T. Optimal detection angle in sub-diffraction resolution photothermal microscopy: application for high sensitivity imaging of bilogical tissues. Opt Express. 2015;22(16):18833–42.
4. De M, Ghosh PS, Rotello VM. Applications of nanoparticles in biology. Adv Mater. 2008;20:4225–41.
5. Wang EC, Wang AZ. Nanoparticles and their applications in cell and molecular biology. Integr Biol. 2014;6:9–26.
6. Salata OV. Applications of nanoparticles in biology and medicine. J Nanobiotechnol. 2004;2(1):1–6.
7. Etoc F, Lisse D, Bellaiche Y, Piehler J, Coppey M, Dahan M. Subcellular control of Rac-GTPase signalling by magnetogenetic manipulation inside living cells. Nat Nanotechnol. 2013;8:193–8.
8. Levet F, Hosy E, Kechkar A, Butler C, Beghin A, Choquet D, Sibarita JB. SR-Tesseler: a method to segment and quantify localization-based super-resolution microscopy data. Nat Methods. 2015;12(11):1065–71.
9. Rizk A, Paul G, Incardona P, Bugarski M, Mansouri M, Niemann A, Ziegler U, Berger P, Sbalzarini IF. Segmentation and quantification of subcellular structures in fluorescence microscopy images using squassh. Nat Protoc. 2014;9(3):586–96.
10. Yang HG, Ahuja N. Automatic segmentation of granular objects in images: combing local density clustering and gradient-barrier watershed. Pattern Recogn. 2014;47(6):2266–79.
11. Bieniek A, Moga A. An efficient watershed algorithm based on connected components. Pattern Recogn. 2000;33(6):907–16.
12. Yang X, Li H, Zhou X. Nuclei segmentation using marker-controlled watershed, tracking using mean-shift, and Kalman filter in timelapse microscopy. IEEE Trans Circuits Syst I. 2006;53(11):2405–14.
13. Gomez O, Gonzalez JA, Morales EF. Image segmentation using automatic seeded region growing and instance-based learning. Lect Notes Comput Sci Prog Pattern Recognit Image Anal Appl. 2007;4756:192–201.
14. Jones TR, Carpenter AE, Lamprecht MR, Moffat J, Silver SJ, Grenier JK, Castoreno AB, Eggert US, Root DE, Golland P, Sabatini DM. Scoring diverse cellular morphologies in image-based screens with iterative feedback and machine learning. Proc Natl Acad Sci U S A. 2008;106(6):1826–31.
15. Mosaliganti KR, Noche RR, Xiong FZ, Swinburne IA, Megason SG. ACME: Automated cell morphology extractor for comprehensive reconstruction of cell membranes. PLoS Comput Biol. 2012;8(12):1–14.
16. Xu C, Su ZC. Identification of cell types from single-cell transcriptomes using a novel clustering method. Bioinformatics. 2015;31(12):1974–80.
17. Liu F, Mackey AL, Srikuea R, Esser KA, Yang L. Automated image segmentation of haematoxylin and eosin stained skeletal muscle cross-sections. J Microsc. 2013;252(3):275–85.
18. Nogueira PA, Teofilo LF. A multi-layered segmentation method for nucleus detection in highly clustered ciroscopy imaging: A practical application and validation using human U2OS cytoplasm-nucleus translocation images. Artif Intell Rev. 2014;42(3):331–46.
19. Dimopoulos S, Mayer CE, Rudolf F, Stelling J. Accurate cell segmentation in microscopy images using membrane patterns. Bioinformatics. 2014;30(18): 2644–51.
20. Valmianski I, Shih AY, Driscoll JD, Mathews DW, Freund Y, Kleinfeld D. Automatic identification of fluorescently labeled brain cells for rapid functional imaging. J Neurophysiol. 2010;104(3):1803–11.
21. Chalfoun J, Majurski M, Dima A, Stuelten C, Peskin A, Brady M. FogBank: a single cell segmentation across multiple cell lines and image modalities. BMC Bioinformatics. 2014;15(431):1–12.
22. Wait E, Winter M, Bjornsson C, Kokovay E, Wang Y, Goderie S, Temple S, Cohen AR. Visualization and correction of automated segmentation, tracking and lineaging from 5-D stem cell image sequences. BMC Bioinformatics. 2014;15(328):1–14.
23. Obara B, Roberts MAJ, Armitage JP, Grau V. Bacterial cell identification in differential interference contrast microscopy images. BMC Bioinformatics. 2013;14(134):1–13.
24. Wang ZZ. A new approach for segmentation and quantification of cells or nanoparticles. IEEE Trans Ind Inform Vol. 2016;12(3):962–71.
25. Wang ZZ. A semi-automatic method for robust and efficient identification of neighboring muscle cells. Pattern Recogn. 2016;53(8):300–12.
26. Pal NR, Pal SK. A review on image segmentation techniques. Pattern Recogn. 1993;26(9):1277–94.
27. Pham DL, Xu CY, Prince JL. Current methods in medical image segmentation. Annu Rev Biomed Eng. 2000;2:315–37.
28. Wang ZZ. A new approach for robust segmentation of the noisy or textured images. SIAM J Imag Sci. 2016;9(3):1409–36.
29. Sezgin M, Sankur B. Survey over image thresholding techniques andquantitative performance evaluation. J Electron Imaging. 2004;13(1):146–65.
30. Li CH, Lee CK. Minimum cross-entropy thresholding. Pattern Recogn. 1993; 26(4):617–25.
31. Cheng HD, Chen YH, Sun Y. A novel fuzzy entropy a proach to image enhancement and thresholding. Signal Process. 1999;75(3):277–301.
32. Ridler TW, Calvard S. Picture thresholding using an iterative selection method. IEEE Trans Syst Man Cybern. 1978;8(8):630–2.
33. Otsu N. A threshold selection method from gray level histogram. IEEE Trans Syst Man Cybern. 1979;SMC-9(1):62–6.
34. Smith LR, Barton ER. SMASH-semi-automatic muscle analysis using segmentation of histology: a MATLAB application. Skelet Muscle. 2014;4(21):1–15.
35. Hodneland E, Kogel T, Frei DM, Gerdes HH, Lundervold A. CellSegm-a MATLAB toolbox for high-throughput 3D cell segmentation. Source Code Biol Med. 2013;8(16):1–24.
36. Broad Bioimage Benchmark Collection-annotated biological image sets for testing and validation. (available at: http://www.broadinstitute.org/bbbc). Accessed 07 Feb 2017.

7

A Bayesian taxonomic classification method for 16S rRNA gene sequences with improved species-level accuracy

Xiang Gao[1], Huaiying Lin[1,2], Kashi Revanna[1,2] and Qunfeng Dong[1,2,3,4*] ⓘ

Abstract

Background: Species-level classification for 16S rRNA gene sequences remains a serious challenge for microbiome researchers, because existing taxonomic classification tools for 16S rRNA gene sequences either do not provide species-level classification, or their classification results are unreliable. The unreliable results are due to the limitations in the existing methods which either lack solid probabilistic-based criteria to evaluate the confidence of their taxonomic assignments, or use nucleotide k-mer frequency as the proxy for sequence similarity measurement.

Results: We have developed a method that shows significantly improved species-level classification results over existing methods. Our method calculates true sequence similarity between query sequences and database hits using pairwise sequence alignment. Taxonomic classifications are assigned from the species to the phylum levels based on the lowest common ancestors of multiple database hits for each query sequence, and further classification reliabilities are evaluated by bootstrap confidence scores. The novelty of our method is that the contribution of each database hit to the taxonomic assignment of the query sequence is weighted by a Bayesian posterior probability based upon the degree of sequence similarity of the database hit to the query sequence. Our method does not need any training datasets specific for different taxonomic groups. Instead only a reference database is required for aligning to the query sequences, making our method easily applicable for different regions of the 16S rRNA gene or other phylogenetic marker genes.

Conclusions: Reliable species-level classification for 16S rRNA or other phylogenetic marker genes is critical for microbiome research. Our software shows significantly higher classification accuracy than the existing tools and we provide probabilistic-based confidence scores to evaluate the reliability of our taxonomic classification assignments based on multiple database matches to query sequences. Despite its higher computational costs, our method is still suitable for analyzing large-scale microbiome datasets for practical purposes. Furthermore, our method can be applied for taxonomic classification of any phylogenetic marker gene sequences. Our software, called BLCA, is freely available at https://github.com/qunfengdong/BLCA.

Keywords: 16S rRNA gene, Taxonomic classification

Background

High-throughput 16S rRNA gene sequencing is widely used in microbiome studies for characterizing bacterial community compositions. A key computational task is to perform taxonomic classification for 16S rRNA gene sequences, with emphasis increasing on species-level classification [1]. The published tools dedicated for 16S rRNA gene classification include the RDP Classifier [2], 16S Classifier [3] and SPINGO [4]. There are also software packages or websites that provide 16S classification options, e.g., QIIME [5] and MG-RAST [6].

Despite the availability of those taxonomic classification tools, species-level classification for 16S rRNA gene sequences still remains a serious challenge for microbiome researchers. Some of the tools simply do not classify at the species level. For example, the standard version of the widely-used software, RDP Classifier, only classifies 16S rRNA gene sequences from the phylum to genus levels,

* Correspondence: qdong@luc.edu
[1]Department of Public Health Sciences, Loyola University Chicago Health Sciences Division, Maywood, IL 60153, USA
[2]Center for Biomedical Informatics, Loyola University Chicago Health Sciences Division, Maywood, IL 60153, USA
Full list of author information is available at the end of the article

although the RDP Classifier can be re-trained for species level classification. Another recently published software, the 16S Classifier, is not capable of classifying sequences at the species level either. For the other tools that can classify at the species level, they suffer from at least one of the two major limitations: i) nucleotide k-mer frequency is used for measuring similarity between query and database sequences, a proxy measurement of true sequence similarity; ii) solid probabilistic-based criteria is lacking for evaluating the confidence of taxonomic assignment results, particularly to evaluate whether the best-matched database sequence is significantly better than other database matches for the taxonomic assignments.

Taxonomic classification of 16S gene sequences typically requires comparing query sequences to annotated database sequences. The k-mer based approaches, e.g., the RDP Classifier and SPINGO, compare the frequency of k-mer nucleotides between query and database sequences. The higher degree of shared k-mer nucleotide frequencies, the more similar the two sequences are. The advantage of k-mer based approaches is its fast computational speed. However, k-mer based approaches rely on two key assumptions: i) the k-mer nucleotides in DNA sequences used as discriminating features among different taxa are independent, and ii) the actual nucleotide position of the k-mers in the DNA sequences is not important. In reality, nucleotides in different positions of a gene sequence can be correlated (e.g., to preserve the secondary or higher-dimensional structure of rRNA folding), and gene sequences with the same set of k-mer in different orders are clearly not the same sequences. Therefore, these two assumptions are the theoretical sources of taxonomic misclassification by k-mer based approaches. There is also a nontrivial practical limitation for a k-mer based approach: it is extremely difficult to determine an optimal size of k-mer for discriminating among different species at different regions of 16S sequences. For example, the accuracy of the RDP Classifier, which uses a k-mer size of eight, varies significantly with different types of bacterial taxa at different 16S gene regions [7]. Therefore, k-mer based approaches rely on a proxy measurement of the sequence similarity between the query and database sequences, which is inherently less accurate than the gold standard sequence-alignment-based method.

As mentioned above, another major limitation for most existing methods is that they lack solid probabilistic-based criteria to evaluate the confidence of their taxonomic assignments. Although all existing methods infer taxonomic classification based on matched database sequences, most of the existing methods do not provide any indication on whether the best-matched database hit sequence is significantly better than other database hits. Since the 16S rRNA gene is highly conserved among different bacterial taxa

and the query sequences in microbiome studies are often only a short fragment of the full-length 16S rRNA gene with sequencing errors, it is common to have several database hits from different taxa that may have comparable sequence similarities to the query sequence. Therefore, it is not reliable to simply transfer the taxonomic annotation associated with the best database hit for the query sequence [8]. Instead, a better method for 16S classification may consider multiple database hits together and evaluate whether the best database hit is significantly better than other database hits.

The Lowest Common Ancestor (LCA) algorithm, implemented in the MEGAN package [9], provides a natural biological framework to integrate taxonomic annotations associated with multiple database hits when classifying query sequences. In MEGAN, all taxa corresponding to the BLAST [10] hits are first mapped to NCBI taxonomic trees and the lowest common ancestor of all mapped taxa is then assigned to the query sequence. For example, if a query sequence has two BLAST hits belonging to two different species, e.g., one from *Lactobacillus acidophilus* and the other one from *L. casei*, the LCA algorithm assigns the query sequence to the genus *Lactobacillus*, which is the lowest common taxonomic level of these two species. However, the LCA algorithm fails to consider the differing degrees of similarity between the query and the database hit sequences. In other words, when inferring the LCA for the query, the algorithm acts as if all the hit sequences, affected by an arbitrary sequence similarity threshold, were equally similar to the query sequence, even though in practice they are often not. Biologically speaking, the greater the degree of sequence similarity between the query and the hit sequences, the more likely they may belong to the same taxon, but the current LCA algorithm lacks a quantitative way to incorporate this important information on sequence similarity in its taxonomic assignment.

To overcome the above limitations of the existing software, we have developed a Bayesian-based LCA method, named BLCA. BLCA can perform species and even sub-species level taxonomic classification. It relies on sequence alignment instead of k-mer frequency for sequence similarity measurement; it considers multiple database hits instead of only the best database hit for taxonomic assignment; it provides a probabilistic-based confidence score for evaluating taxonomic assignments. The novelty of our method is that the contribution of each database hit to the taxonomic assignment of the query sequence is weighted by a Bayesian posterior probability based upon the sequence similarity of the database hit to the query. The calculated Bayesian posterior probability implicitly penalizes dissimilar database hit sequences in a quantitative way, which makes our method insensitive to arbitrary sequence similarity thresholds for selecting candidate database hits for each query sequence. We show that BLCA provides significantly more accurate classification results at the species level when compared to all other existing tools.

Implementation

The BLCA method is implemented as a Python package, which is freely available at https://github.com/qunfeng-dong/BLCA under the GNU General Public License. An overview of the BLCA method is illustrated in Fig. 1. Users start by comparing the query 16S sequences against entries in an annotated 16S database using BLASTN. The taxonomic lineage of each 16S database sequence is extracted from the NCBI taxonomic database (ftp://ftp.ncbi.nih.gov/pub/taxonomy/). As with MEGAN, we chose the 16S rRNA gene collection from NCBI (ftp://ftp.ncbi.nlm.nih.gov/blast/db/16SMicrobial.-tar.gz) as the default database, although users can also use the Greengenes database [11] or adopt any custom collection of 16S sequences provided that the sequence IDs can be mapped to the NCBI or Greengenes taxonomy. Next, the BLAST hits are extracted; by default, BLCA only extracts the BLAST hits from BLAST pairwise alignments with at least 95% identity and 95% coverage with respect to the query, but users can easily change these parameters using the command-line at execution as well as setting an additional criterion to retain only the BLAST hits whose bit scores are

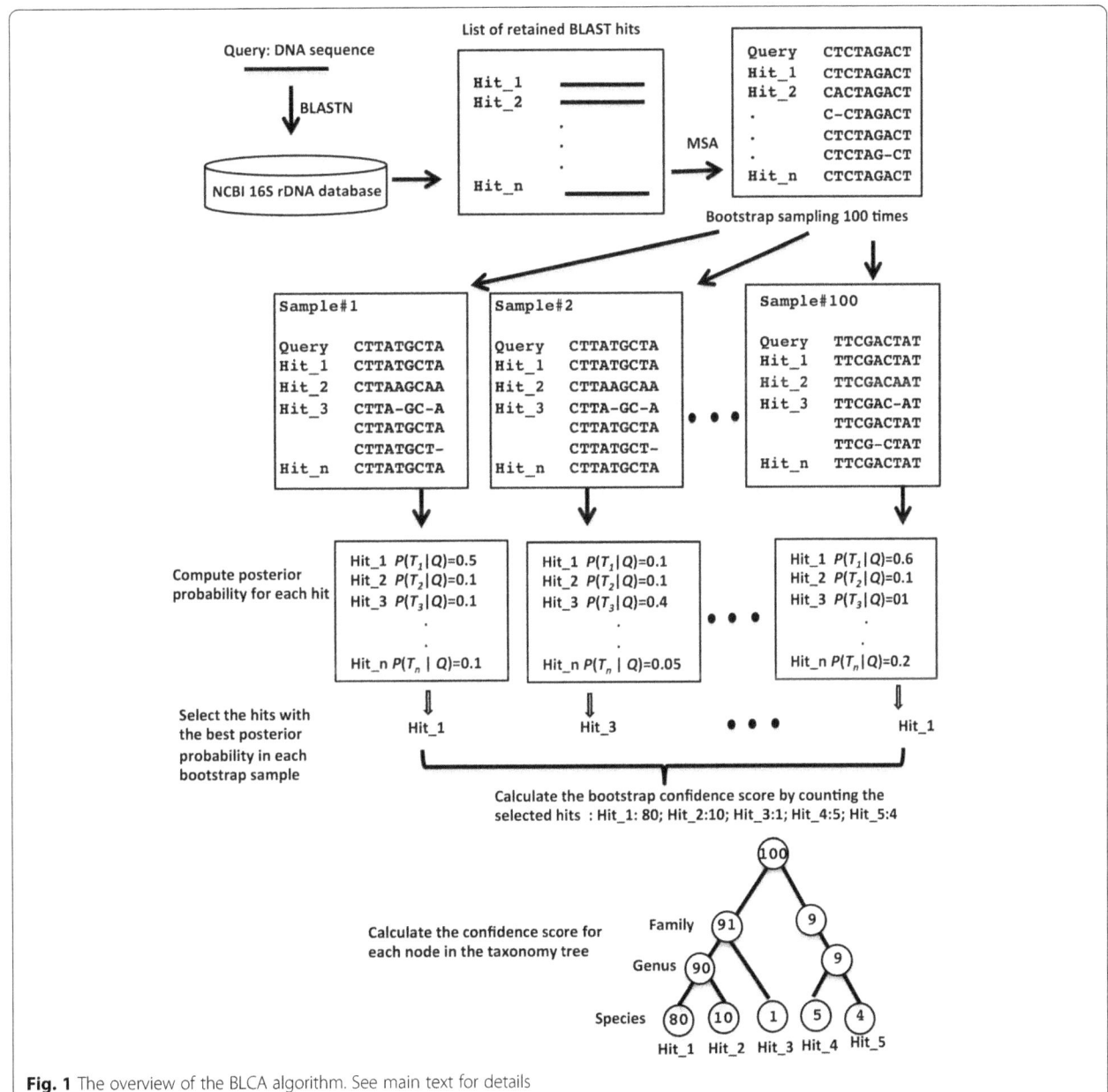

Fig. 1 The overview of the BLCA algorithm. See main text for details

within a certain percentage of difference from the top hits (the same criterion used by MEGAN). Each query sequence and its corresponding BLAST hits are passed as an input to the MUSCLE program [12] for multiple sequence alignment. Because most 16S query sequences are not full-length gene sequences in practice, BLCA only extracts the relevant subsequences of the hits – those that actually align to the query sequences in the BLAST pairwise sequence alignments. An extra 10 nucleotides upstream and downstream relative to the aligned regions from the hit sequences are also included to avoid potential overhangs at the 5' or 3' end of the query sequences in the multiple sequence alignment.

We define $Pr(T_i \mid Q)$ as the Bayesian posterior probability for a taxon T_i being assigned to a given query sequence Q. Based on Bayes' rule, we obtain

$$Pr(T_i|Q) = Pr(Q|T_i)Pr(T_i)/Pr(Q) \qquad (1)$$

wherein $Pr(Q \mid T_i)$ is the likelihood of observing the sequence Q if it were derived from the taxon T_i. The likelihood can be calculated as the pairwise alignment score between the query sequence Q and the database hit sequence annotated as T_i, divided by the pairwise alignment score between the hit sequence T_i to itself. In other words, the likelihood is defined as the similarity score between the query and the database hit normalized by the maximum possible similarity score between any sequences to the hit sequence. The likelihood $Pr(Q \mid T_i)$ is a real number between 0 (i.e., no match between the query Q and the database hit T_i) and 1 (i.e., a perfect match between the query and the database hit). Our definition of $Pr(Q \mid T_i)$ as a likelihood simply reflects the degree of support by the evidence (i.e., similarity between query and the database hit) for the hypothesis (i.e. the query belongs to the taxon of the database hit). In our current implementation, the pairwise alignment score between the query sequence and BLAST hit sequence is computed from the multiple sequence alignment, which tends to be more accurate than the original BLAST pairwise alignment because BLAST alignment performs local alignment, whereas MUSCLE is a global alignment program. Since the alignment is between DNA sequences, the pairwise alignment score can be simply computed with the following criteria: match = +1, mismatch = –2, and gap = –2.5 (these are the exact default scoring criteria used for BLASTN). $Pr(T_i)$ is the prior probability of a particular taxon T_i for the query sequence, which is set to a uniform distribution in our implementation. The uniform prior is a suitable choice for taxonomic classification, since, without knowing the data, we can treat every taxon as equally probable (the same uniform prior is used in the RDP Classifier). If necessary, non-uniform priors can be easily adopted for

specific situations where certain taxa are more likely than others in the same Bayesian framework described in this work. $Pr(Q)$ is the marginal distribution of the query sequence Q, which can be calculated as the summation of the product of likelihoods and priors of all the BLAST hits, i.e., $\sum_{i=1}^{m} Pr(Q|T_i)Pr(T_i)$ for m total BLAST hits, based on the law of total probability. Note that the term $Pr(T_i)$, assumed to be a uniform prior, can be cancelled from the denominator and numerator when calculating $Pr(T_i \mid Q)$. In addition, sequence similarity estimations might be improved by specifying sequencing error models for both query and database sequences (e.g., a Poisson probability distribution of an observed nucleotide in a DNA sequence being incorrect); these can be incorporated in our Bayesian framework by adjusting the likelihood calculation in Eq. (1).

Since T_i corresponds to the taxonomic annotation for an individual BLAST hit sequence, it represents the leaf node in the NCBI taxonomic tree (e.g., at the species or sub-species level). We also need to compute the posterior probability at higher taxonomic levels, i.e., the internal nodes in the taxonomic tree that correspond to the antecedents of T_i (i.e., the common ancestors of all the T_i). Using the addition rule for probability, the posterior probability of any internal node I, $Pr(T_I \mid Q)$, in the taxonomic tree can be computed by a simple summation of those of all the descendant leaf T_i:

$$Pr(T_I|Q) = \sum_{i=1}^{k} Pr(T_i|Q_i) \qquad (2)$$

wherein the internal node I has k total descendant leaf nodes. The Eq. (2) allows us to easily compute the posterior probability of any higher taxonomic level, e.g., from genus to phylum, by simply summing the posterior probabilities associated with all the descendant leaf nodes in the taxonomic trees under any internal nodes. Using the previous example in which a query sequence has one BLAST hit from *L. acidophilus* and the other from *L. casei*, the posterior probability for the genus level of *Lactobacillus* for the query is the sum of the posterior probabilities for *L. acidophilus* and *L. casei*, respectively.

Based on the posterior probabilities calculated for all the nodes in the taxonomic tree, a bootstrap confidence score is derived to evaluate the reliability of the taxonomic assignment for each node. Specifically, aligned nucleotide positions in the multiple sequence alignment between query and BLAST hits are randomly sampled with replacement; the total number of sampled nucleotide positions is the same as the length of the query sequence (i.e., a pseudo multiple-sequence alignment is bootstrapped from the original multiple-sequence alignment). Using the pseudo multiple-sequence alignment, the posterior probability of each leaf node in the taxonomic tree is re-computed by the same procedure as described above and the leaf node with the highest posterior probability is identified and tallied as the "winning" node. The process is repeated 100 times, and the number of times that a leaf node emerged as the winner becomes the confidence score for the taxonomic assignment of the particular node. Similar to the posterior probability calculation, the confidence score for internal nodes can also be obtained by summing up the confidence scores of all their descendent leaf nodes. The RDP Classifier uses a similar bootstrapping strategy to assign confidence scores for its taxonomic classifications. However, unlike the RDP Classifier, which is based on bootstrapping k-mers from query and database sequences, our strategy randomly samples from aligned nucleotides in multiple sequence alignment, a method that is commonly used for evaluating the confidence of branches in molecular phylogenic trees [13].

To assess the accuracy of a classification tool, we must have a benchmark dataset with known taxonomic annotations for each 16S sequence. Therefore, we extracted the V2, V4, V1–V3, V3–V5, and V6–V9 regions of 16S sequences from 1000 randomly selected bacterial species with known taxonomic annotations in the NCBI database as the benchmark dataset. These variable regions were chosen for testing because they represent typical 16S sequences in real-world microbial studies. Instead of using the exact sequences from those regions for testing, we introduced sequencing errors to each sequence, using a customized Python script to generate an average of 1% random mutation based on a Poisson distribution. The 1% mutation rate is based on the reported upper range of the Illumina MiSeq sequencing platform [14]. The test sequences, with sequencing errors, were searched against the 16S sequences from NCBI (downloaded on August 5th, 2016) using BLASTN version 2.5.0. For MEGAN parameters, we set the same default settings (e.g., minimum BLAST bit scores, maximum BLAST expected values, and the percent of BLAST hits) for both BLCA and MEGAN. For BLCA, SPINGO, and the RDP Classifier, two sets of confidence score thresholds were used: (i) 0.8–RDP Classifier's default confidence score and (ii) 0.5–RDP Classifier's confidence score threshold recommended for short-read sequences, as written in the RDP Classifier's documentation. Neither MEGAN nor Kraken [15] have a probabilistic-based parameter for evaluating the assigned taxa, thus we used their default taxonomic assignments for comparison.

For each of the taxa in the benchmark dataset (e.g., a known *E. coli* sequence), we were able to identify whether the classification results from each software represent a true positive (TP, e.g., the predicted taxonomy is also *E. coli*), false negative (FN, e.g., the predicted taxonomy is not *E. coli*), false positive (FP, e.g.,

other non-*E. coli* sequences were incorrectly predicted to be *E. coli*), and true negative (TN, e.g., other non-*E. coli* sequences were correctly predicted to be non-*E. coli* sequences). The total amount of TP, FN, FP, and TN are tallied from the 1000 test sequences from the species to the phylum levels. The rates of TP, FN, FP, and TN were used for computing the F-score, which is a standard measure of a classifier's accuracy by combining both the precision and the recall of the classifier [16]. The procedure above was repeated three times to measure the variability of the classification accuracy.

Besides the above-simulated dataset, we also evaluated the performance of BLCA with a real-world 16S dataset, which was suggested by one of the reviewers of our manuscript. The dataset was originally produced by Pop et al. [17] and is available in the Bioconductor package (referred as the *msd16s* dataset) [18]. The *msd16s* dataset contains 26,044 species-level operational taxonomic unit (OTU) sequences from the V1V2 rRNA gene region. The original authors used the top BLAST hit against the RDP 16S database [19] as the taxonomic annotation for each OTU sequence. Since MEGAN and SPINGO can only use NCBI taxonomy nomenclature, we re-annotated the *msd16s* dataset by using the top BLAST hit against NCBI 16S database (i.e., the same BLAST strategy as in the original study of Pop et al. [17]) in order to ensure that MEGAN and SPINGO can be compared against BLCA and other programs using the same reference taxonomic annotation.

Results

To compare BLCA against other software, we reviewed all recently published 16S taxonomic classification tools. Since BLCA aims to improve species-level classification accuracy compared to existing tools, we excluded the 16S Classifier program since it cannot classify at the species level.

To obtain a fair comparison with MEGAN (version 6.7.1), we used the same default criteria as MEGAN for retaining the BLAST hits. The most important MEGAN parameter for extracting BLAST hits for downstream analysis is the parameter *topPercent*, used to keep only the BLAST hits whose bit scores are within a given percentage of the best BLAST hit. The default value in MEGAN for this parameter is 10%. For example, if the top BLAST hit has a bit score of 1000, we only retain BLAST hits for downstream analysis if their BLAST bit scores are at least 900 (i.e., 1000–1000*10%). As shown in Table 1, BLCA consistently outperforms MEGAN with all the tested 16S variable regions from the species to the family levels of taxonomic classification. From the order to the phylum levels, the accuracies of BLCA, MEGAN and other software are similar and above 98% (data not shown). More importantly, the accuracy of

Table 1 Comparison of the classification accuracies using the simulated dataset

CST = 0.8		V2	V4	V1V3	V3V5	V6V9
Species	BLCA	0.7594 ± 0.0164*	0.5331 ± 0.0208	0.9323 ± 0.0054*	0.8335 ± 0.0072*	0.8690 ± 0.0012*
	Kraken	0.7275 ± 0.0054	0.5326 ± 0.0181	0.8672 ± 0.0072	0.7542 ± 0.0087	0.7572 ± 0.0056
	MEGAN	0.7290 ± 0.0114	0.5238 ± 0.0161	0.7071 ± 0.0053	0.5206 ± 0.0108	0.5227 ± 0.0140
	RDP	0.6102 ± 0.0042	0.3928 ± 0.0292	0.8549 ± 0.0199	0.7307 ± 0.0203	0.7823 ± 0.0124
	SPINGO	0.5700 ± 0.0187	0.3910 ± 0.0106	0.7907 ± 0.0061	0.6900 ± 0.0071	0.7318 ± 0.0116
Genus	BLCA	0.9498 ± 0.0019*	0.8982 ± 0.0107*	0.9965 ± 0.0012*	0.9863 ± 0.0011*	0.9925 ± 0.0012*
	Kraken	0.9072 ± 0.0066	0.8612 ± 0.0189	0.9691 ± 0.0051	0.9463 ± 0.0006	0.9437 ± 0.0034
	MEGAN	0.9334 ± 0.0079	0.8830 ± 0.0115	0.9528 ± 0.0040	0.9002 ± 0.0027	0.8939 ± 0.0041
	RDP	0.8768 ± 0.0065	0.8067 ± 0.0139	0.9629 ± 0.0072	0.9562 ± 0.0065	0.9657 ± 0.0042
	SPINGO	0.8481 ± 0.0002	0.7726 ± 0.0077	0.9333 ± 0.0057	0.9192 ± 0.0034	0.9238 ± 0.0067
Family	BLCA	0.9791 ± 0.0009*	0.9787 ± 0.0018*	0.9984 ± 0.0019*	0.9975 ± 0.0019*	0.9970 ± 0.0014*
	Kraken	0.9594 ± 0.0038	0.9480 ± 0.0028	0.9882 ± 0.0021	0.9850 ± 0.0033	0.9799 ± 0.0032
	MEGAN	0.9495 ± 0.0089	0.9413 ± 0.0015	0.9517 ± 0.0032	0.9397 ± 0.0044	0.9447 ± 0.0034
	RDP	0.9461 ± 0.0093	0.9295 ± 0.0062	0.9818 ± 0.0007	0.9806 ± 0.0054	0.9855 ± 0.0013
	SPINGO	NA	NA	NA	NA	NA
CST = 0.5		V2	V4	V1V3	V3V5	V6V9
Species	BLCA	0.8485 ± 0.0128*	0.6813 ± 0.0115*	0.9629 ± 0.0077*	0.9050 ± 0.0034*	0.9315 ± 0.0045*
	Kraken	0.7275 ± 0.0054	0.5326 ± 0.0181	0.8672 ± 0.0072	0.7542 ± 0.0087	0.7572 ± 0.0056
	MEGAN	0.7290 ± 0.0114	0.5238 ± 0.0161	0.7071 ± 0.0053	0.5206 ± 0.0108	0.5227 ± 0.0140
	RDP	0.7526 ± 0.0107	0.5692 ± 0.0194	0.8997 ± 0.0144	0.8221 ± 0.0105	0.8621 ± 0.0094
	SPINGO	0.6570 ± 0.0124	0.5008 ± 0.0114	0.8256 ± 0.0038	0.7497 ± 0.0041	0.7805 ± 0.0021
Genus	BLCA	0.9722 ± 0.0028*	0.9467 ± 0.0031*	0.9985 ± 0.0019*	0.9947 ± 0.0013*	0.9972 ± 0.0002*
	Kraken	0.9072 ± 0.0066	0.8612 ± 0.0189	0.9691 ± 0.0051	0.9463 ± 0.0006	0.9437 ± 0.0034
	MEGAN	0.9334 ± 0.0079	0.8830 ± 0.0115	0.9528 ± 0.0040	0.9002 ± 0.0027	0.8939 ± 0.0041
	RDP	0.9319 ± 0.0044	0.8960 ± 0.0086	0.9710 ± 0.0049	0.9693 ± 0.0046	0.9729 ± 0.0003
	SPINGO	0.8807 ± 0.0034	0.8354 ± 0.0041	0.9400 ± 0.0030	0.9287 ± 0.0024	0.9317 ± 0.0083
Family	BLCA	0.9870 ± 0.0013*	0.9856 ± 0.0035*	0.9987 ± 0.0021*	0.9991 ± 0.0012*	0.9984 ± 0.0019*
	Kraken	0.9594 ± 0.0038	0.9480 ± 0.0028	0.9882 ± 0.0021	0.9850 ± 0.0033	0.9799 ± 0.0032
	MEGAN	0.9495 ± 0.0089	0.9413 ± 0.0015	0.9517 ± 0.0032	0.9397 ± 0.0044	0.9447 ± 0.0034
	RDP	0.9696 ± 0.0040	0.9674 ± 0.0015	0.9836 ± 0.0017	0.9830 ± 0.0033	0.9868 ± 0.0004
	SPINGO	NA	NA	NA	NA	NA

Each entry in the table shows the average and standard deviation of the F-scores for a particular classifier (i.e., rows) at a specific 16S region (i.e., columns) based on three random sets of 1000 test sequences. Two confidence score thresholds (CST), 0.8 and 0.5, were applied for BLCA, RDP Classifier, and SPINGO as described in the main text. The *indicates that the F-scores of BLCA are significantly higher than those of other software, based on a one-tailed paired t-test with a p-value less than 0.05. Similar statistical significance was also obtained using the one-tailed Wilcoxon signed-rank test. Note that the SPINGO program does not produce family-level classification. In addition, Kraken and MEGAN do not provide any probabilistic-based parameters for evaluating the assigned taxa, thus we used their default taxonomic assignments for comparison

MEGAN drops significantly when the *topPercent* filter was relaxed from 5 to 10% and further to 20% (the recommended range by the original MEGAN publication) at both the species and genus levels (Table 2). For example, using V1–V3 sequences, the species-level accuracy of MEGAN, measured by the F-scores, drops from 0.8394 (with *topPercent* set to 5%) to 0.7071 (with *topPercent* set to 10%), and further down to 0.4673 (with *topPercent* set to 20%). Besides V1–V3, these same trends are observed for all other tested 16S regions

(Table 2). These results are expected because, by relaxing this parameter, more dissimilar BLAST hits (i.e., potentially "bad" BLAST hits) are included in the analysis and the inclusion of bad BLAST hits leads to erroneous taxonomic assignments. This reveals a fundamental limitation of the MEGAN method: its results are sensitive to which BLAST hits are included for analysis and it lacks a probabilistic method to penalize bad BLAST hits. Conversely, the results from BLCA, which showed higher accuracy than MEGAN, remained robust to the

Table 2 BLCA accuracy is insenesitve to the inclusion of dissimilar BLAST hits

Taxonomic levels		Genus		Species	
16S region	*topPercent* Filter	BLCA	MEGAN	BLCA	MEGAN
V2	5%	0.9539 ± 0.0038	0.9531 ± 0.0044	0.7747 ± 0.0150	0.8091 ± 0.0153
	10%	0.9498 ± 0.0019	0.9334 ± 0.0079	0.7594 ± 0.0164	0.7290 ± 0.0114
	20%	0.9487 ± 0.0018	0.8966 ± 0.0080	0.7580 ± 0.0176	0.5983 ± 0.0075
V4	5%	0.9078 ± 0.0078	0.9230 ± 0.0082	0.5597 ± 0.0175	0.6497 ± 0.0058
	10%	0.8982 ± 0.0107	0.8830 ± 0.0115	0.5331 ± 0.0208	0.5238 ± 0.0161
	20%	0.8965 ± 0.0092	0.8016 ± 0.0041	0.5317 ± 0.0189	0.3915 ± 0.0119
V1V3	5%	0.9960 ± 0.0009	0.9778 ± 0.0006	0.9314 ± 0.0058	0.8394 ± 0.0069
	10%	0.9965 ± 0.0012	0.9528 ± 0.004	0.9323 ± 0.0054	0.7071 ± 0.0053
	20%	0.9959 ± 0.0009	0.8609 ± 0.0087	0.9321 ± 0.0053	0.4673 ± 0.0150
V3V5	5%	0.9865 ± 0.0020	0.9550 ± 0.0041	0.8380 ± 0.0064	0.7025 ± 0.0112
	10%	0.9863 ± 0.0011	0.9002 ± 0.0027	0.8335 ± 0.0072	0.5206 ± 0.0108
	20%	0.9863 ± 0.0011	0.7369 ± 0.0094	0.8361 ± 0.0039	0.2880 ± 0.0061
V6V9	5%	0.9933 ± 0.0011	0.9532 ± 0.0050	0.8722 ± 0.0066	0.7258 ± 0.0129
	10%	0.9925 ± 0.0012	0.8939 ± 0.0041	0.8690 ± 0.0012	0.5227 ± 0.0140
	20%	0.9931 ± 0.0017	0.7138 ± 0.0083	0.8701 ± 0.0050	0.2691 ± 0.0255

The parameter *topPercent* is for keeping only the BLAST hits whose bit scores are within a given percentage of the best BLAST hit. The larger the parameter is, the more dissimilar database hits are included for taxonomic classification for the query sequence. The default value in MEGAN for this parameter is 10%. In our comparisons, we set the value of *topPercent* to be 5, 10 and 20% for both BLCA and MEGAN, the recommended range by the original MEGAN publication, to compare the performance of BLCA and MEGAN under different stringencies of retaining BLAST hits. Each table entry shows the average and standard deviation of the F-scores, based on the confidence score threshold of 0.8, for each tested software at the corresponding 16S region. The F-scores of BLCA are much less sensitive to the value of *topPercent* when compared to MEGAN

number of included BLAST hits (Table 2) since bad BLAST hits are penalized using posterior probability scores assigned by the BLCA algorithm. It is worth noting that it is unrealistic to prevent the inclusion of bad BLAST hits in a typical large-scale data analysis since there is no universal cutoff to exclude bad BLAST hits. Any such cutoffs are heuristic in nature, as such, they are inevitably either too stringent or not stringent enough.

The SPINGO program is specifically designed for species-level classification. The authors of SPINGO even showed that SPINGO has superior classification accuracy compared to a customized RDP Classifier and best-matched BLAST hits at species level [4]. Like BLCA and MEGAN, SPINGO uses the NCBI taxonomic database for taxonomic assignments. Unlike those tools, however, SPINGO uses a k-mer based approach instead of sequence alignment to measure the similarity between query and database sequences. The only threshold for SPINGO is its confidence score for taxonomic assignments, which is compatible with the BLCA confidence score. Table 1 shows that the accuracy of BLCA is statistically significantly higher than that of SPINGO in all tested 16S regions at the confidence score thresholds of 0.8 and 0.5, respectively. In addition, SPINGO cannot do subspecies classification, nor can it do family or higher level classification, whereas BLCA can classify reads from any level ranging from subspecies to phylum (though there are not enough annotated subspecies

datasets at NCBI for evaluating BLCA subspecies-level classification accuracy).

Even though the standard release of the RDP Classifier cannot classify 16S sequences at the species level, we obtained the training script from the RDP Classifier's development team (personal communications) and re-trained the RDP Classifier for species-level classification with the same NCBI 16S database that BLCA uses. The NCBI 16S database is used because MEGAN and SPINGO must use NCBI taxonomic database. Therefore, the NCBI database provides a common ground for evaluating the results of all of these tools on the basis of their computational algorithms without being influenced by different taxonomic standards. Similar to SPINGO, the RDP Classifier's confidence score is also compatible with the BLCA confidence score. Although the default threshold for the RDP Classifier's confidence score is 0.8, the developers of the RDP Classifier also recommend a threshold of 0.5 for short read classification. Our results show that BLCA has higher accuracy than the RDP Classifier at the thresholds of 0.8 and 0.5 (Table 1).

Besides these 16S-specific classification tools, there are also metagenomic classification tools that are designed for identifying microbial taxa from whole metagenome shotgun (WMS) sequences. We have chosen Kraken [15] as a representative WMS classification tool to compare with BLCA. Kraken is chosen because of two reasons: i) it has superior or comparable classification accuracy to other existing WMS tools [20] and ii) to our best knowledge, it is the only WMS tool that has been successfully applied in a published 16S study [21]. Kraken's default database incorporates reference genome sequences. To have a fair comparison with BLCA, we have replaced Kraken's default database with the same NCBI 16S database used for BLCA, thus increasing its sensitivity to classify a broader range of bacterial taxa. Kraken, a k-mer based program seeking best database matches, does not provide any confidence score to evaluate the confidence of assigned taxonomies, although Kraken's output can be filtered based on the percent of k-mers matched to each taxa (no guidance is provided by its developer on how to set the filtering threshold). As shown in Table 1, even allowing the maximum sensitivity for Kraken (i.e., without any filtering of Kraken's output), which is the default setting for Kraken, BLCA still significantly outperforms Kraken with all tested 16S regions from the species to the family level.

In addition to using simulated datasets to evaluate BLCA and other software, Table 3 shows that BLCA had either higher or comparable classification accuracies when tested with a real-world 16S dataset. For example, with a confidence score threshold of 0.5 (the recommended threshold for the RDP Classifier for short sequence reads), the species-level classification accuracy of

Table 3 Comparison of the classification accuracies using a real-world dataset

Taxonomy Level	Method	V1V2 Region	
		CST = 0.8	CST = 0.5
Species	BLCA	0.570	0.716
	Kraken	0.589	0.589
	MEGAN	0.544	0.544
	RDP	0.490	0.613
	SPINGO	0.486	0.562
Genus	BLCA	0.729	0.79
	Kraken	0.694	0.694
	MEGAN	0.745	0.745
	RDP	0.643	0.708
	SPINGO	0.605	0.650
Family	BLCA	0.814	0.832
	Kraken	0.777	0.777
	MEGAN	0.869	0.869
	RDP	0.775	0.805
	SPINGO	NA	NA

Each entry in the table shows the F-scores for a classifier (i.e., rows) based on all the OTU sequences in the msd16s dataset, as described in the main text. Two confidence score thresholds (CST), 0.8 and 0.5, were applied for BLCA, RDP Classifier, and SPINGO, the thresholds as in Table 1. Note that the SPINGO program does not produce family-level classification. In addition, Kraken and MEGAN do not provide any probabilistic-based parameters for evaluating the assigned taxa, thus we used their default taxonomic assignments for comparison

BLCA, measured using an F-score, is 0.716, much higher than the classification accuracy of MEGAN (0.544), the RDP Classifier (0.613), and SPINGO (0.562). The same trends were observed when the default confidence score threshold of 0.8 was applied (Table 3). It is worth noting that, as this is a real-world dataset, the true taxonomic classification is unknown. We had to rely on the top BLAST hit as the reference taxonomic classification when we evaluated the classification accuracies of each software. Nonetheless, the results from the real-world dataset were consistent with those from the simulated datasets, showing that BLCA tends to produce higher taxonomic classification accuracies than currently existing software.

Discussion

Despite the importance of species-level classification, the existing tools either do not classify 16S sequences at the species level or their taxonomic assignments are not reliable. As discussed above, k-mer based methods are intrinsically less accurate than an alignment-based sequence similarity measurement. The k-mer based approaches may be sufficient for high level taxonomic classification, since sequences from different higher taxonomic levels tend to be very divergent. For lower level taxonomic classification, however, particularly species-level classification, we have shown that BLCA significantly outperforms k-mer based methods (e.g., SPINGO, the RDP Classifier, and Kraken) in classification accuracy.

In addition, the Bayesian posterior probability of BLCA quantitatively measures the difference between the best database hit and other database hits, and the bootstrapping principle, adopted by BLCA for providing confidence score, has solid statistical foundation for measuring prediction errors [22]. In this study, we have applied 0.5 and 0.8 as thresholds for the BLCA confidence scores for comparison with other software. The confidence score of BLCA is comparable to that of the RDP Classifier and SPINGO. There is no perfect universal threshold that is suitable for all datasets. We recommend that users consider exploring several different thresholds (e.g., 0.6 and 0.8) to examine if their results are consistent under different thresholds. If not, the users need to be wary that their results may be too sensitive based on the particular threshold they have chosen.

It is worth mentioning that BLCA does not require a training process for classification, which can be more convenient for some users when compared to some other software. For example, the 16S Classifier trains a standard machine-learning model, a Random Forest, with k-mer nucleotides from different regions of 16S rRNA genes. We could not even test our V1–V3, V3–V5, and V6–V9 datasets with the 16S Classifier because the published software has not been trained for this region, even though these regions are widely used in microbiome studies. In contrast, our BLCA program requires no training process at all since our algorithm is based on the alignment between query and reference database sequences. Therefore, users only need to download reference 16S database sequences for BLCA and this allows our method to be easily applied to any other DNA marker gene families for taxonomic classification (e.g., rpoB or 18S rRNA gene sequences). The accompanying BLCA package includes instructions on how to replace the default 16S sequences with the user's own customized gene family sequences. For example, to demonstrate the flexibility of alternative database sequences, BLCA provides an option to use the Greengenes 16S database and its associated taxonomy [11] instead of the default NCBI 16S database since many researchers may prefer the Greengenes taxonomy.

We have shown that BLCA has significantly higher accuracy than existing taxonomic classification methods at the species level. This higher accuracy comes with the cost of longer computation time. BLCA is not designed for performing taxonomic classification for raw 16S sequences. Instead, raw 16S sequences should be first clustered into OTUs to eliminate redundant or highly similar sequences before performing taxonomic classification, which is a standard procedure for 16S sequence processing by widely used software packages, e.g., QIIME. With 100,000 OTUs, BLCA can have a run-time of approximately 4 days, which is not unusual for modern-day bioinformatics tasks with large datasets. Considering the significant gains in accuracy with our method, we believe that many researchers will find the time tradeoff to be reasonable. In addition, users can divide the input sequences into multiple files and execute BLCA in parallel on computer clusters to hasten the classification process, if necessary. In addition, not all OTUs require species-level classification in practice. Typically, researchers are only interested in a small subset of OTUs, e.g., a list of OTUs that are differentially abundant in different ecosystems (similar to how molecular biologists are often only interested in detailed gene annotations for a small list of differentially expressed genes instead of all of the genes in an organism). In these cases, BLCA may take only a few minutes to classify a subset of several hundreds of OTUs of interest.

Conclusion

In summary, we have developed a novel computational method that significantly outperforms previously published software for species-level classification accuracy. Its probabilistic-based confidence score helps users evaluate the confidence of the resulting taxonomic assignments based on multiple database hits. In addition,

our methods do not require any training, which makes it easily applicable for different regions of 16S rRNA gene or even different phylogenetic marker genes. Despite its higher computational costs, our method is still suitable for large-scale microbiome datasets, providing a valuable alternative option for microbiome researchers who prefer higher classification accuracy.

Abbreviations
FN: False negative; FP: False positive; LCA: Lowest Common Ancestor; OTU: Operational taxonomic unit; rRNA: ribosomal RNA; TN: True negative; TP: True positive

Acknowledgements
We thank Benli Chai from the RDP research group for providing the training script for the RDP Classifier. We also thank Michael Zhao, Laurynas Kalesinskas and Francis Cocjin for testing the BLCA software package and proofreading the manuscript. We also thank the two anonymous reviewers for their helpful comments and suggestions.

Funding
This work was supported by NIH grants 1R01AI116706-01A1, 1P20DK108268, and U01HL121831. The funding bodies had no role in the design of the study and collection, analysis, and interpretation of data and in writing the manuscript.

Authors' contributions
XG and QD developed the algorithm, designed computational experiments, analyzed the results, and wrote the manuscript. HL and KR independently implemented the initial version of the computer software, conducted testing, developed documentation, and assisted with manuscript preparation. HL produced the final version of the software, and conducted all the computational experiments. All authors have read and approved the final version of the manuscript.

Competing interests
The authors declare that they have no competing interests.

Author details
[1]Department of Public Health Sciences, Loyola University Chicago Health Sciences Division, Maywood, IL 60153, USA. [2]Center for Biomedical Informatics, Loyola University Chicago Health Sciences Division, Maywood, IL 60153, USA. [3]Bioinformatics Program, Loyola University Chicago Lake Shore Campus, Chicago, IL 60660, USA. [4]Department of Computer Science, Loyola University Chicago Water Tower Campus, Chicago, IL 60611, USA.

References
1. Fettweis JM, Serrano MG, Sheth NU, et al. Species-level classification of the vaginal microbiome. BMC Genomics. 2012;13 Suppl 8:S17.
2. Wang Q, Garrity GM, Tiedje JM, et al. Naive Bayesian classifier for rapid assignment of rRNA sequences into the new bacterial taxonomy. Appl Environ Microbiol. 2007;73(16):5261–7.
3. Chaudhary N, Sharma AK, Agarwal P, et al. 16S classifier: a tool for fast and accurate taxonomic classification of 16S rRNA hypervariable regions in metagenomic datasets. PLoS One. 2015;10(2):e0116106.
4. Allard G, Ryan FJ, Jeffery IB, et al. SPINGO: a rapid species-classifier for microbial amplicon sequences. BMC Bioinformatics. 2015;16(1):324.
5. Caporaso JG, Kuczynski J, Stombaugh J, et al. QIIME allows analysis of high-throughput community sequencing data. Nat Methods. 2010;7(5):335–6.
6. Meyer F, Paarmann D, D'Souza M, et al. The metagenomics RAST server–a public resource for the automatic phylogenetic and functional analysis of metagenomes. BMC Bioinformatics. 2008;9(1):1.
7. Vilo C, Dong Q. Evaluation of the RDP classifier accuracy using 16S rRNA gene variable regions. Metagenomics. 2012;1:1–5.
8. Koski LB, Golding GB. The closest BLAST hit is often not the nearest neighbor. J Mol Evol. 2001;52(6):540–2.
9. Huson DH, Auch AF, Qi J, et al. MEGAN analysis of metagenomic data. Genome Res. 2007;17(3):377–86.

10. Altschul SF, Madden TL, Schaffer AA, et al. Gapped BLAST and PSI-BLAST: a new generation of protein database search programs. Nucleic Acids Res. 1997;25(17):3389–402.

11. DeSantis TZ, Hugenholtz P, Larsen N, et al. Greengenes, a chimera-checked 16S rRNA gene database and workbench compatible with ARB. Appl Environ Microbiol. 2006;72(7):5069–72.

12. Edgar RC. MUSCLE: multiple sequence alignment with high accuracy and high throughput. Nucleic Acids Res. 2004;32(5):1792–7.

13. Felsenstein J. Confidence limits on phylogenies: an approach using the bootstrap. Evolution. 1985;39(4):783–91.

14. Quail MA, Smith M, Coupland P, et al. A tale of three next generation sequencing platforms: comparison of Ion Torrent, Pacific Biosciences and Illumina MiSeq sequencers. BMC Genomics. 2012;13(1):1.

15. Wood DE, Salzberg SL. Kraken: ultrafast metagenomic sequence classification using exact alignments. Genome Biol. 2014;15(3):1–12.

16. Powers DM. Evaluation: from precision, recall and F-measure to ROC, informedness, markedness and correlation. 2011.

17. Pop M, Walker AW, Paulson J, et al. Diarrhea in young children from low-income countries leads to large-scale alterations in intestinal microbiota composition. Genome Biol. 2014;15(6):R76.

18. Gentleman RC, Carey VJ, Bates DM, et al. Bioconductor: open software development for computational biology and bioinformatics. Genome Biol. 2004;5(10):R80.

19. Cole JR, Chai B, Farris RJ, et al. The Ribosomal Database Project (RDP-II): sequences and tools for high-throughput rRNA analysis. Nucleic Acids Res. 2005;33(Database issue):D294–6.

20. Lindgreen S, Adair KL, Gardner PP. An evaluation of the accuracy and speed of metagenome analysis tools. Sci Rep. 2016;6:19233.

21. Valenzuela-González F, Martínez-Porchas M, Villalpando-Canchola E, et al. Studying long 16S rDNA sequences with ultrafast-metagenomic sequence classification using exact alignments (Kraken). J Microbiol Methods. 2016;122:38–42.

22. Efron B, Tibshirani RJ. An introduction to the bootstrap. Boca Raton: CRC press; 1994.

8

De novo assembly of highly polymorphic metagenomic data using *in situ* generated reference sequences and a novel BLAST-based assembly pipeline

You-Yu Lin[1,4]* ⓘ, Chia-Hung Hsieh[2], Jiun-Hong Chen[1], Xuemei Lu[3], Jia-Horng Kao[4], Pei-Jer Chen[4], Ding-Shinn Chen[4,5] and Hurng-Yi Wang[4,6,7]*

Abstract

Background: The accuracy of metagenomic assembly is usually compromised by high levels of polymorphism due to divergent reads from the same genomic region recognized as different loci when sequenced and assembled together. A viral quasispecies is a group of abundant and diversified genetically related viruses found in a single carrier. Current mainstream assembly methods, such as Velvet and SOAPdenovo, were not originally intended for the assembly of such metagenomics data, and therefore demands for new methods to provide accurate and informative assembly results for metagenomic data.

Results: In this study, we present a hybrid method for assembling highly polymorphic data combining the partial *de novo*-reference assembly (PDR) strategy and the BLAST-based assembly pipeline (BBAP). The PDR strategy generates *in situ* reference sequences through *de novo* assembly of a randomly extracted partial data set which is subsequently used for the reference assembly for the full data set. BBAP employs a greedy algorithm to assemble polymorphic reads. We used 12 hepatitis B virus quasispecies NGS data sets from a previous study to assess and compare the performance of both PDR and BBAP. Analyses suggest the high polymorphism of a full metagenomic data set leads to fragmentized *de novo* assembly results, whereas the biased or limited representation of external reference sequences included fewer reads into the assembly with lower assembly accuracy and variation sensitivity. In comparison, the PDR generated *in situ* reference sequence incorporated more reads into the final PDR assembly of the full metagenomics data set along with greater accuracy and higher variation sensitivity. BBAP assembly results also suggest higher assembly efficiency and accuracy compared to other assembly methods. Additionally, BBAP assembly recovered HBV structural variants that were not observed amongst assembly results of other methods. Together, PDR/BBAP assembly results were significantly better than other compared methods.

Conclusions: Both PDR and BBAP independently increased the assembly efficiency and accuracy of highly polymorphic data, and assembly performances were further improved when used together. BBAP also provides nucleotide frequency information. Together, PDR and BBAP provide powerful tools for metagenomic data studies.

Keywords: Next generation sequencing, Metagenomics, Hepatitis B virus, Sequence assembly, Assembly pipeline

* Correspondence: youylin@ntu.edu.tw; hurngyi@ntu.edu.tw
[1]Department of Life Science, National Taiwan University, Taipei 106, Taiwan
[4]Graduate Institute of Clinical Medicine, National Taiwan University, Taipei 100, Taiwan
Full list of author information is available at the end of the article

Background

Next-generation sequencing (NGS) has become the mainstream method for obtaining high quantities of genomic data during the past decade, and the increased accessibility of massive datasets has driven up the need for compatible analytic algorithms and software [1]. There are several key components for an assembly algorithm, including the capacity to handle massive data sets, the accuracy and efficiency of the assembly, the nature of the data set itself, and the intended use of the assembly results. The former two are dependent of the hardware and algorithms implemented, whereas the latter two influences the optimization strategy and the type of information to be extracted during assembly. For example, metagenomic studies commonly aim to understand the composition and relative abundances of the data set as well as the intra-species or inter-population heterogeneity, therefore the assembly depth and length as well as accuracy are prioritized for such data sets [2].

A viral quasispecies is a group of highly genetically related viruses found in a single carrier and can be both abundant (viral titer $\approx 10^6$-10^9 ge/ml) and greatly diversified (nucleotide diversity $\approx 10^{-2}$-10^{-3}) within patient carriers [3–5]. Two main NGS platforms, 454/Roche pyrosequencing [6] and Illumina Genome Analyzer [7], have been commonly used for recent quasispecies-related studies. Pyrosequencing has longer sequence reads and typically does not require data set assembly [8–10], although some studies still performed *de novo* assembly [11] or reference sequence assembly [12, 13]. Illumina sequencing generates much larger data sets compared to pyrosequencing, but its shorter read length limits the efficiency for *de novo* assembly [2]. Therefore, Illumina sequenced viral quasispecies data sets are usually assembled using reference sequences as templates [14–17] while *de novo* assembly is applicable but not commonly used [18].

The high throughput Illumina platform, compared to the pyrosequencing platform, is capable of detecting greater amounts of genetic variation within viral quasispecies [15]. However, a major challenge for Illumina quasispecies NGS studies is the sequence assembly of the data sets. Sequence assembly using a reference approach is not only subject to bias of the chosen reference sequence, but also assembles less reads and thus less genetic variation information in the assembly [15]. *De novo* assembly should be able to provide the most complete and accurate genetic information of NGS data, but can be hindered by regions with high levels of diversity. The commonly used *de novo* assembly algorithms, such as Velvet [19], SOAPdenovo [20], CLC Genomics Workbench (CLC, CLC bio, Aarhus, Denmark), and Euler-SR [21], were not originally intended for the assembly of metagenomics data with high diversity and coverage depth. Recent progress have been made in the development of *de novo* assembly algorithms for metagenomes, such as MetaVelvet [22] and Genovo [23].

In this study, we propose a partial *de novo*-reference assembly strategy, PDR, which is a *de novo*-reference hybrid assembly strategy that utilizes the completeness of *de novo* assembly while complementing its low-efficiency with reference assembly. PDR generates an *in situ* reference sequence by *de novo* assembly of a smaller yet less diverse partial data set followed by the reference assembly of the full data set. Results show that the PDR assembly results are more complete and accurate than direct *de novo* or reference assembly of highly polymorphic metagenomic data sets. We also present a novel BLAST-based assembly pipeline, BBAP, capable of both *de novo* and reference assembly specifically designed for assembly of metagenomic data sets. The assembly efficiency and accuracy of both PDR and BBAP were examined using actual NGS data sets as well as *in silico* generated simulated NGS data sets and compared with the assembly results of other assembly methods.

Results

To examine the performance of BBAP and the proposed hybrid assembly strategy, we acquired 12 NGS data sets of HBV viral quasispecies from 7 HBV patient samples [24]. The 12 data sets used for assembly consisted of an average of 21,494,295 101-bp raw reads (RRs), 14,388,844 high quality reads (HQRs, quality score ≥ 20 for all bases; i.e., sequencing error rate = 1%), and 60,228 HRURs (high redundancy unique representative reads; unique representative reads with redundancy ≥ 5, Table 1 and Additional file 1: Table S1). The optimized parameters for BBAP assembly are listed in Additional file 1: Table S2. The same parameters were used for all BBAP assemblies in this study unless mentioned otherwise.

BBAP *de novo* assembly of full and partial data sets

The *de novo* assembly of the full data sets (FD) resulted in an average of 46.0 contigs (minimum length of 150 bp) for each library with an average contig length of 321 bp, suggesting that the assembly results were fragmentized (Table 1 and Additional file 1: Table S3). For *de novo* assembly of partial data sets (PD) of each data set, five partial data sets were initially randomly generated and assembled independently. Because the PD assembly results of the partial data sets from each library were highly similar (data not shown), a single partial data set and its assembly results were used for representation of the sample in further analyses. The PD assembly yielded fewer number of contigs and longer average maximum contig lengths, indicating the PD assembly results were not as fragmentized as FD assembly. Furthermore, PD assembly required fewer contigs than the FD assembly to span the full genome to recover the full length HBV

Table 1 Average assembly statistics of all 12 data sets using BBAP with multiple approaches

	PD[a]	FD[b]	SR[c]	PDR[d]
RRs	214,942	21,494,295	21,494,295	21,494,295
HQRs	143,912	14,388,844	14,388,844	14,388,844
URs	27,150	860,144	860,144	860,144
HRURs	6264	60,228	60,228	60,228
RiHRURs	116,555	13,388,423	13,388,423	13,388,423
Contigs assembled[e]	2.1	46.0	1.0	3.9
Max contig length	3119	1473	3,207	3148
Average contig length	2319	321	3207	1268
% of Mapped HRURs	95.9%	70.3%	67.4%	69.9%
% of Mapped RiHRURs	80.4%	68.7%	82.7%	84.5%

The full data sets were used in the BBAP assembly with FD, SR, and PDR approaches, whereas partial data sets consisting of 1% of randomly selected RRs were used in the BBAP PD assembly approach

[a]Partial data set *de novo* assembly

[b]Full data set *de novo* assembly

[c]Sanger reference assembly

[d]Partial data set reference assembly of the full data set

[e]Only minimum assembled contig length > 150 bp was shown

RRs raw reads, *HQRs* high quality reads (quality score threshold = 20, i.e., sequencing error rate = 1%), *URs* unique representative reads, *HRURs* high redundancy unique representative reads (unique representative reads with redundancy threshold = 5), *RiHRURs* reads included in high redundancy unique representative reads

genome (Fig. 1a, Additional file 2: Figure S1). PD assembly also yielded a higher proportion of mapped HRURs (95.9% vs 70.3%) and RiHRURs (reads included in high redundancy unique representative reads, 80.4% vs. 68.7%) than FD, further demonstrating its better assembly efficiency.

Fragmentation is possibly due to high polymorphic reads from the same genomic regions recognized by BBAP as different haplotypes and subsequently assembled into separate clusters. The proportion of polymorphic sites in overlapping contig regions of D2_1 FD assembly was 10 times higher than that in non-overlapping regions (0.238 vs. 0.022; $p < 10^{-10}$). A similar trend was also found in D2_1 PD assembly (Additional file 1: Table S4). The shorter FD assembled contigs (<300 bp) had a significantly higher proportion of polymorphic sites than the longer FD assembled contigs (Additional file 2: Figure S2, Student's *t*-test, $p < 0.05$). HRURs that were included or excluded in the partial data sets (for PD assembly) had average redundancies of 1,808X (n = 75,173) and 38X (n = 647,561), respectively, within the full data set. Additionally, the redundancies of the included HRURs in the full and partial data sets were highly correlated ($R^2 = 0.9997$). This suggests the random selection partial data sets was unbiased and effectively excluded HRURs of low redundancies, resulting in lower polymorphism levels and, in turn, less fragmented assembly results.

BBAP reference assembly with different reference sequences

To fully represent the full data set, the PD assembled contigs were used as references for the reference assembly of the full data set (PDR). For comparison purposes, a Sanger sequence from each patient sample was chosen as the reference sequence for the reference assembly of the full data set (SR). SR assembly resulted in single contigs with average lengths of 3207 bp, whereas PDR assembly produced an average of 3.9 contigs with maximum and average lengths of 3148 bp and 1268 bp, respectively (Table 1 and Additional file 1: Table S3). Both PDR and SR recovered full HBV genomes and similar levels of

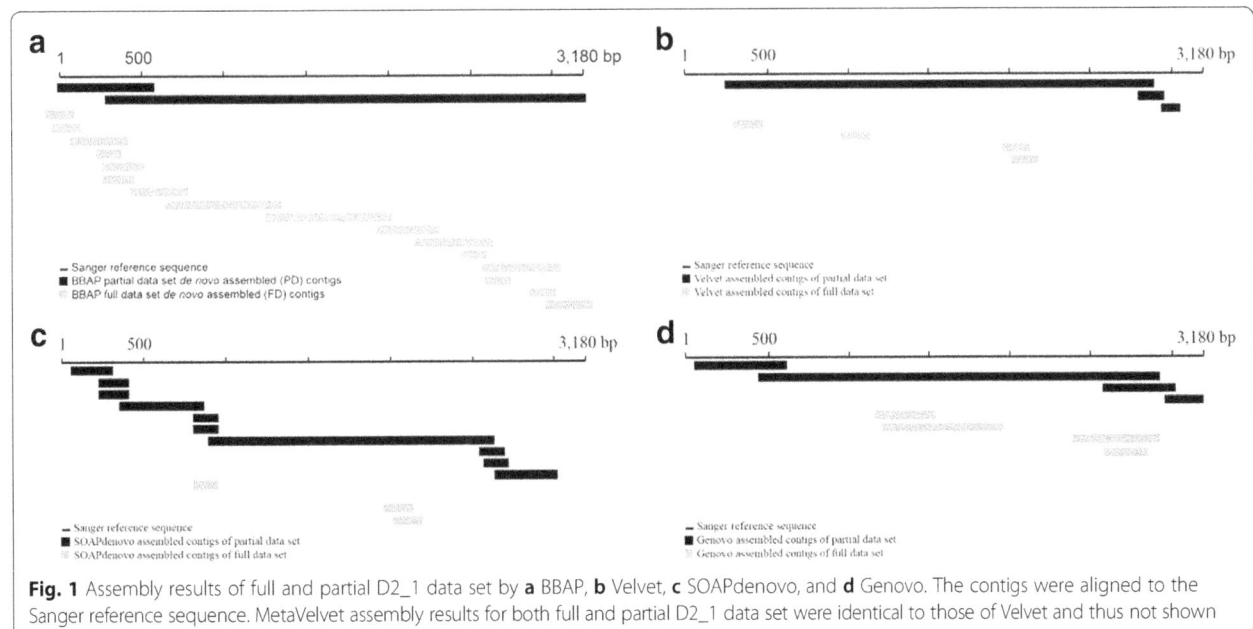

Fig. 1 Assembly results of full and partial D2_1 data set by **a** BBAP, **b** Velvet, **c** SOAPdenovo, and **d** Genovo. The contigs were aligned to the Sanger reference sequence. MetaVelvet assembly results for both full and partial D2_1 data set were identical to those of Velvet and thus not shown

polymorphism in the consensus sequences (Additional file 1: Table S5), but the PDR assembly additionally identified HBV structural variants (Additional file 1: Table S6, Additional file 2: Figure S3-S5 and Additional file 3: SA).

PDR alignment accuracy was also higher than SR. SR assembly of D2_1 resulted in a single contig with 50,587 HRURs, but only 50,211 of the SR assembled HRURs were mapped to the two main PDR assembled contigs (M1 and M2; Additional file 2: Figure S6, 50,396 HRURs) covering the full HBV genome and have identical sequences as the SR contig. Not only did the remaining 376 HRURs all mapped to one of the nine PDR assembled variant contigs, but the SR alignment qualities of those 376 HRURs was less optimal than the 50,211 HRURs, shown by the significantly greater BLAST e-value and lower BLAST alignment score (Wilcoxon rank-sum test, $p < 0.001$), both supporting the higher alignment accuracy of PDR assembly. Overall, results of SR assembly and PDR assembly were similar in recovering sequence variation, but the latter included more HRURs and RiHRURs with increased accuracy due to the additional mapping options of the shorter HBV variant contigs provided by the *de novo* assembly of the partial data set, whereas the lower assembly accuracy of the former resulted in low quality alignments and slightly more polymorphic sites.

We were able to measure the polymorphism level of BBAP assembly results (Additional file 2: Figure S6) by calculating the nucleotide frequencies for each position (Additional file 1: Table S7, Additional file 2: Figure S7 and Additional file 3: SB). Furthermore, the nucleotide frequencies derived from BBAP PDR assembly were validated by pyrosequencing (Additional file 1: Table S8), demonstrating the assembly results of BBAP are reliable.

BBAP assembly results compared with other assembly methods

We next compared the efficiency and accuracy of BBAP to different assembly methods using both full and partial

D2_1 data set. Similar to BBAP FD, the full data set assemblies by Velvet, MetaVelvet, SOAPdenovo, and Genovo resulted in fragmented contigs. *De novo* assembly of full data set with Velvet resulted in 13 contigs with maximum and average lengths of 1102 bp and 303 bp, respectively (Table 2), and recovered only 19% of the HBV genome (Fig. 1b, Additional file 2: Figure S1). MetaVelvet assembly results, which are based on initial Velvet assembly results, did not show any improvement and were completely identical to Velvet assembly results for both full and partial data set. SOAPdenovo generated 8 assembled contigs with maximum and average lengths of 934 bp and 340 bp, respectively, and covered 14% of the HBV genome (Fig. 1c). Genovo assembly for the D2_1 data set resulted in a total of 60 contigs with maximum and average contig lengths of 1352 bp and 395 bp, respectively, but only 44% of the HBV genome were recovered (Fig. 1d, Additional file 2: Figure S1).

We proposed that the high polymorphic nature of virus quasispecies may have hindered the efficiency of sequence assembly, and a randomly extracted yet less polymorphic partial data set may provide a better start for initial assembly as shown in FD vs. PD assemblies. Assembly results of different methods all show that the assembly of the partial data set not only generated longer contigs, but also recovered more than 90% of the full HBV genome, demonstrating that exclusion of low redundant HRURs by random selection of partial data effectively reduced level of polymorphism which, in turn, improved the assembly results as judged by contig length and coverage (Table 2, Additional file 2: Figure S1).

We also noticed that BBAP had better performance in recovering structural variants than the other methods tested. While some of BBAP assembled HBV variants were validated by PCR sequencing (Fig. 2), both Velvet/MetaVelvet and SOAPdenovo did not identify any contigs with HBV structural variation. Although Genovo assembled 34 structural containing contigs, their accuracies were

Table 2 Comparison of D2_1 assembly results with different methods and different data set sizes

	Max length	Average length	Number of contigs	% of HBV genome recovered	Contigs that map to reference HBV genome	Contigs with HBV structural variants
BBAP/FD	998	263	52	100%	16	30
Velvet/Full	1102	303	13	19%	4	0
MetaVelvet/Full	1102	303	13	19%	4	0
SOAPdenovo/Full	934	340	8	14%	3	0
Genovo/Full	1352	395	60	44%	4	34
BBAP/PD	2924	692	6	100%	3	3
Velvet/Partial	2576	973	3	89%	3	0
MetaVelvet/Partial	2576	973	3	89%	3	0
SOAPdenovo/Partial	1723	390	10	95%	10	0
Genovo/Partial	2427	481	12	91%	4	7

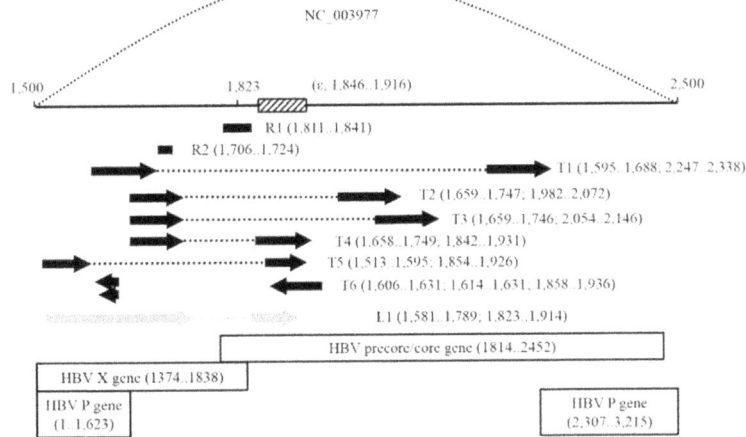

Fig. 2 Schematic summary of corresponding HBV genome (NC_003977) regions for assembled contigs identified as HBV variants. Arrows indicate 5′ to 3′ direction. Only reads containing the sequences spanning the junction regions were assembled separately into variant contigs; reads spanning non-junction regions of the variant contigs (dotted lines) were assembled into the main HBV contig. The L1 sequence, which is similar to T5, resulted from HBV variant validation with PCR using specialized primers followed by Sanger sequencing. Positions are in correspondence with NC_003977, with dotted lines representing the remaining portion of the circular HBV genome, and the boxed section indicating the encapsidation signal (or episilon, ε)

questionable as most of them with non-retraceable junction regions (Additional file 2: Figure S8 and Additional file 3: SC).

Results of *in silico* data set assembly

For a more general assessment and comparison of BBAP performance, *in silico* NGS data sets were generated from the NCBI HBV complete genome and assembled separately using BBAP FD, Velvet, MetaVelvet, SOAPdenovo, and Genovo. Data set sizes were set to 1,726,462 (55,799X), 172,646 (5,579X), 17,264 (557X), and 1726 (55X) HQRs in combination with error rates of 10^{-2}, 10^{-3}, and 10^{-4}/site. Due to computing time considerations, the maximum simulated data set size of 55,799X was approximately 10% of the D2_1 data set size. Five independent data sets were generated for each parameter combination. BBAP assembly results were highly consistent regardless of the data set parameter values. All but one of the 60 assembly results had both perfect coverage and accuracy; the lone standout assembly result had perfect coverage but a 0.9996 (3214/3215) accuracy (Table 3 and Additional file 1: Table S9). The single "inaccurate" nucleotide was not an assembly error, but rather a degenerate nucleotide (Y) representing the reference nucleotide (T, 2/3 or 0.67) and the *in silico* generated erroneous nucleotide (C, 1/3 or 0.33). The corresponding *in silico* data set was generated with the highest error rate (0.01) and smallest data set size (55X), which is the most likely parameter value combination for erroneous nucleotides to exceed the minimum nucleotide frequency threshold (0.2).

Velvet assembly of the *in silico* data sets produced mixed results (Additional file 1: Table S10). Data sets with low error rates and/or small data set sizes were assembled with near perfect coverage and accuracy, whereas both large data sets and high error rates were poorly assembled. As the degree and amount of polymorphism are proportional to the error rate and data set size, respectively, results suggest Velvet is inefficient in assembling highly polymorphic data sets. Unlike the assembly results for D2_1 data sets, MetaVelvet *in silico* data set assembly results, compared to Velvet results, were improved with higher coverage and less fragmentation (Additional file 1: Table S11). MetaVelvet has wider parameter handling range than Velvet, but was still unable to assemble highly polymorphic data sets with high error rates and large data set sizes. Similar to that of Velvet and MetaVelvet, SOAPdenovo could not efficiently assemble data sets of high polymorphism (large data set size and high error rate). In addition, SOAPdenovo also performed poorly when assembling data sets of low polymorphism (low error rate and small data set size). Only data sets of medium sizes and error rates were efficiently assembled by SOAPdenovo (Additional file 1: Table S12). Genovo assembly of smaller data set sizes (55X, 557X, and 5,579X), regardless of the error rate, were highly consistent, with only a single nucleotide assembly error among all 45 assembly results (Additional file 1: Table S13). The assembly result for the largest data sets (55,799X) were slightly fragmentized across all error rates and on average 4 assembly errors were identified among high error rate (0.01) data sets.

Discussion

We developed BBAP, an assembly pipeline designed for the accurate and efficient assembly of highly polymorphic

Table 3 Assembled results of in silico generated data sets from the reference HBV genome by different methods[a]

Data set size	Method	BBAP FD			Velvet			SOAPdenovo			Genovo		
	Error rate	10^{-4}	10^{-3}	10^{-2}	10^{-4}	10^{-3}	10^{-2}	10^{-4}	10^{-3}	10^{-2}	10^{-4}	10^{-3}	10^{-2}
55X	Coverage	1	1	1	1	1	1	0	0	1	1	1	1
	Accuracy	1	1	0.99	1	1	1	0	0	1	1	1	1
	# of contigs	1	1	2	1	1	1	0	0	1	1	1	1
557X	Coverage	1	1	1	1	0.99	1	0	1	**0.27**	1	1	1
	Accuracy	1	1	1	1	1	0.99	0	0.99	**0.99**	1	1	1
	# of contigs	1	1	2	1	1	9	0	1	**5**	1	1	1
5,579X	Coverage	1	1	1	0.99	0.96	**0.03**	1	**0.01**	**0.43**	1	1	1
	Accuracy	1	1	1	1	1	**0.59**	1	**0.20**	**0.99**	1	1	0.99
	# of contigs	1	1	1	1	6	**1**	1	**0**	**11**	1	1	1
55,799X	Coverage	1	1	1	0.98	**0**	**0.11**	**0.02**	**0**	**0.04**	1	1	1
	Accuracy	1	1	1	1	**0**	**0.97**	**0.40**	**0**	**0.80**	1	1	0.99
	# of contigs	1	1	1	3	**0**	**2**	**0**	**0**	**1**	3	2	5

[a]Results represent averages of the assembly results of 5 replicate data sets. Bold areas indicate average assembly results with <80% coverage

metagenomic NGS data sets. BBAP implements a unique BLAST-based greedy algorithm to assemble data set reads and provides multiple intuitive parameters, depending on the nature of the data set, the sequencing platform, and information demands, to adjust the threshold for read alignment, variant retention, and error removal during assembly. BBAP assembly results of both real and simulated NGS data sets were of higher quality than assembly results of other methods compared.

We also introduce a new partial *de novo*-reference (PDR) assembly strategy, which *in situ* generates reference sequences by *de novo* assembly of a randomly extracted partial data set to be subsequently used for the reference assembly of the full data set. Current assembly approaches typically assemble the full data set straightforward with either *de novo* or reference assembly methods, each with their respective advantages and disadvantages. Reference assembly is a much more direct process than *de novo* assembly which reduces alignment ambiguities and low coverage issues. However, the quality of reference assembly is reliant on the representation level of the reference sequence, as the assembly result will be biased towards the reference sequence and sequence variations not represented by the reference sequence will not be captured. *De novo* assembly, which is independent of reference sequences, possesses the potential to generate a more complete assembly result including majority consensus sequences and minor variant sequences, but can be hindered by coverage gaps that lack sequencing information and polymorphic regions with high levels of diversity as shown in Tables 1 and 2.

The partial *de novo*-reference assembly strategy utilizes the advantages of both traditional approaches to contemplate each other. *De novo* assembly of a randomly extracted yet less polymorphic partial data set provides assembly results that are more complete and highly representative of both majority sequence as well as minor variant sequences in the full data set. In turn, the following reference assembly not only assembles more reads due to the accurate representation of the reference sequences, but also has increased assembly accuracy than both straight-up *de novo* and reference assemblies (Table 1). More importantly, the improved quality of assembly resulting from this hybrid PDR approach was not limited to BBAP, as better assembly results using partial data sets were also demonstrated by Velvet, MetaVelvet, SOAPdenovo, and Genovo (Table 2).

The assembly efficiency of metagenomics data sets is also dependent on the algorithms each assembly method employs. Velvet, MetaVelvet, and SOAPdenovo all assemble NGS data sets through the construction of de Bruijn graphs and Eulerian paths. De Bruijn graphs contain overlapping sequence information represented by branching nodes and stemming vertices, and is extremely sensitive and results quickly deteriorate even with the slightest amount of polymorphism [21]. The assembly algorithm of Velvet and SOAPdenovo both manipulate the constructed de Bruijn graph with error removal and simplification to generate optimal assembly results, which effectively excludes the essential polymorphism information vital to metagenomics data sets during assembly. In contrast, MetaVelvet decomposes the de Bruijn graphs into individual subgraphs and assembles each subgraphs into separate contigs. On the other hand, BBAP adopts a greedy assembly approach by incorporating and clustering sequence reads through BLAST results, and Genovo implements a Bayesian-based probabilistic model and takes into account the potential presence of multiple genomes in the data set. Therefore, it was reasonably expected for BBAP, MetaVelvet, and Genovo to have better assembly results than Velvet and SOAPdenovo when assembling metagenomics data sets, and this was consistent with our results that support BBAP, MetaVelvet, and Genovo are better equipped to assemble metagenomics data sets than Velvet or SOAPdenovo.

We compared the average assembly times for *in silico* and NGS data sets on our server (E5310 1.6GHz x4 *x2*, 12GB RAM) between all methods to further assess the performance of both BBAP and PDR. For smaller *in silico* data sets (data set size ≤5,579X or 17.44 Mb) BBAP assembly time was slightly longer than Velvet, MetaVelvet, and SOAPdenovo, but still within a couple minutes (Additional file 1: Table S14). BBAP assembly time for the largest *in silico* data sets tested (data set size = 55,799X or 174 Mb) were similar to the assembly time by the other methods except Genovo, which required considerably much more assembly time than BBAP or the other methods for all *in silico* data sets. The average BBAP PDR assembly time (624 s) for the 12 NGS data sets was drastically faster than the average BBAP FD assembly time (14,347 s). Overall, results suggest not only do both BBAP and PDR individually increase assembly efficiency and accuracy compared to their respective counterparts, but the combination of BBAP and PDR together further improves the overall assembly quality of metagenomic data sets.

Viral pathogens are responsible for the majority of pandemic and epidemic diseases listed by the World Health Organization. Recent studies have utilized the advantages of NGS data sets of the viral quasispecies genome to construct genome-wide diversity profiles for studying the virus-host interactions during infection and, treatment and vaccination [8, 10, 11, 15, 17]. Resistance associated variants and novel variants of the viral quasispecies usually are rare and not detectable by conventional or low depth sequencing, therefore detection of minor variants is clinically important for customizing patient management and treatment strategies [10, 16]. Our results show that BBAP and PDR not only provided an accurate assembly sequence but also generates a high resolution diversity profile of the data set. Additionally, we were able to detect and recover novel variants that were otherwise undetectable to alternative assembly methods.

Conclusions

Assembly of a highly polymorphic NGS data set is a complicated process as it involves multiple steps (such as quality control, read assembly and error removal) and is dependent of several prerequisite factors (data set type, sequencing platform, intended use of results, etc.). In addition, a functional understanding of the algorithms and sufficient parameters are important for the optimization of assembly results. We believe both BBAP and the partial

de novo-reference assembly strategy will provide a powerful tool for future metagenomic and viral quasispecies studies.

Methods
BLAST-based assembly pipeline

The BLAST-based assembly pipeline, BBAP, is divided into four major steps: quality control (QC), blast and cluster (BC), alignment and consensus determination (AC), and contig assembly (CA) (Fig. 3a). BBAP assembles high quality sequences into contigs according to BLAST results. Alignment files of the assembled contigs are generated as a result. The contigs are further assembled into extended-contigs and resulting in contig sequences, a log file, and a statistical analysis of the assembly. All steps, with the sole exception of BLAST, used in-house developed perl scripts.

The QC step excludes sequences with low quality scores, trims sequences from both ends, removes redundant identical sequences, and filters unique representative sequences with low redundancy. First, raw reads (RRs) that include any called base with a quality score less than the given threshold is omitted. The remaining high quality reads (HQRs) are trimmed from both ends for the given length to remove barcodes, artificial sequences such as linker, adapters or vectors, and error-prone regions that are more frequently found in the terminal regions for some sequencing platforms. Identical HQRs are compressed and represented by a single unique representative read (UR) while retaining the redundancy count information. Unique representative reads with redundancy counts greater than or equal to the given threshold, high redundancy unique representative reads (HRURs), are retained for further assembly.

For *de novo* assembly, the BLAST and cluster step (BC) is initiated with the reciprocal BLAST of the HRURs fasta file. The BLAST parameter of repeat masking was set to include repetitive regions into the results (-F ""). BLAST results with gaps or e-value, identity, or BLAST length not meeting the given thresholds were excluded from further assembly. During clustering, if two reads are BLASTed to one another and are both unassigned, then they are assigned to a same new cluster. If only one read has been assigned a cluster, then the unassigned read is added to the cluster of the assigned read. If both have been separately assigned to different clusters, then the two clusters are merged into one single cluster. Finally, clusters with number of assigned reads less than the given threshold sequence number are excluded from further assembly.

The BC step of reference assembly is similar to that of *de novo* assembly but with some minor differences. Instead of reciprocal BLAST, the HRURs fasta file is BLASTed to the reference sequences. If a read has identical e-values for

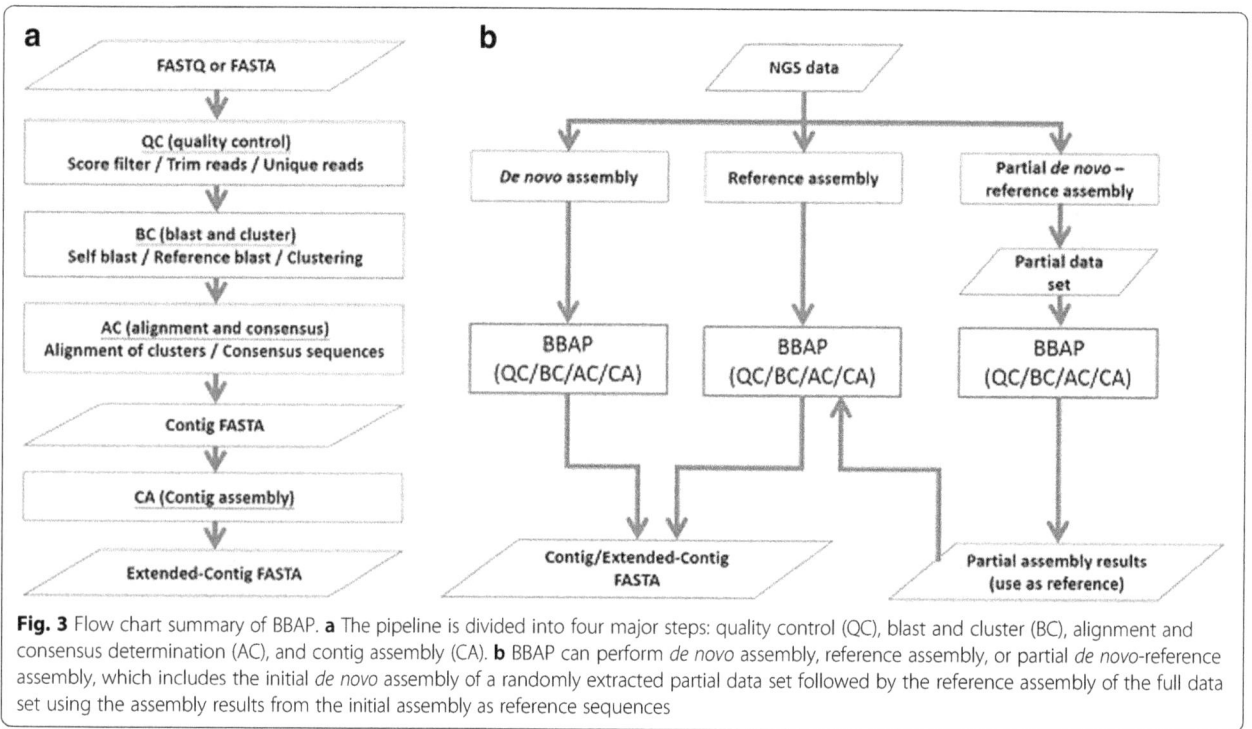

Fig. 3 Flow chart summary of BBAP. **a** The pipeline is divided into four major steps: quality control (QC), blast and cluster (BC), alignment and consensus determination (AC), and contig assembly (CA). **b** BBAP can perform *de novo* assembly, reference assembly, or partial *de novo*-reference assembly, which includes the initial *de novo* assembly of a randomly extracted partial data set followed by the reference assembly of the full data set using the assembly results from the initial assembly as reference sequences

multiple reference sequences, the read will be assigned to the reference sequence with the longest sequence length.

The alignment and consensus determination step (AC) calculates the alignment position for each read of a cluster based on its BLAST results. Only top BLAST results with identity and BLAST length greater than the given thresholds were used for alignment. Consensus sequences were calculated for each base according to the alignment results. Nucleotides with frequencies greater than or equal to the given threshold are retained for polymorphic sites.

Contigs with identical terminal sequences longer than the given threshold are merged together into extended-contigs. Identical terminal sequences were identified by self-BLAST of contigs. This step is optional and dependent on the nature of the data set.

Overall, BBAP uses BLAST results (reciprocal BLAST for *de novo* assembly, and data set to reference sequence BLAST for reference assembly) to cluster reads into contig groups to increase computation efficiency of following steps. The reads in each contig group are then positioned/aligned according to their respective BLAST results into contigs. The grouped reads are then extended into contigs according to positioning/alignment information provided from the BLAST results in a greedy strategy manner. Extension of contigs and prevention of assembly artifacts (such as artificial chimeras) are directly dictated by the BLAST identify and length threshold parameters, and indirectly effected by quality control parameters, including the QC-score threshold and the redundancy threshold.

BBAP can assemble data sets with or without a reference sequence by reference assembly or *de novo* assembly, respectively. We also introduce a third assembly strategy, the partial *de novo*-reference assembly approach (Fig. 3b). A randomly extracted partial data set is first *de novo* assembled, and then the resulting contig sequences are used as reference sequences to assemble the entire data set through reference assembly.

Next generation sequencing data set assembly and statistical analyses

NGS data sets were downloaded from a previous study [24], which consisted of 12 libraries derived from 7 patients chronically infected with HBV within a single family (Additional file 1: Table S15). The full data set was separately assembled with BBAP through full data set *de novo* (FD) assembly, Sanger reference (SR) assembly, and partial *de novo*-reference (PDR) assembly. A single Sanger sequence from each patient sample was chosen and used as the reference sequence for the SR assembly of the corresponding full data set. For the PDR assembly, partial data sets were constructed independently by randomly choosing 1% of the RRs from the full data set and assembled *de novo*, and the results of the partial data set *de novo* (PD) assembly were used as reference sequences for the reference assembly of the full data set. Partial data sets of different ratios were assembled and 1% partial data sets generated the most optimal assembly results (Additional file 1: Table S16 and Additional file 3: SD). Assembly results of different BBAP methods were then compared to each other.

Variant contigs were identified by BLAST against the NCBI HBV complete genome sequence (NC_003977), the Sanger reference sequence, and the NCBI nr/nt database. To verify that the identified variants were not artifacts of incorrect assembly by BBAP, sequences of at least 20 bp and spanning the junction regions of the structural variations were searched for in both the RRs and HQRs fasta files.

The full data set and partial data sets of one library, D2_1 (Additional file 1: Table S15), were also assembled using all methods. Statistical analyses and comparisons between assembly methods were performed with perl scripts.

In silico data set assembly

We also compared the performance of different assembly methods by using simulated data sets. *In silico* data sets were generated by randomly generating 101 bp reads from the reference NCBI HBV complete genome, NC_003977. To mimic observed polymorphism from virus diversity or sequencing error of NGS, different error rates, 10^{-2}, 10^{-3}, and 10^{-4}/site, were applied to the simulated reads. Data set sizes were set to 1,726,462 (55,799X), 172,646 (5,579X), 17,264 (557X), and 1726 (55X) HQRs. Five independent data sets were generated for each parameter combination, error rate and dataset size. Data sets were assembled using BBAP FD assembly, Velvet, MetaVelvet, SOAPdenovo, and Genovo. All *in silico* data sets, except for data sets of high error rate (0.01) coupled with small data set sizes (55X and 557X), used the same BBAP parameter values for NGS *de novo* assembly. For the high error rate-low coverage depth data sets, the redundancy threshold was reduced from 5 to 1 to compensate for its low redundancy. For Velvet, MetaVelvet, and SOAPdenovo assembly, the k-mer size was optimally set to 57, 57, and 63, respectively. For Genovo assembly, different numbers of iterations were used for data sets of different coverage depths because of the extreme long run time for larger data sets; the number of iterations for data sets with coverage depths of 55,799X, 5,579X, 557X and 55X was 10, 2000, 10,000, and 10,000, respectively.

Additional files

Additional file 1: Table S1. Statistics of next generation sequencing data set of HBV genome from patient serum. **Table S2**: Parameters used for *de novo*, reference, and partial *de novo* reference BBAP assembly. **Table S3**: Assembly results of individual data sets using BBAP with multiple approaches. **Table S4**: Comparison of polymorphism between non-overlapping and overlapping regions of D2_1 assembled contigs alignment. **Table S5**: Comparison of polymorphism levels between assembly results of BBAP PDR and SR assemblies. **Table S6**: Summary of assembled contigs from the PDR assembly of D2_1 NGS data set. **Table S7**: Top ten non-synonymous frequency positions of the HBV quasispecies. **Table S8**: Nucleotide frequencies derived from BBAP

PDR assembly and pyrosequencing. **Table S9**: Results of BBAP *de novo* assembled in silico NCBI HBV complete genome (NC_003977) data sets ($n = 5$). **Table S10**: Results of Velvet assembled *in silico* NCBI HBV complete genome (NC_003977) data sets ($n = 5$). **Table S11**: Results of MetaVelvet assembled *in silico* NCBI HBV complete genome (NC_003977) data sets ($n = 5$). Table S12: Results of SOAPdenovo assembled *in silico* NCBI HBV complete genome (NC_003977) data sets ($n = 5$). **Table S13**: Results of Genovo assembled *in silico* NCBI HBV complete genome (NC_003977) data sets ($n = 5$). **Table S14**: Assembly time required for *in silico* data sets by BBAP, Velvet, MetaVelvet, SOAPdenovo, and Genovo. **Table S15**: Summary of study subjects and samples. **Table S16**: Summary of assembly results for D2_1 partial data sets of different size ratio.

Additional file 2: Figure S1. Comparison of HBV recover ratio by BBAP, Velvet, SOAPdenovo, and Genovo assembly of full and partial D2_1 data sets. **Figure S2**: Correlation between assembled scaffold length and scaffold degeneracy for all 12 data sets. **Figure S3**: (a) Nucleotide sequence of the R1 scaffold. (b) Schematic alignment of the R1 scaffold, HBV X gene and HBV precore/core gene. **Figure S4**: Schematic diagram of the T1 scaffold and its corresponding HBV genome regions. **Figure S5**: Schematic diagram of the T6 scaffold and its corresponding HBV genome and Sanger reference sequence regions. **Figure S6**: Alignment of Sanger (SR) and partial D2_1 data set assembled scaffolds reference assembled (PDR) scaffolds to the Sanger reference sequence. **Figure S7**: Diversity profile of D2_1 HBV quasispecies according to assembly results of partial data set reference assembly of the full data set. **Figure S8**: Schematic diagram of two Genovo assembled scaffolds with identified HBV structural variants and its corresponding HBV genome regions.

Additional file 3: A. Variant sequences and human genome sequences. **B**. Diversity profile of D2_1 HBV quasispecies. **C**. Structure variation by Genovo. **D**. Determining optimal size of partial data set.

Abbreviations

AC: Alignment and consensus determination; BBAP: BLAST-based assembly pipeline; BC: Blast and cluster; CA: Contig assembly; CLC: CLC Genomics Workbench; FD: *De novo* assembly of the full data set; HQRs: High quality reads; HRURs: High redundancy unique representative reads; NGS: Next generation sequencing; PD: *De novo* assembly of the partial data set; PDR: Partial *de novo*-reference assembly; QC: Quality control; RiHRURs: Reads included in HRURs; RRs: Raw reads; SR: Reference assembly of the full data set with Sanger generated reference sequences; URs: Unique representative reads

Acknowledgement

We would like to acknowledge Chia-Hua Chen for discussions and suggestions during this study.

Funding

This work was supported by the Ministry of Science and Technology [103-2621-B-002 -003 and 104-2621-B-002 -006] and National Taiwan University [101R7836, 102R7836, and 103R7836]. Both funding bodies listed above (MOST and NTU) did not play any role in the study or conclusions of this study.

Authors' contributions

Conception and design of the study: JHC, JHK, PJC, DSC, HYW. Performed experiments: YYL, XL, HYW. Data analysis: YYL, HYW. Manuscript preparation: YYL, CHH, JHC, HYW. All authors read and approved the final manuscript.

Competing interests

The authors declare that they have no competing interests.

Author details

[1]Department of Life Science, National Taiwan University, Taipei 106, Taiwan. [2]Department of Forestry and Nature Conservation, Chinese Culture University, Taipei 111, Taiwan. [3]Laboratory of Disease Genomics and Individualized Medicine, Beijing Institute of Genomics, the Chinese Academy of Sciences, Beijing 100101, China. [4]Graduate Institute of Clinical Medicine,

National Taiwan University, Taipei 100, Taiwan. [5]Genomics Research Center, Academia Sinica, Taipei 115, Taiwan. [6]Institute of Ecology and Evolutionary Biology, National Taiwan University, Taipei 106, Taiwan. [7]Research Center for Developmental Biology and Regenerative Medicine, National Taiwan University, Taipei 100, Taiwan.

References

1. Miller JR, Koren S, Sutton G. Assembly algorithms for next-generation sequencing data. Genomics. 2010;95(6):315–27.

2. Scholz MB, Lo CC, Chain PS. Next generation sequencing and bioinformatic bottlenecks: the current state of metagenomic data analysis. Curr Opin Biotechnol. 2012;23(1):9–15.

3. Piatak Jr M, Saag MS, Yang LC, Clark SJ, Kappes JC, Luk KC, Hahn BH, Shaw GM, Lifson JD. High levels of HIV-1 in plasma during all stages of infection determined by competitive PCR. Science. 1993;259(5102):1749–54.

4. Wang HY, Chien MH, Huang HP, Chang HC, Wu CC, Chen PJ, Chang MH, Chen DS. Distinct hepatitis B virus dynamics in the immunotolerant and early immunoclearance phases. J Virol. 2010;84(7):3454–63.

5. Picchio GR, Nakatsuno M, Boggiano C, Sabbe R, Corti M, Daruich J, Perez-Bianco R, Tezanos-Pinto M, Kokka R, Wilber J, et al. Hepatitis C (HCV) genotype and viral titer distribution among Argentinean hemophilic patients in the presence or absence of human immunodeficiency virus (HIV) co-infection. J Med Virol. 1997;52(2):219–25.

6. Margulies M, Egholm M, Altman WE, Attiya S, Bader JS, Bemben LA, Berka J, Braverman MS, Chen YJ, Chen Z, et al. Genome sequencing in microfabricated high-density picolitre reactors. Nature. 2005;437(7057):376–80.

7. Bentley DR, Balasubramanian S, Swerdlow HP, Smith GP, Milton J, Brown CG, Hall KP, Evers DJ, Barnes CL, Bignell HR, et al. Accurate whole human genome sequencing using reversible terminator chemistry. Nature. 2008; 456(7218):53–9.

8. Yin L, Liu L, Sun Y, Hou W, Lowe AC, Gardner BP, Salemi M, Williams WB, Farmerie WG, Sleasman JW, et al. High-resolution deep sequencing reveals biodiversity, population structure, and persistence of HIV-1 quasispecies within host ecosystems. Retrovirology. 2012;9:108.

9. Van Loy T, Thys K, Tritsmans L, Stuyver LJ. Quasispecies Analysis of JC Virus DNA Present in Urine of Healthy Subjects. PLoS One. 2013;8(8):e70950.

10. Solmone M, Vincenti D, Prosperi MC, Bruselles A, Ippolito G, Capobianchi MR. Use of massively parallel ultradeep pyrosequencing to characterize the genetic diversity of hepatitis B virus in drug-resistant and drug-naive patients and to detect minor variants in reverse transcriptase and hepatitis B S antigen. J Virol. 2009;83(4):1718–26.

11. Henn MR, Boutwell CL, Charlebois P, Lennon NJ, Power KA, Macalalad AR, Berlin AM, Malboeuf CM, Ryan EM, Gnerre S, et al. Whole genome deep sequencing of HIV-1 reveals the impact of early minor variants upon immune recognition during acute infection. PLoS Pathog. 2012;8(3): e1002529.

12. Prosperi MC, Yin L, Nolan DJ, Lowe AD, Goodenow MM, Salemi M. Empirical validation of viral quasispecies assembly algorithms: state-of-the-art and challenges. Scientific reports. 2013;3:2837.

13. Topfer A, Hoper D, Blome S, Beer M, Beerenwinkel N, Ruggli N, Leifer I. Sequencing approach to analyze the role of quasispecies for classical swine fever. Virology. 2013;438(1):14–9.

14. Zagordi O, Daumer M, Beisel C, Beerenwinkel N. Read length versus depth of coverage for viral quasispecies reconstruction. PLoS One. 2012;7(10):e47046.

15. Borucki MK, Allen JE, Chen-Harris H, Zemla A, Vanier G, Mabery S, Torres C, Hullinger P, Slezak T. The role of viral population diversity in adaptation of bovine coronavirus to new host environments. PLoS One. 2013;8(1):e52752.

16. Kirst ME, Li EC, Wang CX, Dong HJ, Liu C, Fried MW, Nelson DR, Wang GP. Deep sequencing analysis of HCV NS3 resistance-associated variants and mutation linkage in liver transplant recipients. PLoS One. 2013;8(7):e69698.

17. Abolnik C, de Castro M, Rees J. Full genomic sequence of an African avian paramyxovirus type 4 strain isolated from a wild duck. Virus Genes. 2012; 45(3):537–41.

18. Kuroda M, Katano H, Nakajima N, Tobiume M, Ainai A, Sekizuka T, Hasegawa H, Tashiro M, Sasaki Y, Arakawa Y, et al. Characterization of quasispecies of pandemic 2009 influenza A virus (A/H1N1/2009) by de novo sequencing using a next-generation DNA sequencer. PLoS One. 2010;5(4):e10256.

19. Zerbino DR, Birney E. Velvet: algorithms for de novo short read assembly using de Bruijn graphs. Genome Res. 2008;18(5):821–9.

20. Li R, Li Y, Kristiansen K, Wang J. SOAP: short oligonucleotide alignment program. Bioinformatics. 2008;24(5):713–4.

21. Chaisson MJ, Brinza D, Pevzner PA. De novo fragment assembly with short mate-paired reads: Does the read length matter? Genome Res. 2009;19(2):336–46.

22. Namiki T, Hachiya T, Tanaka H, Sakakibara Y. MetaVelvet: an extension of Velvet assembler to de novo metagenome assembly from short sequence reads. Nucleic Acids Res. 2012;40(20):e155.

23. Laserson J, Jojic V, Koller D. Genovo: de novo assembly for metagenomes. J Comput Biol. 2011;18(3):429–43.

24. Lin YY, Liu C, Chien WH, Wu LL, Tao Y, Wu D, Lu X, Hsieh CH, Chen PJ, Wang HY, et al. New insights into the evolutionary rate of hepatitis B virus at different biological scales. J Virol. 2015;89(7):3512–22.

Across-proteome modeling of dimer structures for the bottom-up assembly of protein-protein interaction networks

Surabhi Maheshwari[1] and Michal Brylinski[1,2]* ⓘ

Abstract

Background: Deciphering complete networks of interactions between proteins is the key to comprehend cellular regulatory mechanisms. A significant effort has been devoted to expanding the coverage of the proteome-wide interaction space at molecular level. Although a growing body of research shows that protein docking can, in principle, be used to predict biologically relevant interactions, the accuracy of the across-proteome identification of interacting partners and the selection of near-native complex structures still need to be improved.

Results: In this study, we developed a new method to discover and model protein interactions employing an exhaustive all-to-all docking strategy. This approach integrates molecular modeling, structural bioinformatics, machine learning, and functional annotation filters in order to provide interaction data for the bottom-up assembly of protein interaction networks. Encouragingly, the success rates for dimer modeling is 57.5 and 48.7% when experimental and computer-generated monomer structures are employed, respectively. Further, our protocol correctly identifies 81% of protein-protein interactions at the expense of only 19% false positive rate. As a proof of concept, 61,913 protein-protein interactions were confidently predicted and modeled for the proteome of *E. coli*. Finally, we validated our method against the human immune disease pathway.

Conclusions: Protein docking supported by evolutionary restraints and machine learning can be used to reliably identify and model biologically relevant protein assemblies at the proteome scale. Moreover, the accuracy of the identification of protein-protein interactions is improved by considering only those protein pairs co-localized in the same cellular compartment and involved in the same biological process. The modeling protocol described in this communication can be applied to detect protein-protein interactions in other organisms and pathways as well as to construct dimer structures and estimate the confidence of protein interactions experimentally identified with high-throughput techniques.

Keywords: Protein-protein interactions, Protein docking, Structural bioinformatics, Machine learning, Gene Ontology filters, *e*FindSite[PPI], *e*Rank[PPI]

Background

Protein-protein interactions (PPIs) are ubiquitous and play crucial roles in all biological processes within and between cells by mediating signaling pathways in cellular networks and controlling intracellular communication [1]. Since complex biological systems are governed by sophisticated networks of PPIs, associations between

proteins ultimately determine the behavior of the cell. Genome-sequencing projects provide comprehensive datasets of biological sequences and numerous post-genomic projects are largely focused on the exploration and analysis of PPIs across proteomes [2, 3]. The number of possible PPIs in an organism can be scaled as the square of the total number of monomeric proteins, yielding an estimated number of disparate protein complexes in the order of millions. High-throughput approaches allow the large-scale detection of protein-interaction partners in many organisms. Although the PPI data is being produced at a swift pace, the major issues in using

---

* Correspondence: michal@brylinski.org
[1]Department of Biological Sciences, Louisiana State University, Baton Rouge, LA, USA
[2]Center for Computation & Technology, Louisiana State University, Baton Rouge, LA, USA

the current genome-wide PPI data are a low coverage and high false positive rates [4, 5]. Moreover, inter-study discrepancies between different experimental approaches applied to the same biological system are not uncommon [6]. Last but not least, while these high-throughput methods identify proteins interacting with one another, they do not provide structural information on biologically relevant protein complexes.

On the other hand, interaction details, which can only be obtained from three-dimensional structures, are crucial to fully comprehend interaction mechanisms at the atomic level. Unfortunately, despite ongoing efforts in structural genomics projects to determine complex structures, structural biology is lagging behind in the current trends of high-throughput methods. While the repertoire of monomeric protein structures solved by X-ray crystallography and NMR spectroscopy is increasing exponentially, the structural space of interacting proteins is still far from complete. In fact, there is an increasing gap between the number of identified interactions and the number of 3D structures of these associations. Thus, it is imperative to develop and continuously improve computational techniques to accurately identify interacting proteins and the corresponding complex structures.

A number of computational approaches have been developed to discover and model new interactions at a system level. Modeling complex structures can be accomplished using two distinct types of techniques, template-free and template-based. The former methods, also known as protein docking, construct a complex model by assembling the monomeric structures of target proteins through a conformational search followed by the selection of high scoring binding orientations. In contrast, template-based approaches build complex structures by mapping monomeric targets to experimentally solved template complexes often followed by the refinement of the initial structural framework. Both methods have advantages and disadvantages. Template-based approaches can construct dimeric models directly from target sequences, therefore, monomer structures may not be required. Further, these techniques select templates based on sequence [7, 8], sequence-to-structure [9] and structure alignments [10, 11] often yielding more accurate results than template-free docking [12, 13]. Although dimer templates are available in the Protein Data Bank (PDB) [14] to model all complexes in which the monomer structures are either known or can independently be modeled [15], the success rate of template-based docking is only about 23% when no closely homologous templates with a sequence identity to the target of >40% can be found for at least one monomer chain. Analogous interaction templates cannot be identified in the current PDB to effectively guide template-based docking in those failed cases [16]. The fact that suitable templates are available only for a limited number of

interactions significantly lowers the coverage of proteome-scale datasets.

In contrast, template-free methods are, in principle, applicable to those protein targets whose monomer structures are either solved experimentally or can be generated with homology modeling. These techniques do not require the structures of related complexes to model the association between targets proteins. Consequently, template-free approaches provide a higher coverage in large-scale applications focusing on the construction and analysis of PPI networks. Although template-free modeling is often applied to a pair of proteins known to interact with one another, several studies have successfully employed the exhaustive rigid-body protein docking and post-docking analysis to predict PPIs and PPI networks [17–19]. For instance, a docking experiment comparing the distribution of docking scores collected for proteins known to interact to those between putatively non-interacting proteins was reported [20].

Another study attempted to predict the protein-protein interaction network of the bacterial chemotaxis signaling pathway using an all-to-all docking approach [21]. Here, two docking tools, MEGADOCK [18] and ZDOCK [22], were employed to conduct rigid-body docking of all possible combinations of 101 proteins belonging to 13 families, which are known to be part of the chemotaxis signaling pathway. Based on a previous observation that the decoys of interacting proteins form dense clusters as opposed to the lack of dense clusters formed by non-interacting proteins [17, 18], clustering high-scoring decoys was used to evaluate protein binding affinity and to predict the PPI network. Encouragingly, combining positive predictions from both docking tools correctly identified almost all core-signaling interactions in bacterial chemotaxis. Although the aforementioned methods were shown to discriminate true protein interactions from likely non-interacting pairs, the native complexes of interacting proteins have not been recovered mainly due to an insufficient ranking accuracy of docking algorithms. Further, the reported benchmarking calculations conducted using relatively small datasets of experimental structures may not be indicative of the performance of the proteome-scale identification of molecular interactions.

In that regard, we developed a new approach to discover and model PPIs across proteomes employing an exhaustive all-to-all docking strategy. This pipeline comprises six major steps including protein threading and homology modelling, the prediction of binding interfaces, a rigid body docking, the flexible refinement and scoring of the modeled interfaces, and a series of function annotation filters. Our approach was carefully benchmarked on a large and representative dataset of experimental structures and computer-generated models of target proteins. In order to demonstrate its utility in large-scale projects, we modeled dimer structures and predicted PPIs across the proteome of *Escherichia coli*. Interaction data generated for *E. coli* is primed for experimental validation and further computational analyses. In addition, we validated our method against the human immune disease pathway. Encouragingly, our results demonstrate that protein docking can be used not only to identify near-native complexes but also to predict interaction partners. Overall, this study shows that combining computational modeling, structural bioinformatics, machine learning, and function annotation provides a powerful methodology for the bottom-up assembly of protein-protein interaction networks.

Methods
Datasets
The pipeline to model PPIs is benchmarked on the BM1905 dataset (available at http://www.brylinski.org/content/efindsiteppi-datasets), which was previously compiled to evaluate the accuracy of interface residue prediction and the re-ranking of docked models [23, 24]. This dataset contains experimental target structures (BM1905C) as well as high-quality computer-generated models (BM1905H). The quality of monomer models was assessed by the root-mean-square deviation (RMSD) and the Template Modeling score (TM-score) [25]. The latter ranges from 0 to 1 with values >0.4 indicating a significant structural similarity to the native conformation. BM1905H comprises models whose mean Cα-RMSD is 6.94 Å ±4.61 and mean TM-score is 0.72 ± 0.15.

The algorithm to predict binary interactions is trained and validated against a non-redundant and representative dataset of 18,162 protein dimers selected from the PDB. First, all dimers having at least 20 interface residues were categorized as either homo-dimers whose individual chains share at least 85% sequence identity or hetero-dimers when the sequence identity was below 85%. Next, each subset was clustered with CD-HIT [26] at 80% sequence identity. Finally, redundant dimers that have similar interfaces with the Matthews correlation coefficient (MCC) calculated over interface residues of >0.5 were removed from each cluster. This procedure resulted in a set of 14,944 homodimers (HOM14944) and a set of 3,519 heterodimers (HET3519). In addition, the algorithm to predict binary interactions is tested on 1,688 non-interacting protein pairs derived from the Negatome 2.0 database [27]. Computer models of individual proteins in Negatome 2.0 were built with Modeller [28] using templates identified by *e*Thread [29], followed by a high-resolution structure refinement with ModRefiner [30].

The developed pipeline to predict PPI networks is validated using *E. coli* as a model organism. Protein interaction data for *E. coli* consisting of 13,374 known interactions formed by 2,994 bacterial proteins were

downloaded from the Database of Interacting Proteins (DIP) [31] in March 2016. We removed from the original dataset redundant proteins as well as those targets longer than 600 residues, which may be difficult to model with threading, and shorter than 50 residues because these molecules are likely peptides. The final *E. coli* dataset consists of 2,300 proteins forming 6,341 interactions. DIP provides the sequences of interacting proteins, therefore, we constructed monomer structures with Modeller [28] using templates identified by *e*Thread [29], followed by a high-resolution structure refinement with ModRefiner [30].

Finally, the protocol to predict and model protein interactions is validated against the human immune disease pathway associated with the Toll-Like Receptor (TLR) signaling cascade. Information on proteins involved in this pathway as well as experimentally detected interactions were obtained from the Reactome database [32] in June 2016. The human immune pathway comprises 26 proteins connected through 112 interactions; protein monomer structures are constructed with the same protocol as that used to model DIP proteins.

Protein docking, ranking and refinement
For a given pair of protein targets, a collection of docking solutions is generated with the FFT-based rigid body docking program ZDOCK version 3.02 [33]. We use the default parameters to exhaustively search the 3D grid space around the receptor by rotating and translating the ligand. Subsequently, the top 2,000 conformations reported by ZDOCK are re-ranked with eRank^{PPI} [23], a recently developed algorithm to identify near-native conformations from the high-scoring hits. The scoring function implemented in eRank^{PPI} employs multiple features including residue-level interface probability estimates, protein docking potentials, and energy-based scores. Surface residues in target receptors are annotated with interface probability estimates by eFindSite^{PPI} [24], a structure/evolution-based approach to detect interface residues. eFindSite^{PPI} builds on a strong conservation of the location and geometry of binding sites in evolutionarily related dimers and employs meta-threading, structural alignments, and machine learning to predict interfacial residues for a target protein. The top 10 models selected by eRank^{PPI} are finally subjected to a flexible refinement with FiberDock [34]. FiberDock mimics the induced fit by accounting for both side-chain and backbone flexibility. The side-chain flexibility is modeled using a rotamer library, whereas a normal mode procedure is used to model the backbone flexibility.

Assessing the quality of protein complex models
The accuracy of dimer models is primarily assessed with iAlign [35] against experimental complex structures retrieved from the PDB. iAlign evaluates the quality of

structural models with the Interface Similarity score (IS-score) combining Cartesian distances with the overlap of interfacial contact patterns [36]. IS-score ranges from 0 to 1 with values greater than 0.210, 0.311 and 0.473 indicating a statistically significant interface similarity at p-values of 10^{-2}, 10^{-5} and 10^{-10}, respectively. In addition, the quality of dimer models is assessed with iRMSD, a standard evaluation measure in the Critical Assessment of PRedicted Interactions (CAPRI) [37] and the Pairwise Contact Score (PCS) [23]. iRMSD is the interfacial Cα-RMSD between ligands in the predicted and experimental complexes upon the superposition of receptor structures. In iRMSD calculations, interface residues are defined as those having at least one atom within 10 Å from any atom in the other protein chain. The PCS employs the Matthews correlation coefficient to evaluate the overlap between predicted and the actual interfacial contacts; it ranges from about 0 (random prediction) to 1 (perfect prediction). The docking success rate is defined as the percentage of targets for which at least one correct model is ranked within the top 10 conformations. The acceptance criteria for correct predictions are an iRMSD of ≤2.5 Å and a PCS of ≥0.65 for experimental structures, and an iRMSD of ≤8.5 Å and a PCS of ≥0.30 for computer-generated models, as described in [23].

Protein-protein interaction prediction with supervised learning

The scoring function to identify biologically relevant assemblies was trained and cross-validated against the HET3519 dataset of experimental hetero-dimers used as positives and a simulated dataset of 14,944 likely non-interacting pairs used as negatives. The negative dataset was constructed by randomly swapping ligands within the HOM14944 dataset. Since HOM14944 proteins share less than 80% sequence identity, this procedure resulted in a random set of hetero-dimers referred to as RND14944. Uniformly choosing random protein pairs excluding experimental interactions produces an unbiased estimate of the distribution of negatives in the prediction of protein-protein interactions [38]. Hence, this procedure is a common practice to generate negative datasets containing at most a negligible fraction of interacting proteins [39–41]. FiberDock calculates several binding energy scores, including attractive and repulsive van de Waals forces, the atomic contact energy, partial electrostatics, hydrogen and disulfide bonds, π-stacking, and aliphatic interactions. These scores were used as a feature vector to train a Random Forest Classifier (RFC) returning a single probabilistic score to assess whether two interacting proteins are biologically relevant. The machine learning model was 10-fold cross-validated against the positive set HET3519 and the negative set RND14944.

Annotation filters

Positive predictions are further subjected to filtering with Gene Ontology (GO) terms. GO is a hierarchically organized database providing a controlled vocabulary to characterize gene products, divided into three sub-ontologies: cellular component (CC), biological process (BP) and molecular function (MF) [42]. Here, we use GO slims, which are cut-down versions of the GO ontologies without the detail of the specific fine grained terms. GO slims were extracted from the PANTHER classification system [43], whereas annotations for *E. coli* proteins were obtained from the EcoCyc database [44] in May 2016. We tested whether CC, BP and MF slims can be used to refine prediction results by considering proteins localized in the same cellular component, assigned to the same biological process, and having different molecular functions.

Performance evaluation metrics

PPI prediction is assessed using standard evaluation metrics for classification problems:

True positive rate:

$$TPR = \frac{TP}{TP + FN} \tag{1}$$

False positive rate:

$$FPR = \frac{FP}{FP + TN} \tag{2}$$

Accuracy:

$$ACC = \frac{TP + TN}{TP + FP + TN + FN} \tag{3}$$

Matthews correlation coefficient:

$$MCC = \frac{TP \times TN - FP \times FN}{\sqrt{(TP + FP)(TP + TN)(FP + FN)(TN + FN)}} \tag{4}$$

where TP (True Positives), FN (False Negatives) and FP (False Positives) are the number of correctly predicted, under-, and over-predicted PPIs, respectively. TN (True Negatives) is the number of correctly predicted non-interacting partners. The MCC quantifies the strength of the correlation between predicted and actual classes; by heavily penalizing both over- and under-predictions, it provides a convenient assessment measure that balances the sensitivity and specificity.

Results and discussion

The goal of this study was to develop and test a new protocol to model putative protein complex structures across proteomes that can subsequently be used to assemble protein-protein interaction networks. The modeling procedure for a pair of proteins is presented in Fig. 1. The

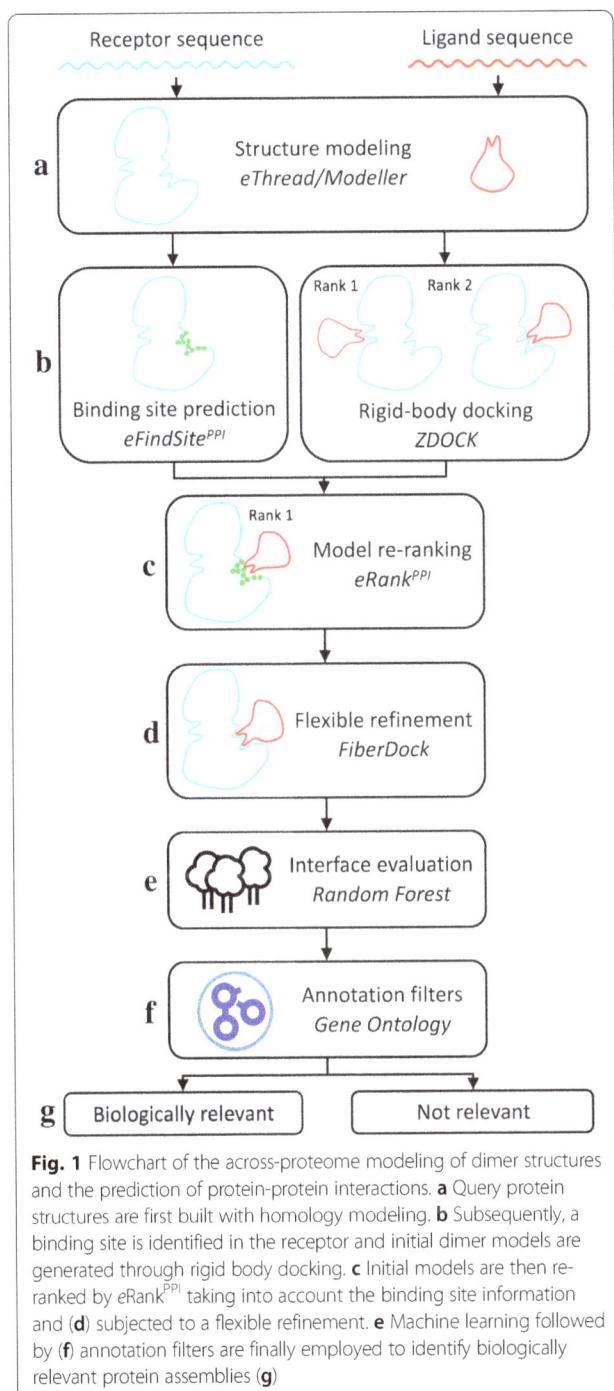

Fig. 1 Flowchart of the across-proteome modeling of dimer structures and the prediction of protein-protein interactions. **a** Query protein structures are first built with homology modeling. **b** Subsequently, a binding site is identified in the receptor and initial dimer models are generated through rigid body docking. **c** Initial models are then re-ranked by eRank^{PPI} taking into account the binding site information and (**d**) subjected to a flexible refinement. **e** Machine learning followed by (**f**) annotation filters are finally employed to identify biologically relevant protein assemblies (**g**)

docking conformations are filtered and re-ranked with eRank^{PPI} utilizing the binding interface predicted by eFindSite^{PPI} (Fig. 1c). The identified putative dimers are then subjected to a flexible refinement with FiberDock (Fig. 1d) followed by the evaluation of binding energies with the RFC in order to select the final model (Fig. 1e). A probability score reported by the RFC is used together with annotation filters according to Gene Ontology terms (Fig. 1f) to make the final decision whether or not the constructed dimer is biologically relevant (Fig. 1g).

Although the comprehensive benchmarks of eFindSite^{PPI} and eRank^{PPI} have been already reported [23, 24], we found that a flexible refinement improves the accuracy of dimers assembled from experimental as well as computer-generated monomer structures. In addition, using machine learning to evaluate the refined interfaces is shown to reliably detect biologically relevant protein complexes. Finally, we demonstrate that annotation filters can successfully be employed in genome-wide projects to further refine the classification results and more accurately identify putative pairs of interacting proteins.

Sampling and scoring in template-free docking

In this work, the structures of protein complexes are modeled via a protocol utilizing template-free docking with ZDOCK. Template-free docking consists of two successive tasks, sampling and scoring. Sampling employs a rigid-body search over different rotational-translational degrees of freedom, whereas the purpose of scoring is to rank the sampled poses in order to identify near-native configurations. Consequently, sampling and scoring failures are two major reasons for the lack of success in protein docking. The former are caused by an insufficient sampling, *viz.* near-native conformations are not generated by a sampling algorithm, therefore, reliable dimer models cannot be constructed. These errors can frequently be corrected simply by increasing the sampling exhaustiveness. Scoring failures are unsuccessful docking calculations, in which at least one near-native conformation is generated, however, it is not selected by a scoring function as a feasible solution; correcting these errors is more challenging compared to sampling failures. eRank^{PPI} was developed specifically to address scoring failures by improving the accuracy of dimer ranking in protein docking [23].

Here, we assess docking success rates, sampling and scoring failures for crystal structures as well as computer-generated models for the BM1905 dataset. The results are shown as IS-score spectrum plots in Fig. 2. For instance, at an IS-score of 0.210 corresponding to a p-value of 10^{-2}, the success rate of ZDOCK against crystal structures is 73.4%, with the remaining 26.6% cases classified as scoring failures (Fig. 2a). Re-ranking of the docked poses with eRank^{PPI} increases the success rate to 88.1%, decreasing

construction of a hetero-dimer starts with the prediction of 3D structures of individual monomer chains using eThread and Modeller (Fig. 1a). Here, the larger monomer is the receptor and the smaller monomer is the ligand; the size is proportional to the number of amino acid residues. Subsequently, eFindSite^{PPI} is employed to predict a protein binding site in the receptor structure and, simultaneously, a rigid-body docking of the ligand to the receptor is performed with ZDOCK (Fig. 1b). In the next step,

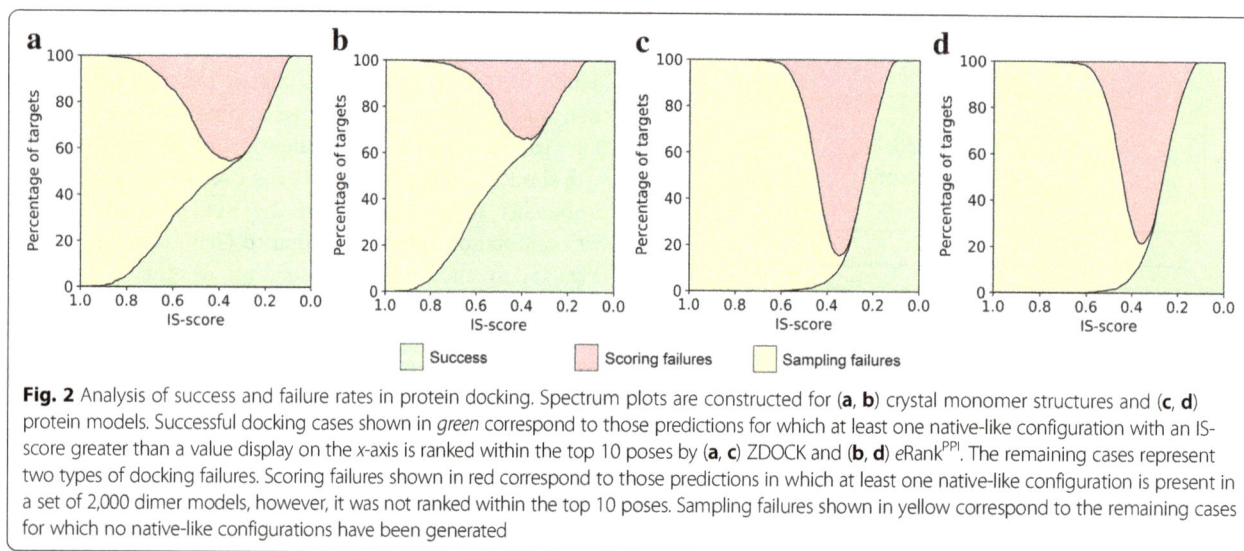

Fig. 2 Analysis of success and failure rates in protein docking. Spectrum plots are constructed for (**a**, **b**) crystal monomer structures and (**c**, **d**) protein models. Successful docking cases shown in *green* correspond to those predictions for which at least one native-like configuration with an IS-score greater than a value display on the x-axis is ranked within the top 10 poses by (**a**, **c**) ZDOCK and (**b**, **d**) eRank^{PPI}. The remaining cases represent two types of docking failures. Scoring failures shown in red correspond to those predictions in which at least one native-like configuration is present in a set of 2,000 dimer models, however, it was not ranked within the top 10 poses. Sampling failures shown in yellow correspond to the remaining cases for which no native-like configurations have been generated

the rate of scoring failures to only 11.9% (Fig. 2b). For computer-generated models, the success rates (scoring failures) are 64.4% (35.6%) for ZDOCK and 71.9% (28.1%) for eRank^{PPI} (Fig. 2c and d, respectively). Note that the lack of sampling failures at an IS-score of 0.210 suggests that rigid-body docking successfully samples the conformational space of dimers assembled with experimental as well as computer-generated models of monomer proteins. Sampling failures come into sight only at higher IS-score values, for example, conformations with an IS-score of at least 0.473 corresponding to a p-value of 10^{-10} are not constructed by ZDOCK for 19.1 and 61.1% of the cases when experimental monomer structures and computer-generated models are used, respectively. However, one should keep in mind that the models of individual monomers may already contain significant inaccuracies, thus interfaces highly similar to those in experimental structures simply cannot be constructed by rigid-body docking. Overall, this analysis shows that scoring failures are responsible for the majority of unsuccessful docking calculations and that eRank^{PPI} improves the success rate by reducing the number of scoring failures by 14.7% for crystal structures and 7.5% for protein models.

Dimers constructed from experimental monomer structures
Interface quality in the modeled dimer structures is assessed in Fig. 3 by the distribution of IS-scores [36] across the BM1905 dataset. Figure 3a shows the accuracy of complex models constructed from experimental monomeric structures with ZDOCK alone, ZDOCK followed by FiberDock, eRank^{PPI}, and eRank^{PPI} followed by FiberDock. For each receptor-ligand pair, we first selected the top 10 highest scoring ZDOCK models and picked the model with the best IS-score. At least one model with a statistically highly significant IS-score of 0.473 is found in 34.9% of the cases. This percentage

increases to 42.4% when the initial dimers are refined by FiberDock. Next, we re-ranked the top 2,000 models from ZDOCK with eRank^{PPI} in order to more reliably identify near-native structures. Encouragingly, in 50.5% of the cases, at least one model having an IS-score higher than 0.473 is now found within the top 10 dimers re-ranked by eRank^{PPI}. Further refinement with Fiber-Dock increases this fraction to as high as 57.5%. In addition to the IS-score, Table 1 shows that success rates measured with iRMSD as well as PCS increase when eRank^{PPI} and FiberDock are included in the modeling protocol.

Altogether, eRank^{PPI} and FiberDock generate the most accurate dimers in these benchmarking calculations. Figure 3a and Table 1 show that re-ranking with eRank^{PPI} places more near-native structures within the top-ranked models compared to ZDOCK, which is in accordance with our previous studies [23] reporting ~10% improvement in the success rate. In general, the refinement by FiberDock considering both backbone and sidechain flexibility consistently improves the model accuracy, however, the improvement clearly depends on the quality of the top-ranked dimers. Most significant improvement for models selected by eRank^{PPI} is achieved when the IS-score of the initial dimers is in the range of 0.4-0.8.

Dimers constructed from computer-generated monomer structures
The unavailability of experimentally determined structures for a vast majority of gene products necessitates using computer-generated models for genome-wide determination of PPIs. On that account, we investigate how protein docking, and dimer re-ranking and refinement are affected when computer-generated models are used instead of experimental structures. Figure 3b

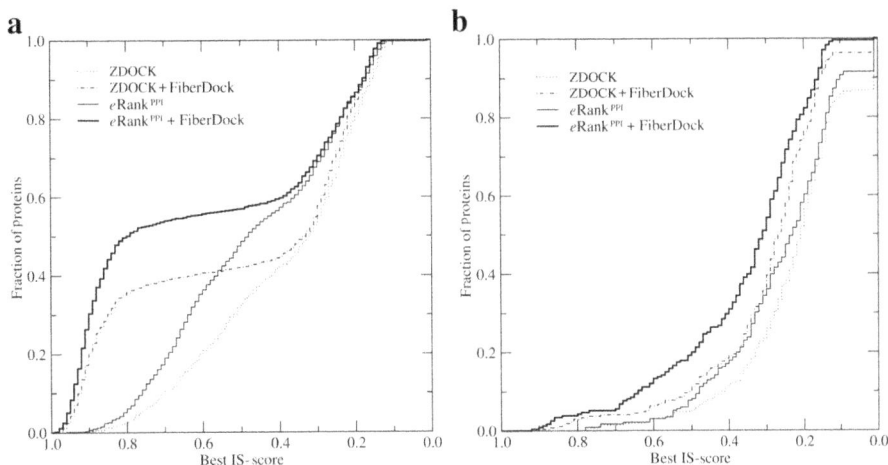

Fig. 3 Performance of ZDOCK, eRank^{PPI} and FiberDock on the BM1905 dataset. Dimer complexes are constructed using (**a**) experimentally solved monomer structures (BM1905C) and (**b**) computer generated monomer models (BM1905H). The results are presented as the cumulative fraction of proteins with the IS-score between predicted and experimental complex structures larger than or equal to the value displayed on the x-axis

shows the accuracy of dimer models constructed using four protocols described above. Since monomers are weakly homologous models containing structural inaccuracies, the modeling results are evaluated with a lower, yet still statistically significant IS-score threshold of 0.311. We find that in 22.3 and 31.0% of the cases, at least one model with an IS-score of ≥0.311 is found within the top 10 conformations ranked by ZDOCK and eRank^{PPI}, respectively. Furthermore, a flexible refinement with FiberDock increases the percentage of successful cases to 32.2% for ZDOCK and to 48.7% for eRank^{PPI}. Table 1 shows that similar results are obtained with the iRMSD and PCS used to measure the success rate. Therefore, not only dimer models reranked by eRank^{PPI} and additionally refined by FiberDock are the most accurate, but also the refinement procedure yields better improvements for eRank^{PPI} compared to ZDOCK. Despite the fact that protein docking using weakly homologous monomer structures is a difficult task and the dimer accuracy cannot be expected to be higher than the accuracy of the monomers, our analysis demonstrates that, in many cases, using a protocol combining eRank^{PPI} and FiberDock constructs reliable complexes as assessed by the IS-score, iRMSD, and PCS.

Predicting biologically relevant interactions

Macromolecular complexes are stabilized by a variety of interactions including solvation effects, changes in the internal energy upon binding, electrostatics, van der Waals interactions, hydrogen bonds, π-stacking, and hydrophobic contacts across the interface. These interactions are prevalently found in the crystal structures of protein assemblies deposited in the PDB. Given that protein crystals mimic the actual interactions in an aqueous solution, biologically relevant complex structures can be predicted based on these contributions to the binding energy. Figure 4 shows the distribution of various energy terms calculated by FiberDock for the positive dataset HET3519 and the negative dataset RND14944. Note a clear distinction in the distribution of most energies between interacting and non-interacting protein pairs suggesting that these scores can be utilized to identify biologically relevant interactions. For example, the median attractive (repulsive) van der Waals energy is -0.230 (-0.187) and 0.214 (-0.195) for interacting and non-interacting pairs, respectively. Another highly discriminatory term is the hydrogen bond energy with the median value of -0.068 for interacting and 0.418 for non-interacting pairs, which is consistent with other studies reporting that the hydrogen bond potential greatly improves the

Table 1 Comparison of the success rates for protein dimers assembled from the crystal structures and computer-generated models of monomers

Protocol	Crystal structures		Protein models	
	iRMSD ≤2.5 Å	PCS ≥0.65	iRMSD ≤8.5 Å	PCS ≥0.30
ZDOCK	51.5%	52.1%	28.1%	23.2%
ZDOCK + eRank^{PPI}	58.3%	59.6%	43.7%	39.3%
ZDOCK + eRank^{PPI} + FiberDock	72.8%	73.2%	52.4%	48.7%

The acceptance criteria for correct predictions are an iRMSD of ≤2.5 Å and PCS ≥0.65 for crystal structures, and an iRMSD of ≤8.5 Å and PCS ≥0.30 for protein models. The best of top 10 dimer models is considered

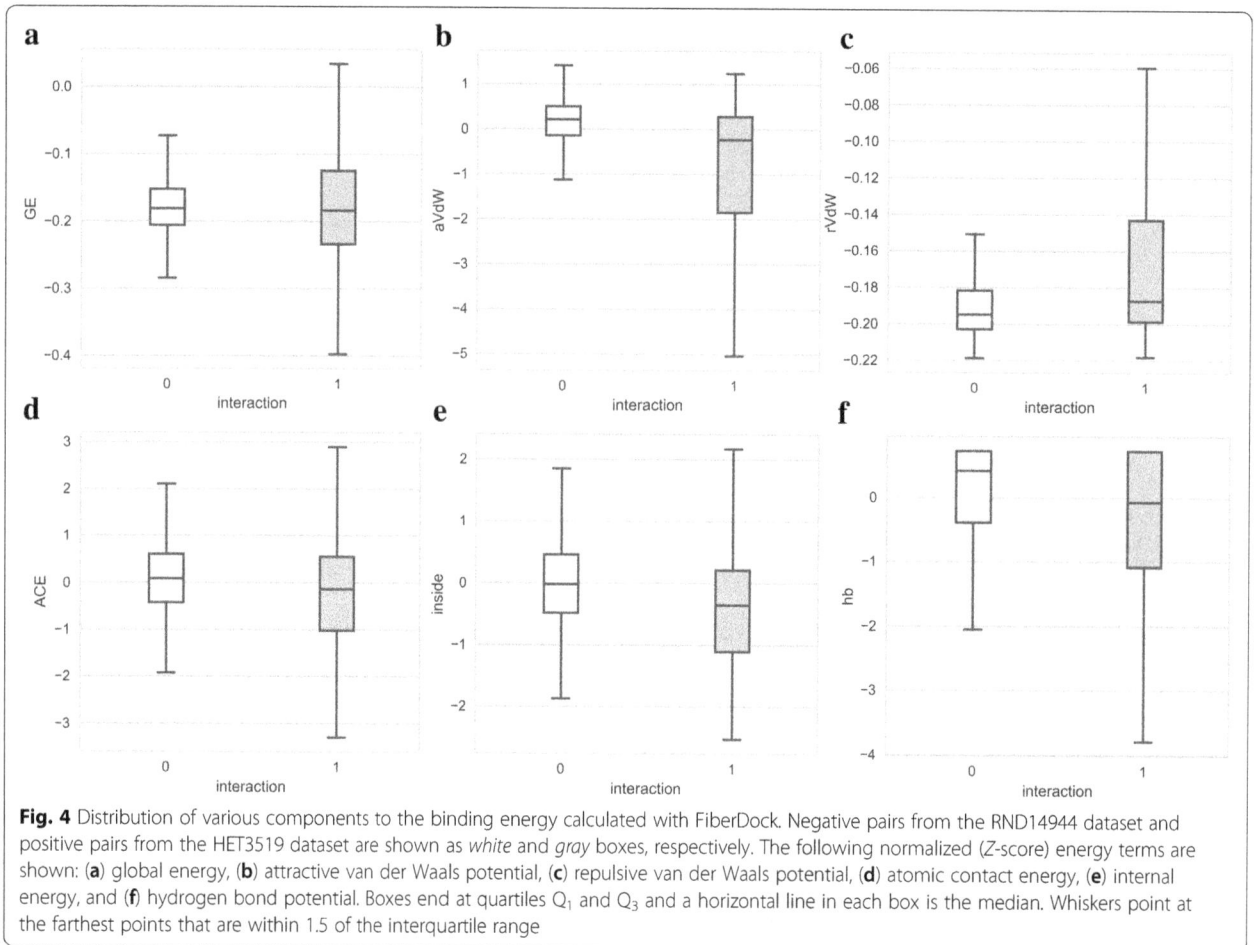

Fig. 4 Distribution of various components to the binding energy calculated with FiberDock. Negative pairs from the RND14944 dataset and positive pairs from the HET3519 dataset are shown as *white* and *gray* boxes, respectively. The following normalized (*Z*-score) energy terms are shown: (**a**) global energy, (**b**) attractive van der Waals potential, (**c**) repulsive van der Waals potential, (**d**) atomic contact energy, (**e**) internal energy, and (**f**) hydrogen bond potential. Boxes end at quartiles Q_1 and Q_3 and a horizontal line in each box is the median. Whiskers point at the farthest points that are within 1.5 of the interquartile range

recognition of correctly docked protein-protein complexes from large sets of alternative structures [45].

Next, we combine various interactions at the interface for the top 3 refined models in order to evaluate the complex stability and to predict whether the interaction is biologically relevant or not. Specifically, the RFC is employed to estimate a probability that a given complex model represents a true interaction. Figure 5 shows a receiver operating characteristic (ROC) plot evaluating the performance of a classifier separating true interactions within the HET3519 dataset from negative pairs present in the RND14944 dataset. Using the top-ranked model, the area under the curve for the prediction of biologically relevant interactions is 0.72. The probability threshold of 0.13 (a solid triangle in Fig. 5) maximizes the MCC to a value of 0.43 at a true positive rate of 0.51 and a false positive rate of 0.14. Essentially, this threshold corresponds to a point in the ROC space farthest from the diagonal representing the performance of a random classifier (gray area in Fig. 5).

Next, we improved the classification procedure by employing up to top 5 ranked models constructed for a given pair of receptor and ligand proteins. A pair is

predicted to represent a true interaction if a positive predictive score is greater than the optimized probability threshold of 0.13 for at least one out of top *n* models. Table 2 shows that this strategy indeed enhances the discriminatory power. Considering the top 3 models maximizes the MCC to a value of 0.61 with a true positive rate of 0.81 and a false positive rate of 0.19 (a solid circle in Fig. 5). Finally, we independently test our classification protocol against the Negatome 2.0 database, which provides a collection of protein pairs unlikely to physically interact with each other [27]. We obtained a false positive rate of 0.23, i.e. 23% of non-interacting pairs included in Negatome 2.0 are predicted as interacting proteins. This false positive rate is similar to that calculated for the HET3519 and RND14944 datasets suggesting that the RFC classifier is robust and its performance is independent on the validation dataset. Overall, the classifier performance is sufficiently high to be applicable at a proteome scale.

Modeling protein-protein complex structures for *E. coli*

All-against-all docking of 2,300 proteins in *E. coli* produced 2,643,850 possible binary PPIs with 3 putative dimer

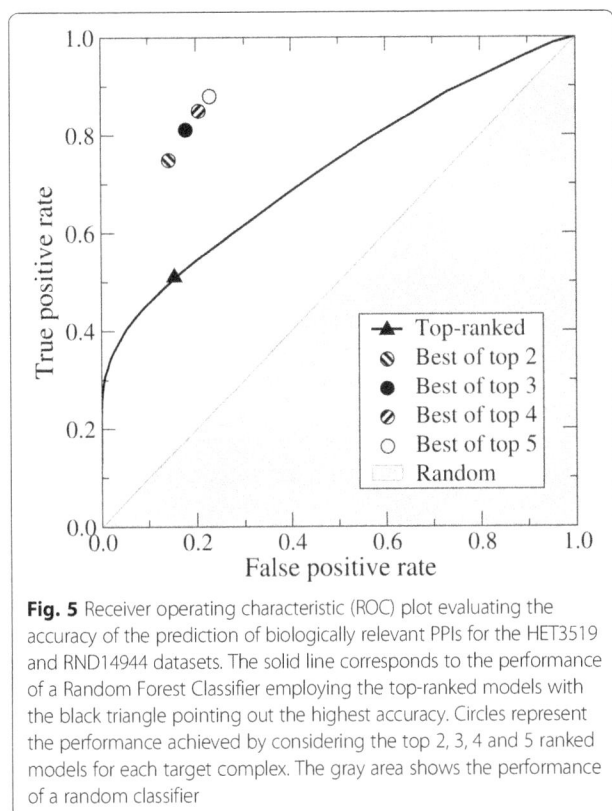

Fig. 5 Receiver operating characteristic (ROC) plot evaluating the accuracy of the prediction of biologically relevant PPIs for the HET3519 and RND14944 datasets. The solid line corresponds to the performance of a Random Forest Classifier employing the top-ranked models with the black triangle pointing out the highest accuracy. Circles represent the performance achieved by considering the top 2, 3, 4 and 5 ranked models for each target complex. The gray area shows the performance of a random classifier

models generated for each unique receptor-ligand pair, totaling 7,931,550 3D complex structures of bacterial proteins. Applying the RFC trained on the HET3519 and RND14944 datasets predicted 425,412 biologically relevant interactions corresponding to 18.2% of all possible PPIs (Additional file 1). Note that although the experimentally covered PPI space provided by DIP [31] is very limited with only 6,341 validated interactions, our structure-based pipeline correctly identified 3,930 (62%) of these true PPIs. According to the BioGRID Database Statistics, an estimated number of 164,717 non-redundant interactions are present in *E. coli*, suggesting that that additional filters are required to further refine the set of predicted interactions. On that account, we added annotation filters from Gene Ontology

Table 2 Accuracy of the prediction of biologically relevant PPIs for the HET3519 and RND14944 datasets

Number of models	MCC	TPR	FPR
1	0.43	0.53	0.11
2	0.58	0.74	0.14
3	0.61	0.81	0.19
4	0.58	0.85	0.20
5	0.58	0.88	0.22

Here, we consider up to top 5 ranked models constructed for a given pair of receptor and ligand proteins
MCC Matthews correlation coefficient, *TPR* true positive rate, *FPR* false positive rate

to support the identification of biologically relevant dimers constructed for the *E. coli* proteome.

Integrating structure-based prediction with Gene Ontology

First, we tested whether CC, BP and MF slims can be used as filters to identify interacting proteins by comparing GO annotations in positive and negative protein pairs. Here, the positive set contains known protein interactions according to the DIP database, whereas the negative set is compiled by randomly pairing *E. coli* proteins included in the DIP database. Those protein pairs having at least one common GO slim pass the annotation filter. About 82% of positives pass the CC filter that requires two proteins to co-localize in order to form a physical interaction. In contrast, only 58% of negatives are located in the same cellular component. Further, as many as 93% of positives are part of the same biological process, whereas 66% of negatives pass the BP filter. These results are in line with previous studies demonstrating that proteins localized in the same cellular compartment are more likely to interact than those residing in spatially distant compartments [46, 47]. Similarly, proteins involved in the same biological process have on average a higher chance to interact compared to molecules functioning in different biological processes. Thus, both CC and BP filters retain the majority of true interactions and reject a number of non-interacting protein pairs leading to a better classification performance. In contrast, molecular function cannot be used to improve the identification of biologically relevant interactions because a similar percentage of positives (48%) and negatives (52%) pass the MF filter. To further corroborate these results, we applied both CC and BP filters to the HET3519 and RND14944 datasets. Encouragingly, as many as 91 and 93% of HET3519 complexes passed CC and BP filters, respectively. In contrast, significantly fewer pairs from the random dataset RND14944 passed CC (63%) and BP (44%) filters. The discriminatory performance of GO filters applied to HET3519 and RND14944 is consistent with that obtained for the *E. coli* dataset.

Assembly and analysis of PPI network in E. coli

In order to assemble the network of protein-protein interactions in *E. coli*, we first applied the CC filter to 425,412 putative hetero-dimers identified by the RFC bringing this number down to 253,230 interactions between proteins localized in the same cellular compartment. Next, we selected only those protein pairs involved in the same biological process further reducing the number of putative hetero-dimers to 81,280. Although the BP filter is highly sensitive correctly identifying 93% of true interactions, this significant reduction of the number of positive predictions is mainly attributed to the fact that BP annotations are available for only 1,294 out of 2,300 proteins. Combining structure-based prediction of PPIs with both annotation filters results in 61,913 biologically relevant interactions. Note that GO filters are frequently employed to automatically refine large sets of protein interactions. For instance, the F-measure assessing the accuracy of PPI prediction for the bacterial chemotaxis signaling pathway increased from 0.52 to 0.69 when the protein localization was taken into consideration [21]. Our final set of protein interactions with confidently modeled dimer conformations provide a tremendous source of structural data relating to the network of protein-protein interactions in E. coli.

Subsequently, we investigated several properties of the PPI network constructed for E. coli in comparison with a random network comprising the same number of nodes and edges. The only difference between the predicted and random networks is that the latter is built on interactions randomly assigned to pairs of proteins. For the PPI network predicted for E. coli by the structure-based approach, the degree, diameter, and clustering coefficient [48] are 110.5, 6, and 0.30, respectively. Although the random network has a similar degree of 111.4, its diameter is 3 and the clustering coefficient is only 0.11. This analysis reveals that the global topology of the constructed network significantly differs from that of a random network. Specifically, the predicted PPIs tend to cluster together forming functional units around highly connected hubs, whereas PPIs are distributed more uniformly in a random network. In order to further corroborate these findings, we constructed a PPI network from experimental interactions included in the DIP database and the corresponding random network having the same number of nodes and edges. Here the degree, diameter and clustering coefficient calculated for the DIP (random) network are 6.9 (6.8), 12 (7), and 0.08 (0.004), respectively. The differences between the network predicted by a structure-based approach and that built on interaction data from DIP result from the incompleteness of the latter, i.e. the DIP network is sparse, having about 17 times less connections per node than the predicted network. Nonetheless, the deviations of both networks from their random counterparts are qualitatively similar showing a notable tendency to form clusters and sub-networks.

Figure 6 shows hive plots [49] generated for the predicted (Fig. 6a) and random (Fig. 6b) networks of PPIs in E. coli. In both plots, true positives and false positives with respect to experimentally validated interactions from the DIP database are colored in green and red, respectively. First, the structure-based approach including GO filters correctly identifies the majority of experimental interactions (green lines), whereas these connections are largely missed in the random network (red lines). Second, the axes in both hive plots are sorted by the clustering coefficient of individual nodes and the axis

scales in Fig. 6a and b are significantly different. Third, considering the global network topology, the majority of nodes in the random network are assigned to a medium-degree group (y-axis) forming extensive connections to themselves as well as to low- (x-axis) and high-degree (z-axis) groups. In contrast, extensive connections between all groups are present in the network predicted by the modeling of quaternary structures. These hive plots effectively visualize differences between the predicted and random networks described above.

Examples of dimer models selected from the E. coli network

Since the PPI network for the E. coli proteome is assembled by the modeling of interactions between proteins, we discuss a couple of representative examples of the modeled dimer structures. Note that experimentally solved structures are unavailable for these proteins, therefore, the presented molecular assemblies have been constructed solely from the primary sequences of individual monomers. Although monomer models are built on templates whose sequence identity to the target protein is less than 40%, the estimated Global Distance Test (GDT) [50] is greater than 0.7 indicating that these computer-generated structures are highly confident. The first example is a hetero-dimer assembled from fadJ and fadI proteins involved in the fatty acid beta oxidation pathway, which is part of lipid metabolism. This interaction was proposed to increase the efficiency of anaerobic beta-oxidation by favoring substrates of different chain length [51]. Even though there is experimental evidence that these two proteins interact with one another [52], no structural data is available for the individual proteins nor the complex. The modeling procedure developed in this study correctly identified these proteins to be interaction partners with the putative fadJ/fadI hetero-dimer shown in Fig. 7. A protein binding site confidently predicted by eFindSite^{PPI} on fadJ comprises 11 residues, out of which 9 are also found at the interface in the modeled fadJ/fadI complex. Moreover, fadJ has a NAD binding domain according to the Pfam database [53]. Interestingly, we were able to not only identify a binding pocket for NAD in the fadJ structure model with eFindSite [54], but also to dock a NAD molecule to this pocket using our in-house ligand docking software eSimDock [55].

The second example is glutaminase 2 (glsA2), an amidohydrolase enzyme responsible for generating glutamate from glutamine, demonstrated to be a self-assembling protein [56]. The GDT of the glsA2 monomer estimated by eThread is 0.78 indicating a confident structure model. Next, we predicted the structure of glsA2 homo-dimer as a symmetric complex shown in Fig. 8. A unique feature of eFindSite^{PPI} is that it not only

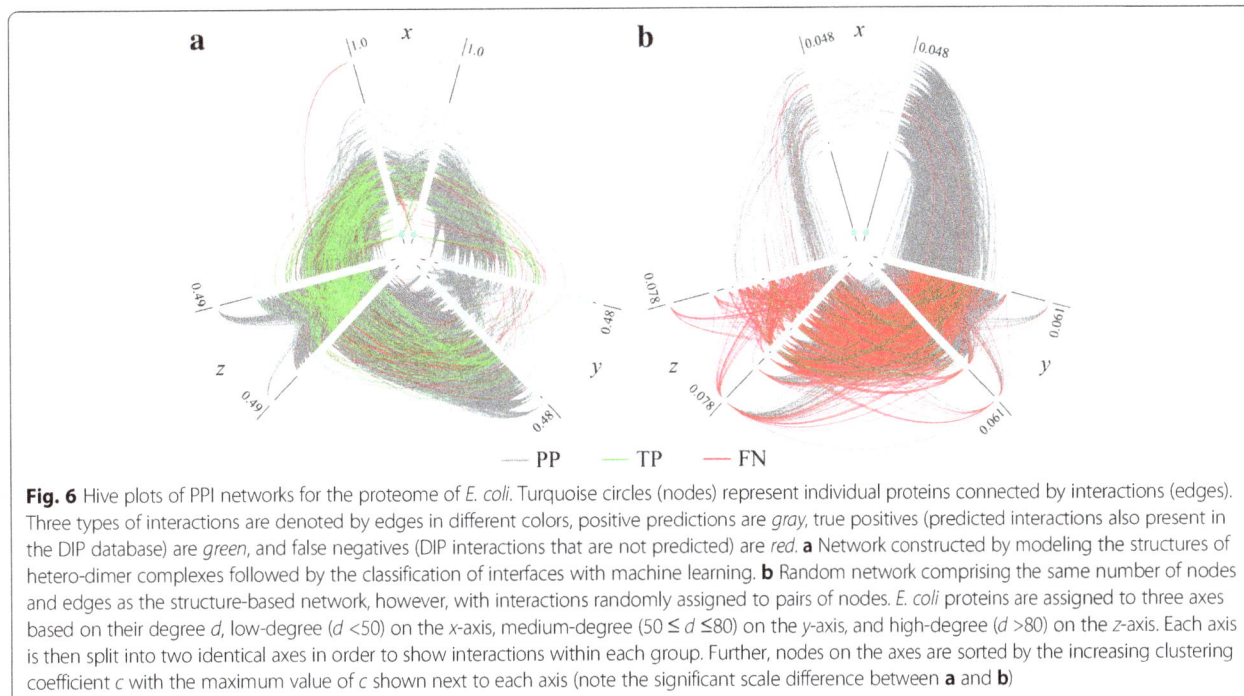

Fig. 6 Hive plots of PPI networks for the proteome of *E. coli*. Turquoise circles (nodes) represent individual proteins connected by interactions (edges). Three types of interactions are denoted by edges in different colors, positive predictions are *gray*, true positives (predicted interactions also present in the DIP database) are *green*, and false negatives (DIP interactions that are not predicted) are *red*. **a** Network constructed by modeling the structures of hetero-dimer complexes followed by the classification of interfaces with machine learning. **b** Random network comprising the same number of nodes and edges as the structure-based network, however, with interactions randomly assigned to pairs of nodes. *E. coli* proteins are assigned to three axes based on their degree d, low-degree ($d < 50$) on the *x*-axis, medium-degree ($50 \leq d \leq 80$) on the *y*-axis, and high-degree ($d > 80$) on the *z*-axis. Each axis is then split into two identical axes in order to show interactions within each group. Further, nodes on the axes are sorted by the increasing clustering coefficient c with the maximum value of c shown next to each axis (note the significant scale difference between **a** and **b**)

detects interaction sites, but also points out specific molecular interactions that stabilize a putative complex. Molecular interactions predicted by $e\text{FindSite}^{\text{PPI}}$ for glsA2 include a salt bridge between the side chains of R232 (chain A) and E82 (chain B) as well as aromatic contacts between W252 (chain A) and W252 (chain B),

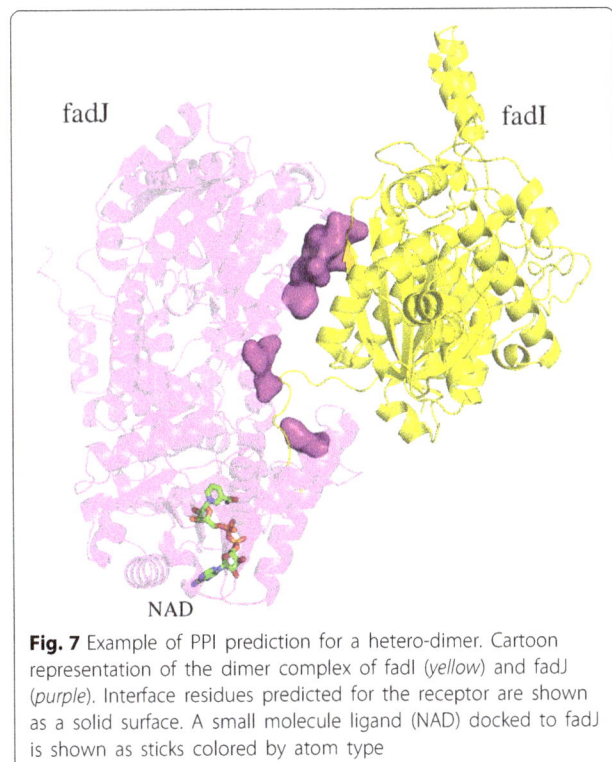

Fig. 7 Example of PPI prediction for a hetero-dimer. Cartoon representation of the dimer complex of fadI (*yellow*) and fadJ (*purple*). Interface residues predicted for the receptor are shown as a solid surface. A small molecule ligand (NAD) docked to fadJ is shown as sticks colored by atom type

which are found in the top-ranked complex model selected by $e\text{Rank}^{\text{PPI}}$.

Analysis of PPIs in the human immune disease pathway

Finally, based on experimental data provided by the Reactome database, we modeled protein complex structures for the human immune disease pathway associated with the TLR signaling cascade. TLRs are sensors of the innate immune system recognizing pathogen-associated molecular patterns [57, 58]. These molecular sensors participate in the first line of defense against invading pathogens by promoting the activation and nuclear translocation of certain transcription factors to induce the secretion of inflammatory cytokines. Out of 26 gene products involved in this pathway, we included the following 17 proteins whose 3D structures have been modeled (estimated GDT values are given in parentheses): P58753 (0.64), Q15399 (0.45), Q9Y2C9 (0.46), P08571 (0.48), P16671 (0.59), O15111 (0.56), O14920 (0.54), Q99836 (0.48), Q9NWZ3 (0.65), O60602 (0.49), Q15653 (0.71), Q00653 (0.32), Q04206 (0.52), P25963 (0.70), P19838 (0.33), Q9BXR5 (0.41), and Q9Y6Y9 (0.77). The remaining 9 structures have not been modeled due to either their large size, the unavailability of reliable templates, or a significant content of transmembrane regions. Although the total number of possible interactions for this dataset is 153, only 58 are confirmed experimentally according to the Reactome database. Figure 9 shows the network structure and a binary interaction matrix for PPIs predicted for this pathway. The structure-based approach predicted a total of 90 unique interactions (dashed blue

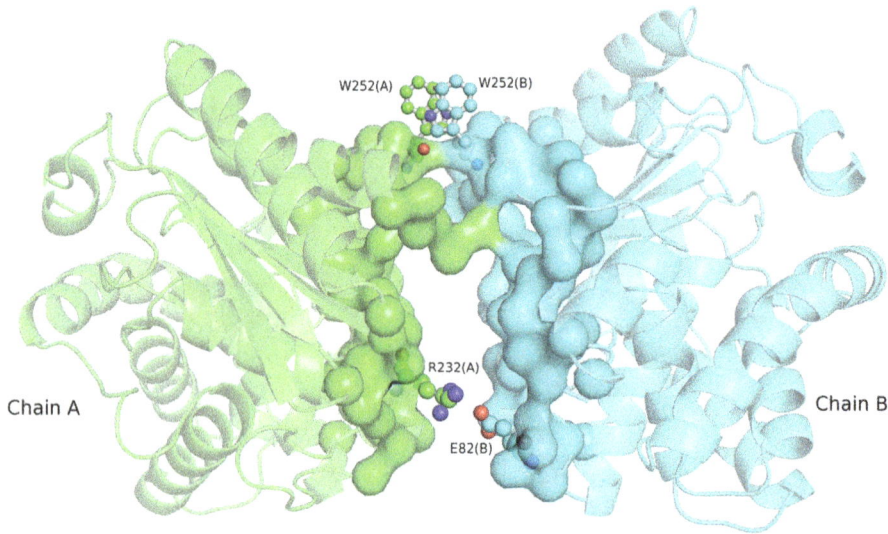

Fig. 8 Example of PPI prediction for a homo-dimer. Cartoon representation of the dimer complex of YneH with chains A and B colored in *green* and *blue*, respectively. Protein interfaces predicted for the monomers are shown as a solid surface. Residues predicted to be involved in a salt bridge R32(A)-E28(B) and aromatic contact W525(A)-W525(B) are shown as balls and sticks

connections in Fig. 9a) including 38 known interactions (solid green connections in Fig. 9a). Only 20 known interactions have not been predicted by the quaternary structure modeling (dotted red connections in Fig. 9a). Therefore, about two-thirds of true PPIs were correctly

recovered by the modeling of the complex structures of proteins involved in the human immune disease pathway. These results are in line with the analysis of the interaction network in *E. coli*, where our protocol correctly identified 62% of known PPIs.

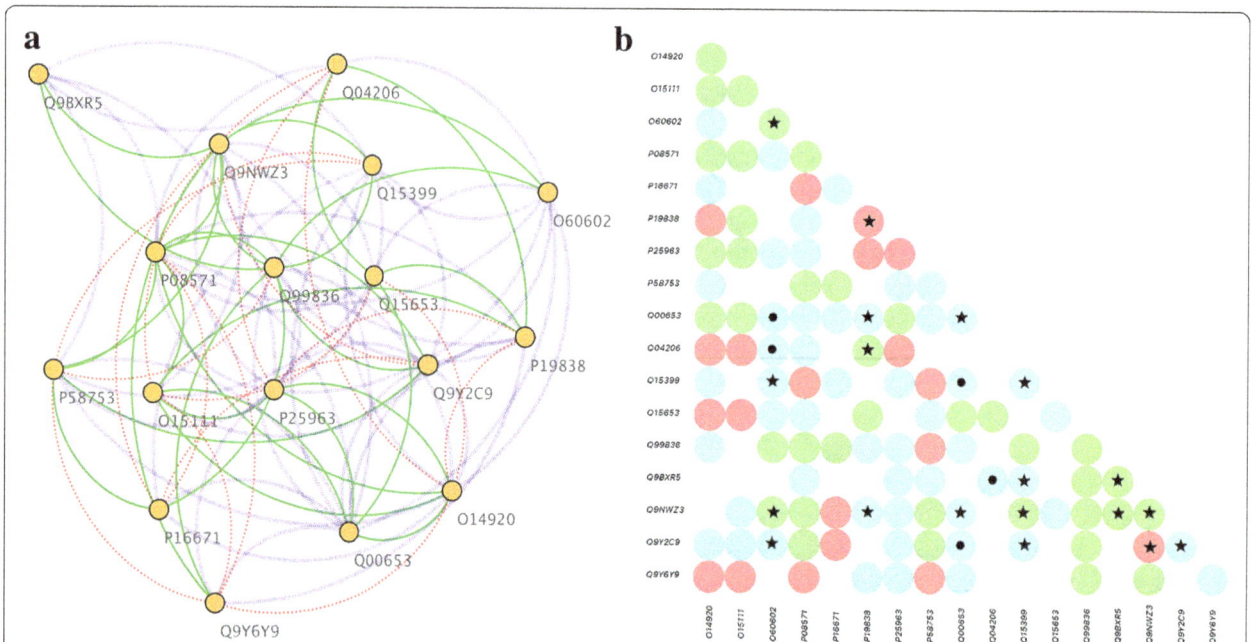

Fig. 9 Structure-based prediction of PPIs for the human immune disease pathway. **a** Network diagram of the human immune disease pathway. *Yellow* circles (nodes) represent individual proteins connected by interactions (edges). Three types of interactions are denoted by edges in different colors, positive predictions are *blue*, true positives (predicted interactions also present in the Reactome database) are *green*, and false negatives (interactions from Reactome that are not predicted) are *red*. **b** Matrix of binary interactions including positive predictions (*blue*), true positives (*green*), and false negatives (*red*). Circles marked with a star and a dot show those protein pair that pass and fail the CC filter, respectively. UniProt IDs of proteins involved in this pathway according to the Reactome database are shown in both **a** and **b**

Across-proteome modeling of dimer structures for the bottom-up assembly of protein-protein interaction...

99

In addition, positive predictions, true positives and false negatives are shown as a binary interaction matrix in Fig. 9b. Here, we also mapped GO Slims for the cellular component to individual proteins in order to improve the PPI prediction accuracy by including function annotation filters. Since GO annotations were available only for 8 proteins, the CC filter was applied to 17 hetero-dimer models constructed by our structure-based approach. Encouragingly, 12 of the predicted complexes passed the CC filter (black stars in Fig. 9b), while only 5 failed (black dots in Fig. 9b). Although, the GO annotation filter can be applied only to a fraction of structure-based predictions for this pathway, it turns out to be quite accurate. Therefore, we expect that new function annotations available in the future will selectively reduce the number of positive predictions leading to more accurate PPI prediction results.

Conclusions

In this work, we developed a new method combining molecular modeling, structural bioinformatics, machine learning, and functional annotation data to predict PPIs across proteomes. We first comprehensively tested this protocol on representative datasets of experimental structures and computer-generated models of protein dimers and then we applied this methodology to predict PPIs across the proteome of *E. coli* and within the human immune disease pathway. Our results indicate that protein docking supported by evolutionary restraints and machine learning can be used to reliably identify and model biologically relevant protein assemblies. Furthermore, the accuracy of the identification of interaction partners can greatly be improved by including only those protein pairs co-localized in the same cellular compartment and involved in the same biological process. The proposed method can be applied to detect PPIs in other organisms and pathways as well as to construct structure models and estimate the confidence of interactions experimentally identified with high-throughput techniques. Finally, with the growing volume of structural data, experimentally confirmed protein interactions, and functional annotation, we expect the coverage and accuracy of our approach to increase over time.

Abbreviations

BP: Biological process; CAPRI: Critical Assessment of PRedicted Interactions; CC: Cellular component; DIP: Database of Interacting Proteins; GDT: Global Distance Test; GO: Gene Ontology; IS-score: Interface Similarity score; MCC: Matthews correlation coefficient; MF: molecular function; PCS: Pairwise Contact Score; PDB: Protein Data Bank; PPIs: Protein-protein interactions; RFC: Random Forest Classifier; RMSD: Root-mean-square deviation; ROC: Receiver operating characteristic; TLR: Toll-Like Receptor; TM-score: Template Modeling score

Acknowledgements

The Authors acknowledge computing resources provided by Louisiana State University.

Funding

Research reported in this publication was supported by the National Institute of General Medical Sciences of the National Institutes of Health under Award Number R35GM119524. The content is solely the responsibility of the authors and does not necessarily represent the official views of the National Institutes of Health.

Authors' contributions

SM and MB designed the study. SM prepared datasets, developed protocols, and performed calculations. SM and MB analyzed data. SM and MB wrote the manuscript. All authors read and approved the final manuscript.

Competing interests

The authors declare that they have no competing interests.

References

1. Jones S, Thornton JM. Principles of protein-protein interactions. Proc Natl Acad Sci U S A. 1996;93(1):13–20.
2. Gandhi TK, Zhong J, Mathivanan S, Karthick L, Chandrika KN, Mohan SS, Sharma S, Pinkert S, Nagaraju S, Periaswamy B, et al. Analysis of the human protein interactome and comparison with yeast, worm and fly interaction datasets. Nat Genet. 2006;38(3):285–93.
3. Sanderson CM. The Cartographers toolbox: building bigger and better human protein interaction networks. Brief Funct Genomic Proteomic. 2009;8(1):1–11.
4. Rual JF, Venkatesan K, Hao T, Hirozane-Kishikawa T, Dricot A, Li N, Berriz GF, Gibbons FD, Dreze M, Ayivi-Guedehoussou N, et al. Towards a proteome-scale map of the human protein-protein interaction network. Nature. 2005; 437(7062):1173–8.
5. Yu H, Braun P, Yildirim MA, Lemmens I, Venkatesan K, Sahalie J, Hirozane-Kishikawa T, Gebreab F, Li N, Simonis N, et al. High-quality binary protein interaction map of the yeast interactome network. Science. 2008;322(5898):104–10.
6. Bjorklund AK, Light S, Hedin L, Elofsson A. Quantitative assessment of the structural bias in protein-protein interaction assays. Proteomics. 2008;8(22):4657–67.
7. Kundrotas PJ, Lensink MF, Alexov E. Homology-based modeling of 3D structures of protein-protein complexes using alignments of modified sequence profiles. Int J Biol Macromol. 2008;43(2):198–208.
8. Launay G, Simonson T. Homology modelling of protein-protein complexes: a simple method and its possibilities and limitations. BMC Bioinformatics. 2008;9:427.
9. Mukherjee S, Zhang Y. Protein-protein complex structure predictions by multimeric threading and template recombination. Structure. 2011;19(7):955–66.
10. Sinha R, Kundrotas PJ, Vakser IA. Protein docking by the interface structure similarity: how much structure is needed? PLoS One. 2012;7(2):e31349.
11. Tuncbag N, Keskin O, Nussinov R, Gursoy A. Fast and accurate modeling of protein-protein interactions by combining template-interface-based docking with flexible refinement. Proteins. 2012;80(4):1239–49.
12. Kundrotas PJ, Vakser IA. Global and local structural similarity in protein-protein complexes: implications for template-based docking. Proteins. 2013;81(12):2137–42.
13. Vreven T, Hwang H, Pierce BG, Weng Z. Evaluating template-based and template-free protein-protein complex structure prediction. Brief Bioinform. 2014;15(2):169–76.
14. Berman HM, Westbrook J, Feng Z, Gilliland G, Bhat TN, Weissig H, Shindyalov IN, Bourne PE. The Protein Data Bank. Nucleic Acids Res. 2000;28(1):235–42.
15. Kundrotas PJ, Zhu Z, Janin J, Vakser IA. Templates are available to model nearly all complexes of structurally characterized proteins. Proc Natl Acad Sci U S A. 2012;109(24):9438–41.
16. Szilagyi A, Zhang Y. Template-based structure modeling of protein-protein interactions. Curr Opin Struct Biol. 2014;24:10–23.
17. Matsuzaki Y, Matsuzaki Y, Sato T, Akiyama Y. In silico screening of protein-protein interactions with all-to-all rigid docking and clustering: an application to pathway analysis. J Bioinform Comput Biol. 2009;7(6):991–1012.
18. Ohue M, Matsuzaki Y, Uchikoga N, Ishida T, Akiyama Y. MEGADOCK: an all-to-all protein-protein interaction prediction system using tertiary structure data. Protein Pept Lett. 2014;21(8):766–78.

19. Tsukamoto K, Yoshikawa T, Hourai Y, Fukui K, Akiyama Y. Development of an affinity evaluation and prediction system by using the shape complementarity characteristic between proteins. J Bioinform Comput Biol. 2008;6(6):1133–56.

20. Wass MN, Fuentes G, Pons C, Pazos F, Valencia A. Towards the prediction of protein interaction partners using physical docking. Mol Syst Biol. 2011;7:469.

21. Matsuzaki Y, Ohue M, Uchikoga N, Akiyama Y. Protein-protein interaction network prediction by using rigid-body docking tools: application to bacterial chemotaxis. Protein Pept Lett. 2014;21(8):790–8.

22. Chen R, Li L, Weng Z. ZDOCK: an initial-stage protein-docking algorithm. Proteins. 2003;52(1):80–7.

23. Maheshwari S, Brylinski M. Predicted binding site information improves model ranking in protein docking using experimental and computer-generated target structures. BMC Struct Biol. 2015;15:23.

24. Maheshwari S, Brylinski M. Prediction of protein-protein interaction sites from weakly homologous template structures using meta-threading and machine learning. J Mol Recognit. 2015;28(1):35–48.

25. Zhang Y, Skolnick J. Scoring function for automated assessment of protein structure template quality. Proteins. 2004;57(4):702–10.

26. Li W, Jaroszewski L, Godzik A. Clustering of highly homologous sequences to reduce the size of large protein databases. Bioinformatics. 2001;17(3):282–3.

27. Blohm P, Frishman G, Smialowski P, Goebels F, Wachinger B, Ruepp A, Frishman D. Negatome 2.0: a database of non-interacting proteins derived by literature mining, manual annotation and protein structure analysis. Nucleic Acids Res. 2014;42(Database issue):D396–400.

28. Sali A, Blundell TL. Comparative protein modelling by satisfaction of spatial restraints. J Mol Biol. 1993;234(3):779–815.

29. Brylinski M, Lingam D. eThread: a highly optimized machine learning-based approach to meta-threading and the modeling of protein tertiary structures. PLoS One. 2012;7(11):e50200.

30. Xu D, Zhang Y. Improving the physical realism and structural accuracy of protein models by a two-step atomic-level energy minimization. Biophys J. 2011;101(10):2525–34.

31. Xenarios I, Rice DW, Salwinski L, Baron MK, Marcotte EM, Eisenberg D. DIP: the database of interacting proteins. Nucleic Acids Res. 2000;28(1):289–91.

32. Croft D, O'Kelly G, Wu G, Haw R, Gillespie M, Matthews L, Caudy M, Garapati P, Gopinath G, Jassal B, et al. Reactome: a database of reactions, pathways and biological processes. Nucleic Acids Res. 2011;39(Database issue):D691–7.

33. Mintseris J, Pierce B, Wiehe K, Anderson R, Chen R, Weng Z. Integrating statistical pair potentials into protein complex prediction. Proteins. 2007;69(3):511–20.

34. Mashiach E, Nussinov R, Wolfson HJ. FiberDock: Flexible induced-fit backbone refinement in molecular docking. Proteins. 2010;78(6):1503–19.

35. Gao M, Skolnick J. iAlign: a method for the structural comparison of protein-protein interfaces. Bioinformatics. 2010;26(18):2259–65.

36. Gao M, Skolnick J. New benchmark metrics for protein-protein docking methods. Proteins. 2011;79(5):1623–34.

37. Janin J, Henrick K, Moult J, Eyck LT, Sternberg MJ, Vajda S, Vakser I, Wodak SJ, Critical Assessment of PI. CAPRI: a Critical Assessment of PRedicted Interactions. Proteins. 2003;52(1):2–9.

38. Ben-Hur A, Noble WS. Choosing negative examples for the prediction of protein-protein interactions. BMC Bioinformatics. 2006;7 Suppl 1:S2.

39. Barman RK, Jana T, Das S, Saha S. Prediction of intra-species protein-protein interactions in enteropathogens facilitating systems biology study. PLoS One. 2015;10(12):e0145648.

40. Chang JW, Zhou YQ, Ul Qamar MT, Chen LL, Ding YD. Prediction of protein-protein interactions by evidence combining methods. Int J Mol Sci. 2016;17(11): E1946.

41. You ZH, Chan KC, Hu P. Predicting protein-protein interactions from primary protein sequences using a novel multi-scale local feature representation scheme and the random forest. PLoS One. 2015;10(5):e0125811.

42. Ashburner M, Ball CA, Blake JA, Botstein D, Butler H, Cherry JM, Davis AP, Dolinski K, Dwight SS, Eppig JT, et al. Gene ontology: tool for the unification of biology. The Gene Ontology Consortium. Nat Genet. 2000;25(1):25–9.

43. Thomas PD, Campbell MJ, Kejariwal A, Mi H, Karlak B, Daverman R, Diemer K, Muruganujan A, Narechania A. PANTHER: a library of protein families and subfamilies indexed by function. Genome Res. 2003;13(9):2129–41.

44. Keseler IM, Mackie A, Peralta-Gil M, Santos-Zavaleta A, Gama-Castro S, Bonavides-Martinez C, Fulcher C, Huerta AM, Kothari A, Krummenacker M, et al. EcoCyc: fusing model organism databases with systems biology. Nucleic Acids Res. 2013;41(Database issue):D605–12.

45. Kortemme T, Morozov AV, Baker D. An orientation-dependent hydrogen bonding potential improves prediction of specificity and structure for proteins and protein-protein complexes. J Mol Biol. 2003;326(4):1239–59.

46. De Bodt S, Proost S, Vandepoele K, Rouze P, Van de Peer Y. Predicting protein-protein interactions in Arabidopsis thaliana through integration of orthology, gene ontology and co-expression. BMC Genomics. 2009;10:288.

47. Qi Y, Bar-Joseph Z, Klein-Seetharaman J. Evaluation of different biological data and computational classification methods for use in protein interaction prediction. Proteins. 2006;63(3):490–500.

48. Assenov Y, Ramirez F, Schelhorn SE, Lengauer T, Albrecht M. Computing topological parameters of biological networks. Bioinformatics. 2008;24(2):282–4.

49. Krzywinski M, Birol I, Jones SJ, Marra MA. Hive plots–rational approach to visualizing networks. Brief Bioinform. 2012;13(5):627–44.

50. Zemla A, Venclovas C, Moult J, Fidelis K: Processing and analysis of CASP3 protein structure predictions. Proteins 1999, Suppl 3:22-29.

51. Campbell JW, Morgan-Kiss RM, Cronan Jr JE. A new Escherichia coli metabolic competency: growth on fatty acids by a novel anaerobic beta-oxidation pathway. Mol Microbiol. 2003;47(3):793–805.

52. Butland G, Peregrin-Alvarez JM, Li J, Yang W, Yang X, Canadien V, Starostine A, Richards D, Beattie B, Krogan N, et al. Interaction network containing conserved and essential protein complexes in Escherichia coli. Nature. 2005;433(7025):531–7.

53. Finn RD, Mistry J, Tate J, Coggill P, Heger A, Pollington JE, Gavin OL, Gunasekaran P, Ceric G, Forslund K, et al. The Pfam protein families database. Nucleic Acids Res. 2010;38(Database issue):D211–22.

54. Brylinski M, Feinstein WP. eFindSite: improved prediction of ligand binding sites in protein models using meta-threading, machine learning and auxiliary ligands. J Comput Aided Mol Des. 2013;27(6):551–67.

55. Brylinski M. Nonlinear scoring functions for similarity-based ligand docking and binding affinity prediction. J Chem Inf Model. 2013;53(11):3097–112.

56. Marino-Ramirez L, Minor JL, Reading N, Hu JC. Identification and mapping of self-assembling protein domains encoded by the Escherichia coli K-12 genome by use of lambda repressor fusions. J Bacteriol. 2004;186(5):1311–9.

57. Kawai T, Akira S. The role of pattern-recognition receptors in innate immunity: update on Toll-like receptors. Nat Immunol. 2010;11(5):373–84.

58. Pasare C, Medzhitov R. Toll-like receptors: linking innate and adaptive immunity. Adv Exp Med Biol. 2005;560:11–8.

10

Genetic sequence-based prediction of long-range chromatin interactions suggests a potential role of short tandem repeat sequences in genome organization

Sarvesh Nikumbh[1][*] and Nico Pfeifer[1,2]

Abstract

Background: Knowing the three-dimensional (3D) structure of the chromatin is important for obtaining a complete picture of the regulatory landscape. Changes in the 3D structure have been implicated in diseases. While there exist approaches that attempt to predict the long-range chromatin interactions, they focus only on interactions between specific genomic regions — the promoters and enhancers, neglecting other possibilities, for instance, the so-called structural interactions involving intervening chromatin.

Results: We present a method that can be trained on 5C data using the genetic sequence of the candidate loci to predict potential genome-wide interaction partners of a particular locus of interest. We have built locus-specific support vector machine (SVM)-based predictors using the oligomer distance histograms (ODH) representation. The method shows good performance with a mean test AUC (area under the receiver operating characteristic (ROC) curve) of 0.7 or higher for various regions across cell lines GM12878, K562 and HeLa-S3. In cases where any locus did not have sufficient candidate interaction partners for model training, we employed multitask learning to share knowledge between models of different loci. In this scenario, across the three cell lines, the method attained an average performance increase of 0.09 in the AUC. Performance evaluation of the models trained on 5C data regarding prediction on an independent high-resolution Hi-C dataset (which is a rather hard problem) shows 0.56 AUC, on average. Additionally, we have developed new, intuitive visualization methods that enable interpretation of sequence signals that contributed towards prediction of locus-specific interaction partners. The analysis of these sequence signals suggests a potential general role of short tandem repeat sequences in genome organization.

Conclusions: We demonstrated how our approach can 1) provide insights into sequence features of locus-specific interaction partners, and 2) also identify their cell-line specificity. That our models deem short tandem repeat sequences as discriminative for prediction of potential interaction partners, suggests that they could play a larger role in genome organization. Thus, our approach can (a) be beneficial to broadly understand, at the sequence-level, chromatin interactions and higher-order structures like (meta-) topologically associating domains (TADs); (b) study regions omitted from existing prediction approaches using various information sources (e.g., epigenetic information); and (c) improve methods that predict the 3D structure of the chromatin.

Keywords: Long-range interactions prediction, Support vector machines, Multitask learning, Hi-C, Visualizations

*Correspondence: snikumbh@mpi-inf.mpg.de
[1]Computational Biology & Applied Algorithmics, Max Planck Institute for Informatics, Saarland Informatics Campus, Building E1.4, D-66123 Saarbruecken, Germany
Full list of author information is available at the end of the article

Background

It is well known that chromatin, a complex of DNA and proteins, is packed in three-dimensional (3D) space inside the nucleus of the cell in a highly regulated fashion. The spatial conformation of chromosomes is governed by certain principles [1–3]. The structure of chromatin depends on the functional state of the cell (viz. normal/diseased) and gene activity among other cellular properties. Thus, a better understanding of 3D chromatin structure and the underlying mechanisms determining this structure helps in gaining an enhanced comprehension of many genomic functions. With the advent of chromosome conformation capture (3C)-based technologies in the last decade, starting with 3C itself in 2002, chromosome conformation capture-on-chip and circular chromosome conformation capture (both abbreviated as 4C), and 3C-carbon copy (5C) in 2006, chromatin interaction analysis by paired-end tag sequencing (ChIA-PET), 2009 [4–8], more recently Hi-C [9] and in situ high-resolution Hi-C [10] which is still quite expensive, genome-wide analysis of the interaction profiles is now possible [11]. Studies have revealed a correlation between long-range chromatin interactions and the functional state of the cell, e.g., in [12] and more generally, cell-type specificity as evidenced by [11]. These long-range interactions comprise pairs of loci that are close in space, but not necessarily close in sequence. The spatial co-localization of different chromosomal regions (*cis* as well as *trans*) can be due to a mix of factors, for example specific, direct contacts between two loci, nonspecific binding as a result of the packing of the chromatin fibre or co-localization due to functional association or having the same subnuclear structure [13].

Any long-range interaction (i.e., interaction between genomic loci separated by >1 or 2 mega base pairs) can typically occur to bring about or increase the likelihood of a certain activity at either of these loci itself (e.g., between an enhancer and a promoter region) or so that they can trigger or play an important role in any activity (e.g., facilitating binding of a protein) taking place at these loci or in their neighborhood on the genome. Knowledge of which loci interact over a long-range and evaluating the effect of such interactions can help us further our understanding of genome regulation and organization. Thus, it is of general interest to be able to predict whether a given pair of loci lying very far apart on the chromosome would interact. There exist machine learning-based approaches for predicting such long-range interactions between enhancer and promoter regions, for example, [14]. They combine the contact information output by a chromatin interaction experiment with various information sources, for example, epigenetic information [14], to make these predictions, but these approaches leave out genomic regions for which such information is not available. A sequence-level model,

in addition to primarily furthering our understanding of chromatin interactions at the most basic level, can also be useful to study any genomic region including the ones omitted by other approaches. Having a model that can predict, based on sequence information alone, whether two regions are likely to interact has several potential applications. One is to use the predicted label as additional information for the prediction of boundaries of topologically associating domains (TADs) [15]. Another is to assist methods that predict the 3D structure of the chromosome from Hi-C data [16].

As a word of caution, since the genetic sequence is only the primary level at which genomic function and organization information is encoded, it is apparent that higher levels of modifications will have the final say towards these chromatin interactions, more so for cell line-specificity. In other words, one would not expect a model using sequence information alone to outshine one that (also) utilizes additional information sources in terms of prediction accuracy. But, a sequence-level model has its advantages as already stated. Thus, we would like to stress upon our aim in performing this study:

> (a) Answer the question: To what extent can the genetic sequence alone predict these long-range chromosomal interactions? We report on various computational experiments, using our genetic-sequence based prediction method, to establish that the DNA sequence is informative to identify potential interaction partners of a given genomic locus, and (b) Understand the characteristic sequence features underlying such long-range interactions. This is achieved with the help of our two new visualization methods that aid in interpreting the sequence signals that contributed towards predicting locus-specific interaction partners and reveal interesting biological connections.

In general, we believe that such an approach using sequence-level information could be useful to study sequence peculiarities among the interaction partners of a particular locus. Our approach could augment existing methods for prediction of 3D chromatin structure and also TAD boundary predictions methods.

Approach

In this study we built a method based on support vector machines (SVMs) [17] to predict which genomic loci potentially interact with a given locus based on the genetic sequence. In a nutshell, we do the following: given a contact matrix delineating interactions between various genomic loci, we build a predictor for a locus of interest (LoI) from the contact matrix. This predictor learns the characteristics of the genomic loci that happen to significantly interact with the LoI as against the set of loci that do not. Thus, we build a predictor per locus. Such locus-specific predictors that use the genetic sequence information at these loci have the potential to uncover peculiarities of the interacting partners of this particular locus which can be useful to understand interactions at the sequence level. Such an understanding can guide us in our efforts to know the role-players at the genetic level and comprehend mechanisms of higher levels of chromatin organization viz. TADs and their hierarchies, and compartments. When dealing with contact matrices output from a chromatin interaction experiment where a large population of non-synchronized cells are studied, such an approach can still give us a holistic view.

We analyzed 5C contact matrices for three *human* cell lines — GM12878, K562 and HeLa-S3 — and demonstrated that the genetic sequence is predictive of the long-range interactions. Additionally, we utilized these locus-specific models, that were trained on the 5C data, to independently predict potential interaction partners across the chromosome for the same LoI. This computational validation is done on high-resolution Hi-C datasets from Rao et al. [10]. Our new visualization methods help to intuitively visualize the sequence features that proved useful for discerning the interaction partners of a LoI from those that do not interact with it, consequently rendering our models to be more than black boxes. Due to the models being locus-specific, one is also able to compare the sequence features found useful by a model (using our visualization) for a locus in one cell line to those found useful by the model for the same locus in another cell line. This is discussed in "Identifying cell-line specific characteristic signals among (non-)interactors of the same locus in different cell lines" section.

Results and discussion

Applicable to information on long-range contacts facilitated by a 4C, 5C or a Hi-C experiment, we describe our pipeline and the corresponding computational experiments performed on data from a 5C experiment [18] that detects interactions between a group of transcription start site (TSS)-containing regions (TCRs [18]) and distal enhancers in the three cell lines GM12878, K562 and HeLa-S3. Here, for each cell line, we built a separate classifier per TCR. Given the set of loci, for which the contact frequency with the TCR of interest (ToI) is known (from the contact matrix), we trained an SVM [17] which, when presented with a new, unseen locus, can classify it as positive or negative (i.e., interacting with the ToI or not). We use string kernels, which provide a measure of similarity between sequences, in conjunction with the SVM.

The aim was to build a pipeline with the best possible locus-specific classifiers (a separate classifier for each LoI/ToI), and also be able to determine subsequently, which sequence features were most important for any classifier to distinguish between the positive and the negative set of genomic loci corresponding to the LoI. Our pipeline is shown in Fig. 1 and described in the "Methodical details" section.

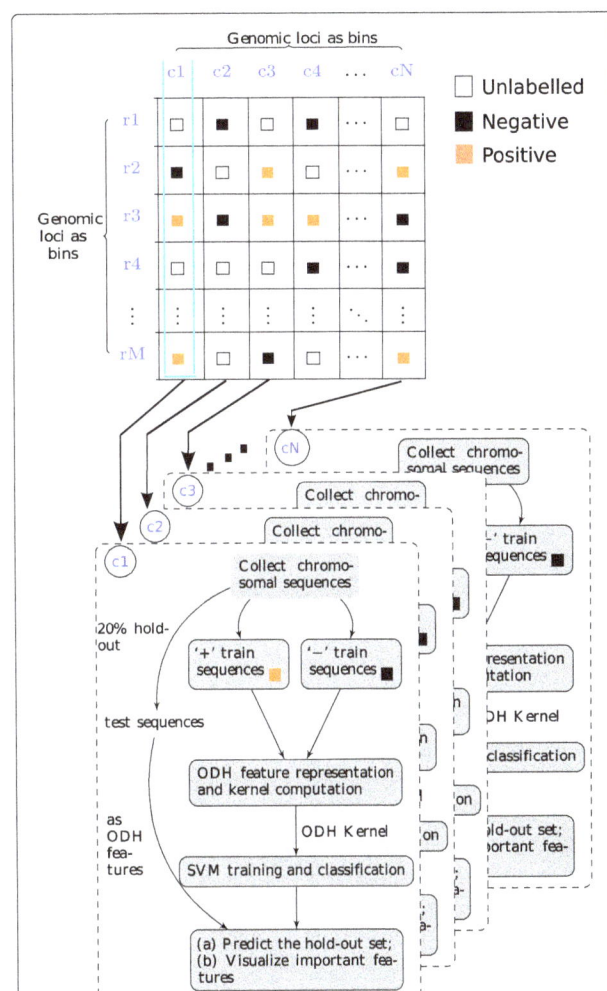

Fig. 1 Pipeline for predicting locus-specific long-range chromatin interactions using the genetic sequence. In the contact matrix, cells denoted by filled *orange boxes* correspond to loci that are called significantly interacting with the LoI in all replicates of any experiment profiling chromatin interactions. This constitutes the positive set of sequences for the corresponding classifier. Those denoted by filled *black boxes* correspond to loci that are not called significantly interacting in any of the replicates. This constitutes the negative set of sequences for the corresponding classifier. This leaves those loci which are called significantly interacting in at least one, but not in all of the replicates. They are visualized by *unfilled boxes* and are not used by the classifier. The genomic loci along the columns of the contact matrix (c1, c2, c3,...,cN) are the LoI for which we build locus-specific classifiers

Prediction of long-range chromatin interactions is possible from the sequence alone using non-linear SVMs

To evaluate the potential of the DNA sequence to serve as the sole information source in predicting the long-range interactions, we selected ten regions per cell line. For each cell line, these are the top 10 regions when ranked based on the number of positive examples available for them (see Supplementary Table S1 in Additional file 1). In each model, the varied-length sequences were represented as fixed-length feature vectors using the oligomer distance histograms (ODH) [19] representation. This represents any sequence by the histograms of distances between K-mers in the sequence (see 'Methodical details" section for more details). We performed experiments with K-mer values 3 and 5 and the maximum distance between K-mers as 100. Intuitively, K-mer value 5 encodes more specificity towards the set of sequences in a collection for a model while K-mer value 3 maintains relative generality. Once these are fixed, the ODH kernel has no other hyper-parameters to be tuned.

Table 1 summarily shows good test AUC (area under the ROC curve) values for all studied regions in all the three cell lines resulting from our 5-fold nested cross validation. Furthermore, our pipeline is also capable of handling imbalances in the data. For all the regions in our computational experiments, the positive class is in minority. We report performances with data imbalance handled (see "Methodical details").

The average test AUC values for the individual tasks are as follows. Oligomer length 3: {GM12878, K562, HeLa-S3}: {0.7251, 0.7534, 0.6782}; Oligomer length 5: {GM12878, K562, HeLa-S3}: {0.7443, 0.7716, 0.7153}. Box plots of all the test performances for different regions in all three cell lines are given in Fig. 2, and Additional file 1 (Supplementary Figures S3, S4 and S5 in Additional file 1). Owing to small sample sizes, the model test performances mostly show high variance (Fig. 2, and Supplementary Figures S3, S4 and S5 in Additional file 1).

For any interaction the complete length of the fragment may not be causal for the interaction, but only part(s) of it. However, this information is not available from the chromatin interaction experiments due to the length distribution of the fragments. Our locus-specific models are able to work around this situation and capture the features from different parts of the locus. This is due to the nature of ODH feature representations which capture the relative structure spread across the sequence rather than occurrences at different absolute positions in the sequence. Section "Tandem repeat motifs are an important feature distinguishing interaction partners" discusses how our visualizations help bring out this aspect of our models.

Tandem repeat motifs are an important feature distinguishing interaction partners

Figures 3 and 4 show our new visualizations of the set of K-mer pairs that influenced the prediction most. In both these visualizations, any K-mer pair is represented as an adjoined {$2K$}-mer separated by '|', e.g., 3-mer pairs as 6-mers, and we loosely address these K-mer pairs as 'motifs', although they are not contiguous. Figure 3 shows the 'Absolute Max Per Distance' (AMPD) visualization for a region (*region* 9) in cell line GM12878. The AMPD visualization shows, at each distance value (plotted on vertical axis), the K-mer pair that contributes the most in predicting a locus as positive and negative. The weights of these K-mer pairs (fetched from the SVM weight vector) are plotted on the horizontal axis. In the visualization, the K-mer pairs at even and odd distance values are segregated from each other to improve legibility. In the left panel, one sees 6-mers consisting of the 3-mer pairs separated by '|' (see Fig. 3), and in the right panel are 10-mers consisting of the 5-mer pairs. Owing to the high dimensionality of the 5-mer case, we observe that the magnitudes of the weights quickly shrink in this case. We filter this information further and visualize only the top few high-scoring features in the 'TopN' visualization shown in Fig. 4. At any distance value, all motifs that exceeded the threshold (shown as an inner dashed circle) are collected along with their weight magnitudes and stacked one over the other to finally represent them with a consensus motif (refer to "Visualizing the important features for each prediction model" section for more details). These consensus motifs are visualized radially.

Across various regions, among many motifs, tandem repeat sequences are prominently observed, especially di- and trinucleotide repeats, at various distances. Our 'AMPD' visualizations facilitate spotting of patterns spread over distances while the 'TopN' visualizations, due to the consensus motifs, can help spot possibly hidden shorter K-mer signals. Refer to Fig. 3 for the following discussion. The dinucleotide pattern 'GT' being repeated is observed in both cases, 3-mers and 5-mers, for distances up to 26 and 34 respectively, to have a maximal contribution among the various K-mer pairs towards predicting a locus as a potential interacting partner of locus chr21:34819525-34821921 (*region* 9) in GM12878. The 3-mer case shows patterns prominently containing more 'T's, from distance ∼30-60 as compared to the smaller distance values, among negatively contributing pairs, while the maximal positive contributors are devoid of them. Various such patterns are observed for different regions across cell lines.

Our literature search revealed some relevant studies on tandem repeat sequences and their potential biological roles. A 1990 review by Vogt [20] provides a very comprehensive and extensive account of

Table 1 Locus information for regions and prediction performances

GM12878

R	TCR	#TP	#NP	Test AUC A	B	C	D
0	chr7:115847372-115857098	63	226	0.7417	0.7538	0.8979	0.9042
1	chr7:115890993-115892266	56	234	0.7141	0.7341	0.8876	0.8960
2	chr7:115861595-115870968	52	252	0.7346	0.7763	0.9152	0.9376
3	chr5:131722317-131724751	39	91	0.6122	0.6547	0.8666	0.8286
4	chr5:131892428-131895867	34	80	0.5971	0.6343	0.8889	0.8543
5	chr7:90224881-90229046	34	122	0.8078	0.8307	0.9221	0.9118
6	chr7:116434729-116454408	33	292	0.7785	0.7787	0.7308	0.7036
7	chr7:90337078-90341001	32	158	0.8163	0.8275	0.9286	0.9324
8	chr22:32162110-32166713	31	127	0.7779	0.7832	0.7789	0.7738
9	chr21:34819525-34821921	30	201	0.6704	0.6694	0.7157	0.6901

K562

R	TCR	#TP	#NP	Test AUC A	B	C	D
0	chr22:32764253-32784733	46	105	0.8163	0.8121	0.9308	0.9382
1	chr22:32920308-32927723	45	109	0.6808	0.7242	0.7744	0.7972
2	chr22:32012966-32043914	42	104	0.7145	0.7324	0.8378	0.8599
3	chr21:35242603-35256847	39	150	0.7321	0.725	0.7251	0.7407
4	chr7:115847372-115857098	37	238	0.7521	0.7756	0.7765	0.7908
5	chr7:89787744-89795672	35	118	0.8546	0.8648	0.8566	0.8727
6	chrX:153625659-153635385	34	46	0.8501	0.8495	0.8044	0.8184
7	chr22:32170492-32188129	32	97	0.7456	0.7146	0.8003	0.8228
8	chr22:32740683-32750950	32	112	0.7167	0.7582	0.8836	0.9166
9	chr11:5721056-5732713	31	85	0.671	0.76	0.7345	0.7545

HeLa-S3

R	TCR	#TP	#NP	Test AUC A	B	C	D
0	chr7:115847372-115857098	98	207	0.6914	0.7111	0.8007	0.8228
1	chr7:116434729-116454408	71	211	0.73	0.7674	0.8573	0.8738
2	chr22:32920308-32927723	53	109	0.644	0.6369	0.7338	0.7091
3	chr7:115890993-115892266	50	243	0.6817	0.7225	0.907	0.9162
4	chr7:89787744-89795672	49	108	0.8108	0.8007	0.8005	0.8084
5	chr7:115861595-115870968	40	284	0.6624	0.732	0.8964	0.9114
6	chr22:32170492-32188129	40	102	0.677	0.755	0.8245	0.8590
7	chr22:32053085-32061138	37	115	0.6018	0.6420	0.7886	0.7991
8	chr22:33262063-33266567	37	112	0.5634	0.6564	0.8449	0.8491
9	chr21:34750664-34761738	37	147	0.7194	0.7294	0.7053	0.7273

#TruePeaks (#TP) and #NonPeaks (#NP) for all the studied genomic regions (column 'R') for the three cell lines (GM12878, K562 and HeLa-S3). Columns marked 'A', 'B', 'C' and 'D' show the mean test AUC values with oligomer length 3 and 5 respectively for two settings: Individual tasks ('A' and 'B') and Multiple tasks ('C' and 'D'). Refer "Pipeline for predicting long-range chromatin interactions", "Prediction of long-range chromatin interactions is possible from the sequence alone using non-linear SVMs" and "Multitask learning (MTL) helps mitigate issue of having too few interacting partners per locus" sections for more information

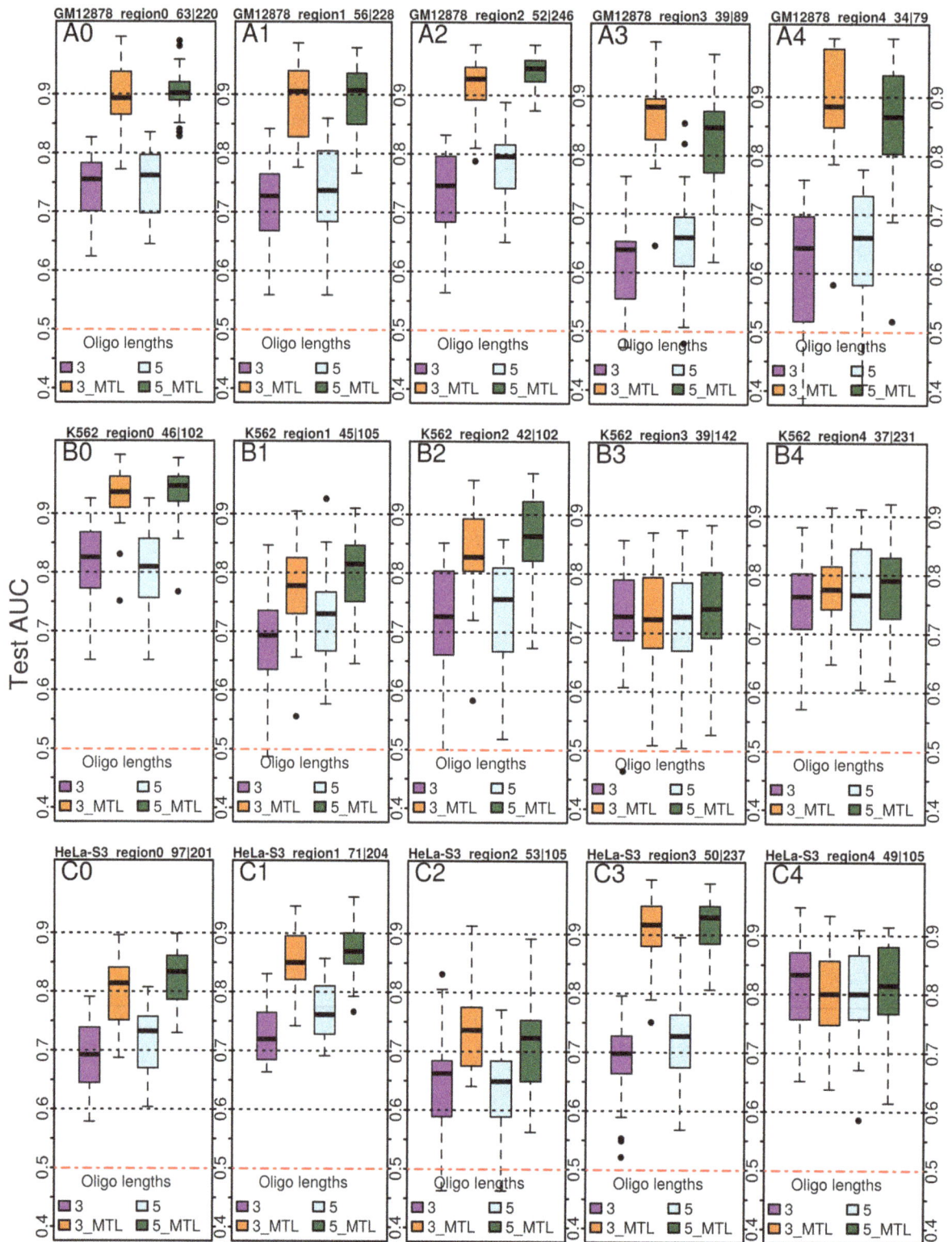

Fig. 2 *Box-plots* of SVC performances for cell lines GM12878, K562 and Hela-S3. Five *regions* (numbered 'A0-A4', 'B0-B4' and 'C0-C4' for GM12878, K562 and Hela-S3 respectively) out of 10 are shown. Individual tasks setting, oligomer lengths = {3,5} in *purple* and *light blue* respectively. MTL with 10 tasks, oligomer lengths = {3, 5} in *orange* and *green*. Distances between *K*-mer pairs upto $D = 100$. *Box-plots* for the other five regions among the 10 are given in the Supplementary Figures S3, S4 and S5 in Additional file 1

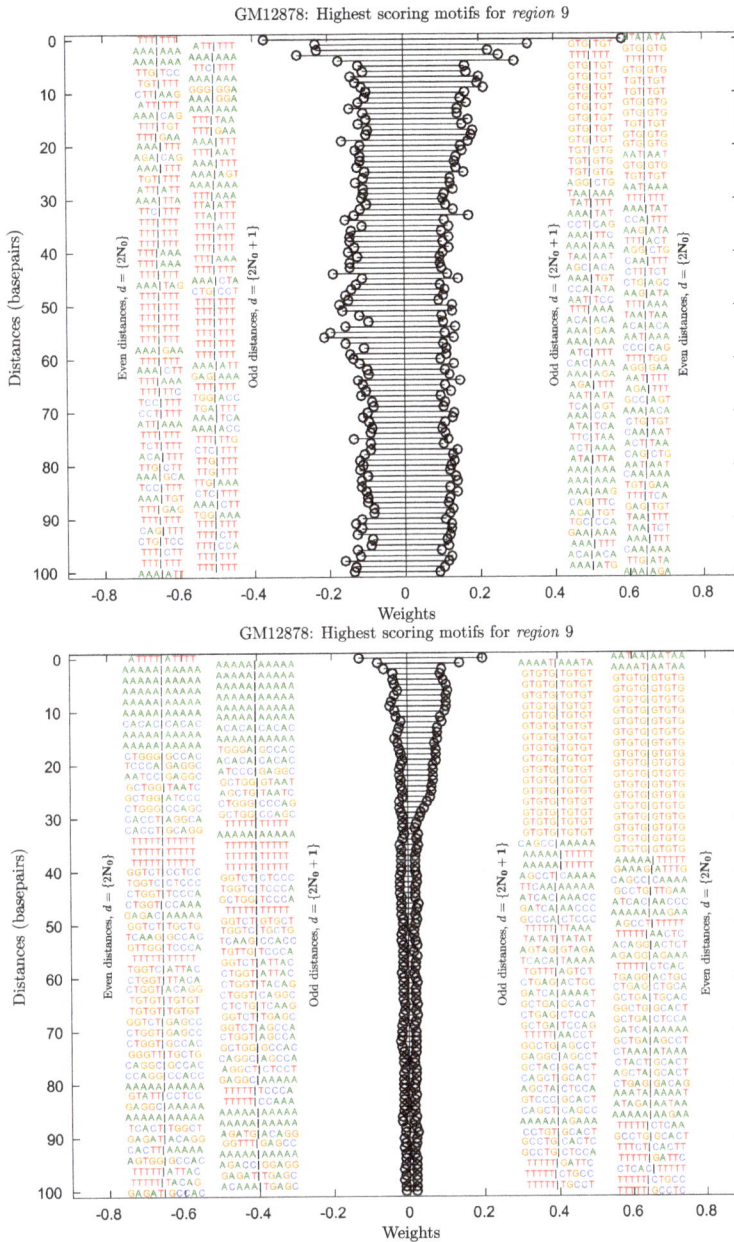

Fig. 3 'AMPD' visualization of the informative K-mer pairs from the predictor for *region* 9 in GM12878 (Refer Table 1 for *region* details). *Top*: At distances in $\{0, \ldots, 100\}$ basepairs, the 3-mer pair that maximally contributes towards positive and negative classification of a given locus is shown. Weights are shown on the *horizontal axis*, distances on the *vertical axis*. *Bottom*: 'AMPD' visualization for the 5-mer case

the potential functions of tandem repeat sequences in the human genome [20]. Among many other things, it includes an exhaustive discussion of the various repeat sequences, viz. mono-, di-, tri-, tetranucleotides and beyond, and the postulates of their association with a multitude of nuclear proteins that help them assume specific chromosomal structures. The author terms this ability of the tandem sequence repeat blocks to render locus-specific higher order structure and play a role in

organization as the 'chromatin folding code' [20]. In the review [20], the author also points to a specific case of the dinucleotide 'TG' as a simple repeating block, which has already been shown to have an enhancer function in vitro [21] in as early as 1984. More recently, a 2014 study [22] identified dinucleotide repeat motifs (DRMs) as general features that can render a nonfunctional sequence into an active enhancer element. Another comprehensive study of the simple sequence repeats in 2014 [23] suggests

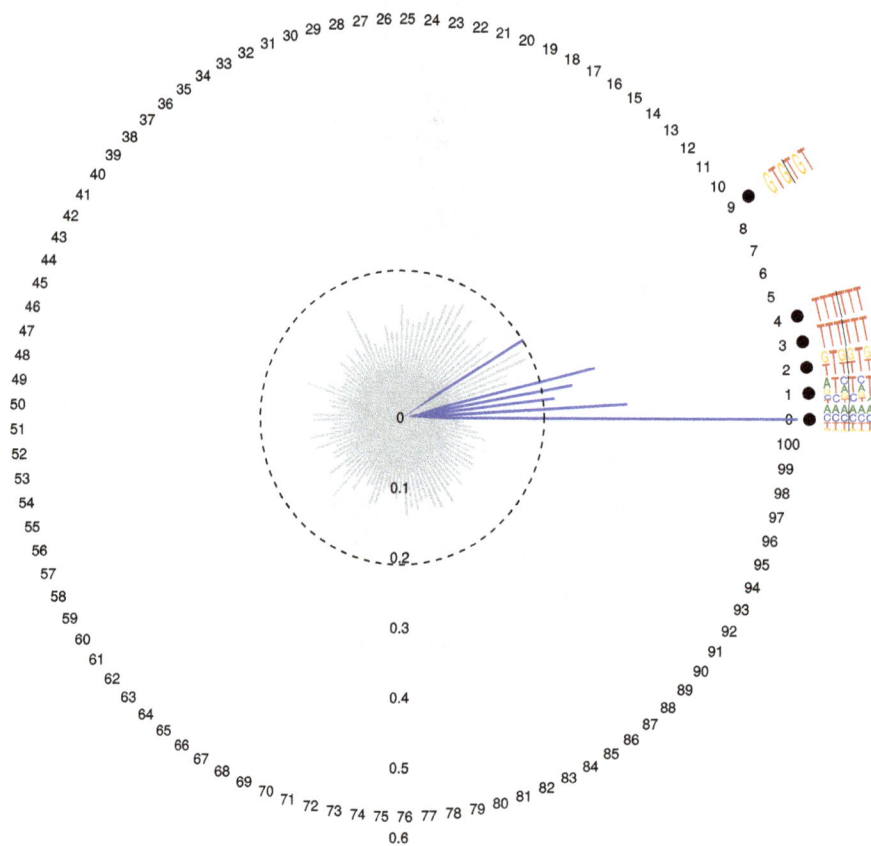

Fig. 4 'Top25' visualization of the informative 3-mer pairs separated by various distances and their magnitudes from the predictor for *region* 9 in GM12878 (Refer Table 1 for *region* details). Top-25 3-mer pairs, with weight magnitudes higher than the threshold (*dashed inner circle*), for the positive class (*blue*). The *dashed inner circle* is the threshold to select the top-25 entries of the averaged SVM weight vector

their potential role in genome regulation and organization. Variable number tandem repeats (VNTRs), as these sequence repeats are broadly termed, have already been implicated in many complex neurological disorders (e.g., Huntington disease [24]) and are generally known to be polymorphic [25].

With this backdrop, it is interesting that, enabled by the visualizations, our models using sequence-level information also reveal such tandem repeat motif signals (at times, even lengths of their tracts) as distinguishing characteristics between potential locus-specific interaction partners, suggesting a potentially important role of such sequence repeats in genome organization and regulation.

Identifying cell-line specific characteristic signals among (non-)interactors of the same locus in different cell lines

As discussed in "Prediction of long-range chromatin interactions is possible from the sequence alone using non-linear SVMs" section, an advantage of studying locus-specific interactions at the sequence-level is realized when our models can reveal the characteristic signals among interaction partners of the same locus in two different cell lines. Consider the locus `chr22:32170492-32188129` which is, both, *region* 6 and *region* 7 among our models for HeLa-S3 and K562 respectively (see Table 1). Refer to their 'AMPD' visualizations with 3-mers in Fig. 5. For K562, the 'CA' dinucleotide repeat sequence stretch of length ~20 markedly denotes a non-interacting partner while this same repeat sequence seems to be interrupted with a short stretch of 'T's in HeLa-S3. Also, another repeat sequence, 'AGA', is notable beyond distance values 50 among the non-interacting partners for this locus in K562 as compared to HeLa-S3 where it is only intermittently observed. These signals are, similarly, also picked up by our 5-mer models. The 3-mer and 5-mer 'AMPD' visualizations for *region* 7 in cell line K562 and *region* 6 in HeLa-S3 are given in Supplementary Figures S9 and S12 respectively in Additional file 1. The corresponding 'Top25' visualizations for these regions are given in Supplementary Figures S10, S11, S13 and S14 in Additional file 1.

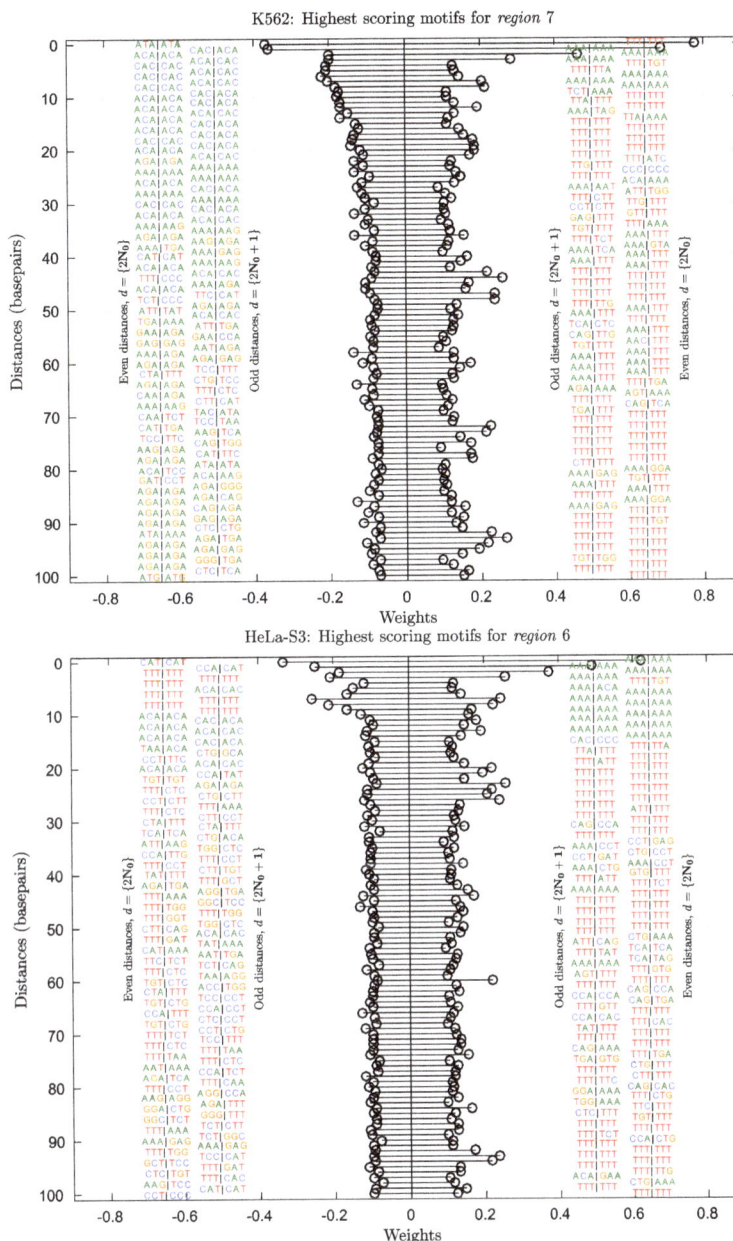

Fig. 5 'AMPD' visualization of the informative 3-mer pairs from the classifiers for locus `chr22:32170492-32188129` which is, both, *region* 7 in K562 and *region* 6 in HeLa-S3 (Refer Table 1 for *region* details). *Top panel*: At distances in (0-100) basepairs, the 3-mer pair that maximally contributes towards positive and negative classification of a given locus is shown. Weights are shown on the horizontal axis, distances on the vertical axis. *Bottom panel*: 3-mer 'AMPD' visualization of the same locus in HeLa-S3

Multitask learning (MTL) helps mitigate issue of having too few interacting partners per locus

Each locus-specific prediction problem in our scenario is termed as a *task* in the MTL setting. The small sample sizes in the single-task setting can be mitigated with the help of the so-called 'multitask' setting (see "Methodical details" for more details). In order to evaluate the efficacy of MTL for this problem, we used the available 10 individual tasks. Here, to compute the task similarity,

we used the 'model-defining' locus (the LoI) information. The locus sequence of every 'model-defining' region was represented as an ODH feature vector using the K-mer values 3 and 5, separately, and maximum distance 100. The similarities between these regions, the *tasks*, were given by the resulting dot products. For models that used oligomer length 3 and 5 representations for the sample sequences, we used the corresponding task similarities also with oligomer length 3 and 5 respectively. The mean test AUC values for the multitask setting with 10 tasks are shown in columns marked 'C' and 'D' (oligomer length 3 and 5 respectively) of Table 1. Mean performance increase across all regions: Oligomer length 3: {GM12878, K562, HeLa-S3}: {0.13, 0.06, 0.13}; Oligomer length 5: {GM12878, K562, HeLa-S3}: {0.09, 0.06, 0.11}. Their box plots are shown in Fig. 2 and Supplementary Figures S3, S4 and S5 (Additional file 1). Performances in the MTL setting mostly show reduced variance as compared to the single-task performances.

Thus, our pipeline in the MTL setting can (a) mitigate the issue of having either too few interacting partners per locus, or (b) in the extreme case, identify putative interaction partners of a locus not profiled in the 5C experiment provided that at least some regions from the same cell line have been profiled in a chromatin interaction experiment, for example, 4C or 5C.

Computational validation with high-resolution Hi-C

Rao et al. performed Hi-C experiments resulting in contact matrices at very high-resolution e.g. 1k, 5k, 10k, 25k base pairs (bps), etc. for various cell lines including GM12878, K562 [10]. Corresponding to the 'model-defining' *regions*, we picked relevant columns from the 5k Hi-C *cis*-contact matrix of the relevant chromosome. For example, if the 'model-defining' genomic *region* was 12,000 bps long, we collected candidate regions (across the rows) corresponding to three column loci. The candidate regions are those which have a non-zero KR-normalized [26] interaction frequency with the LoI. After normalizing, to identify significantly interacting partners at any given resolution, we computed their observed/expected (O/E) values and used an ad-hoc cut-off of 2.5 (i.e., a locus with a normalized O/E value ≥ 2.5 was considered significantly interacting with the LoI), as used earlier in [9]. This criterion is made more stringent as follows. The final set of loci that are considered sig-

nificantly interacting with any individual 'model-defining' *region* are only those that are significant at 5k resolution and also at 10k or 25k resolutions (all using the same cut-off). In other words, if a locus was deemed significant only at 5k resolution but not at 10 or 25k, then we did not consider it a true positive.

These *cis*-interacting genomic loci from the high-resolution contact maps are treated as unseen test sequences for the classifiers built for each *region* using the 5C data. In the pipeline, these are thus treated similarly to the 20% hold-out set: their ODH feature representations are fed to the classifier to predict their labels. We performed this experiment for cell lines GM12878 and K562.

When evaluating performances of our models regarding predictions on unseen loci from Hi-C data, we did so for two scenarios: (a) all chromosome-wide loci together; and (b) considering only loci lying beyond 1M bps from the 'model-defining' locus, i.e., excluding the regions probed in the 5C experiment [18] for the evaluation. Using this stringent criterion, the mean AUC values and their standard deviations are as follows. For prediction with oligomer length 3 models (a) chromosome-wide interaction partners: {GM12878, K562} : {0.5358±0.025, 0.5122± 0.084}; (b) interaction partners beyond 1M bps: {0.5327 ± 0.019, 0.5304±0.057}. And, with oligomer length 5 models (a) chromosome-wide interaction partners: {GM12878, K562} : {0.5278 ± 0.028, 0.5238 ± 0.081}; (b) interaction partners beyond 1M bps: {0.5220 ± 0.026, 0.5294 ± 0.064}. For both cell lines, when considering only the first five regions, the average performance was ~0.55 test AUC (see Table 2). Models for K562 show higher variance than models for GM12878.

We observed that performances of models predicting interaction partners for some LoI are comparatively poorer than those of other models. These 'model-defining' LoI either have very few negative samples to

Table 2 Computational validation with high-resolution Hi-C data

Cell-type	Oligomer length 3 (mean±s.d.)	Oligomer length 5 (mean±s.d.)
Chromosome-wide interaction partners		
GM12878 (regions 0-4)	0.5552 ± 0.009	0.5503 ± 0.006
GM12878 (regions 0-9)	0.5358 ± 0.025	0.5279 ± 0.028
K562 (regions 0-4)	0.5508 ± 0.091	0.5650 ± 0.088
K562 (regions 0-9)	0.5122 ± 0.084	0.5239 ± 0.081
Interaction partners beyond 1M bp		
GM12878 (regions 0-4)	0.5468 ± 0.005	0.5419 ± 0.007
GM12878 (regions 0-9)	0.5327 ± 0.019	0.5220 ± 0.026
K562 (regions 0-4)	0.5593 ± 0.062	0.5646 ± 0.064
K562 (regions 0-9)	0.5304 ± 0.058	0.5294 ± 0.064

(s.d.: standard deviation)

learn from (refer to Table 1) or are themselves rather long loci (refer to column 'length (bp)' in Supplementary Table S1 in Additional file 1). In general, from the perspective of training on 5C data and predicting contacts chromosome-wide, the issues of having few negative samples to learn from and having a rather long model-defining *region* (both, in 5C data) make the problem harder. This could be due to following reasons: (a) the experiments give no information on the potential causal portion(s)(causal for the said interaction), if any, along the complete restriction fragment; (b) the interacting as well as non-interacting partners of a rather long 'model-defining' locus could have varying characteristics in them which may not be comprehensively captured by the available few samples in the 5C data; and (c) the 5C experiments are performed on selected promoter regions and distal enhancers [18] while we make these models trained on such restricted 5C data to predict a potential interaction partner anywhere on the genome not just promoter or distal enhancer regions. Thus, learning on 5C data for a very small subset of the chromosome and then predicting interactions chromosome-wide is a very hard problem (see for example [14], and "Related work" section).

Related work
Recently, Roy et al. [14] developed a model for predicting cell-line specific interactions between only enhancers and promoters using various regulatory genomic datasets. Their predictive model learns from interacting and non-interacting pairs, also from 5C data [18], where the participating promoter and enhancer (of a contact-pair) are encoded as a real or binary vector marking information from 23 datasets including histone marks and transcription factor binding for various cell lines. Additionally, they also attempt at building a minimal classifier that uses information from 11 datasets out of the 23. They achieved a performance (area under precision-recall curve (auPRC)) of ~0.75-0.78 when training and predicting on the same experiment (5C) data. They also performed tasks of training on 5C data [18] and predicting interactions in high-resolution Hi-C data [10]. For this task, they consider an interaction involving a 5k bps locus pair as a true interaction if it is called a peak in any one of the three resolutions 5k, 10k and 25k, and achieved comparatively modest performances (auPRCs) of 0.643 (K562) and 0.687 (GM12878).

In comparison to the literature for prediction of promoter-enhancer interactions, we have used the term long-range chromatin interactions in a broader sense that includes possible interactions between intervening chromatin regions in addition to those (significant looping interactions) between specific genomic (functional) elements such as the enhancers and promoters. We hypothesize that the intervening chromatin could play an important role in maintaining a favorable landscape for the loci to interact, as also observed in more recent capture-C experiments data [27], where there is a possibility of weaker interactions due to putative low-affinity binding sites (e.g., [28]) which, in general, have been largely unexplored still. In our work we have focused on characterizing the long-range chromatin interactions pertaining to a particular genomic locus and investigating the capability of genomic sequence alone in characterizing them. Also, for the task of learning on 5C data and predicting on high-resolution Hi-C data, we have used a comparatively more stringent criterion for considering an interaction a true one. Approaches that use various additional information sources, e.g., epigenetic information [14], typically leave out genomic regions for which these are not available. Our sequence-based approach can be especially helpful in such scenarios. Furthermore, we expect that our models can be further strengthened or supported by utilizing the additional regulatory (epi)genomic information wherever available.

Conclusion
To the best of our knowledge, from the point of view of understanding chromatin interactions at the sequence level, ours is the first approach to do so. In this study, we have taken a broader view of these interactions and based on the hypothesis that the sequence at the intervening chromatin and the loci could also play a part in these interactions given the possibility of such 'interfacing' taking place via various mechanisms, like direct contact or formation of mini-loops or via diffusion after mere juxtaposing in physical vicinity [2], and for various reasons as motivated in the "Background" section. Our computational experiments using data from 5C experiments, for three cell lines GM12878, K562 and HeLa-S3 from [18] achieve good performances of ~0.75 (with oligomer length 5, as average test AUC values across various regions evaluated in this study from the three cell lines) in the single-task setting.

We developed two new, intuitive visualization methods that are suited for our problem scenario namely dealing with varied-length sequences and an appropriately chosen ODH feature representation. Aided by these visualizations, notwithstanding the very high-dimensionality of the feature space (e.g., the 5-mer case), our per-locus models shed light on the potential sequence signals that can characterize the interacting vs. the non-interacting partners of a LoI. We discussed how this can help understand which sequence features in the given region made it interact with one LoI and not with another LoI. Analysis of the various sequence signals from our models suggests a potential functional and organizational role for tandem repeat sequence stretches in the genome.

We also demonstrated how knowledge of individual models could be transferred to those of other regions (those having too few examples to learn from) via multitask learning. Mean performance for the multitask setting, performances of models for oligomer length 3 and 5 combined together, is 0.83. We already observed that several models show less variance in their prediction performances than their single-task counterparts.

Furthermore, we made our models trained on 5C data predict interactions between 5k bps long loci from the recent high-resolution Hi-C [10] data for cell lines where the Hi-C data was available. Even with a very stringent criterion to identify true positives in the high-resolution Hi-C data, we showed that our approach is capable of predicting interesting loci that could interact although lying very far away, even further than 1–2M bps, on the genome using features learned from 5C data that is limited to this 1–2M bps distance. This ability to identify potentially interacting loci lying very far away on the genome could be useful from the point of view of understanding topologically associating domains at the sequence level.

An important point to note here is that since our models do not require any locus to be either a TCR or an enhancer region per se, in principle, it can be seamlessly applied to contact matrices output by any 5C-based or even high resolution Hi-C-based experiments (as training data). At places, we have used the terms TCR and enhancers for the interacting regions because the contact matrices we use in this study come from 5C experiments involving these loci. So, when given a Hi-C contact matrix, any locus therein could be used to learn corresponding models in a similar fashion and it need not necessarily be an enhancer or a promoter region. In comparison, earlier approaches focus only on promoter-enhancer interactions and exclude all other genomic loci from their analysis. Thus, we have preferred to call these genomic loci as simply *regions* in this study. The models in this work are not specific to particular properties of any genomic region and do not make use of supplementary epigenetic information at the locus; we have only used the sequence information. Even with this much harder premise, we still achieved a good performance of ~0.75.

As of today, high resolution Hi-C data is still very expensive. Therefore, our prediction method could also be used in a setting where high-resolution 5C data, but only low-resolution Hi-C data is available to predict additional interaction partners for any regions of interest. These additional predicted contacts could augment methods for predicting the 3D structure of the chromatin as well as methods for predicting boundaries of TADs. Thus, we envisage that our approach of using only sequence-based models can, most importantly, be helpful in (a) understanding higher-order structures like (meta-) TADs at the sequence-level; and (b) giving additional input to methods that estimate the 3D structure of the chromatin for different organisms from the interaction data.

Methods
Materials
We use the 5C contact matrices from experiments published by Sanyal et al. [18]. They probed a collection of regions for two tier-I cell lines (GM12878 and K562) and a tier-II cell line (HeLa-S3) from ENCODE (The ENCODE Project Consortium, 2012). In these experiments involving two biological replicates, for each replicate, upon filtering to exclude certain primers owing to outlier fragments, the contact frequencies are normalized for the trans signal in turn correcting for detection biases per restriction fragment [18]. The intra-chromosomally interacting restriction fragments are then tested for significance, accounting for the inverse relationship between contact frequencies and the genomic distance between the restriction fragments, and peaks are called, conservatively, at a false discovery rate (FDR) cutoff of 1%. [18] term the interactions that are called peaks in both replicates as 'TruePeaks' and those not called peaks in either replicate as 'NonPeaks'. Consequently, in our study, positive examples for any classifier are 'TruePeaks' and negative examples, 'NonPeaks'. We considered different FDR cutoff values (1%, 10% and 15%) and selected an FDR cutoff of 10% (discussed in "Relaxation of FDR cutoff to enable studying of putative 'bystander' or structural interactions" section). Table 1 gives information on the number of 'TruePeaks' (#TP) and the number of 'NonPeaks' (#NP) for the genomic regions for which we built our models in this study to evaluate whether the DNA sequence is informative in predicting the long-range interactions (Refer to Supplementary Tables S1–S4 for additional details about the studied genomic regions). These are the 'model-defining' *regions* for our study. All genomic coordinates are w.r.t. hg19, GRCh37 assembly. The 'model-defining' loci are among the TSS-containing regions (by GENCODE v7 [29]) and the sets of loci in the positive and negative class for the individual classifiers are restriction fragments corresponding to enhancers (also by GENCODE v7 [29]) [18]. All values of #TruePeaks and #NonPeaks in Table 1 are for FDR 10%. For the computational validation with high-resolution Hi-C data, we used the data from Rao et al. [10] deposited at GEO [30], namely 'GSE63525_GM12878_combined_contact_matrices.tar.gz' and 'GSE63525_K562_intrachromosomal_contact_matrices.tar.gz'.

Relaxation of FDR cutoff to enable studying of putative 'bystander' or structural interactions

From a biological point of view, we attempted to take a more broader view and defined an interaction that takes into account not just the significant 'looping interactions' but also the possibility of so-called 'bystander' or structural interactions between intervening chromatin [18, 27]. Thus, in all computational experiments, in order to distinguish significant interactions from non-interactions in the 5C data, we relaxed the FDR cutoff to 10%, instead of 1% as in [18]. In other words, we traded off between being very conservative (which would allow only significant 'looping interactions' as prevalently defined in the community) and comparatively liberal in considering TruePeaks at FDR cutoff 10%. At the same time, this relaxation still maintained a significantly higher mean z-score of the interactions for TruePeaks in comparison to NonPeaks for all the cell lines, similar to the 1% cutoff case (see Supplementary Figure S1 in Additional file 1). While, 15% FDR also shows a significant difference, it did not provide much benefit in the number of additional TruePeaks per region (i.e., positive examples per classification problem in our study) in comparison to relaxing the FDR from 1 to 10%, consistently across all three cell lines.

Methodical details

The genomic loci we study in this work are the restriction fragments reported in the 5C experiments in [18] (see "Materials" section for details). We use string kernels, which provide a measure of similarity between sequences, in conjunction with an SVM as a classifier. Because these loci have highly diverse lengths (Supplementary Figure S2 in Additional file 1), we could not directly use position-aware string kernels like the oligo kernel [31] or weighted degree (WD) kernels [32, 33] for representing the loci.

A feature representation based on oligomer distance histograms (ODH) and the ODH kernel

In 2006 Lingner and Meinicke introduced the ODH feature representation and the corresponding ODH kernel [19]. It provides a fixed-length feature space representation of any arbitrary length sequence based on histograms of distances between short oligomers in the sequence. For alphabet $\sum$, consider all oligomers (or interchangeably, K-mers) $m_i \in \sum^K, i = 1, \ldots, M$. For any sequence s of length $|s| := L_{\max}$, let $D = L_{\max} - K$, the maximum distance between any two K-mers, with distance between a pair of K-mers defined as the difference in their starting positions in the sequence s. The distance histogram vector of s corresponding to the K-mer pair (i, j) is given by $\mathbf{h}_{ij}(s) = [h_{ij}^0(s), h_{ij}^1(s), \ldots, h_{ij}^D(s)]^T$ where T denotes transpose. For all such K-mer pairs over $\sum$, the corresponding distance histogram vectors are concatenated together giving a complete feature space transformation $\Phi(s)$.

$$\Phi(s) = \left[\mathbf{h}_{11}^T(s), \mathbf{h}_{12}^T(s), \ldots, \mathbf{h}_{MM}^T(s) \right]^T \quad (1)$$

The set of feature vectors for N training samples is: $\mathbf{X} = [\Phi(s_1), \ldots, \Phi(s_N)]$ and the $N \times N$ kernel matrix is given by:

$$\mathbf{K} = \mathbf{X}^T \mathbf{X} \quad (2)$$

with k_{ij}, the entries of matrix $\mathbf{K}$, being proportional to the similarity between sequence s_i and s_j. Lingner and Meinicke used this kernel for remote homology detection in protein sequences [19].

Multitask learning (MTL)

Often, for various reasons across domains, one has to deal with the issue of having very few training samples for a given prediction problem also called task. This can affect the generalization ability of any standard machine learning technique such as an SVM [34]. When multiple related tasks are to be learnt, MTL attempts to mitigate this issue by sharing information across these multiple related tasks. From a different perspective, it can be advantageous to leverage information from multiple related tasks to improve the prediction performance of a single task [34]. Depending upon the problem at hand, a suitable measure of task-relatedness (how similar are two given tasks) needs to be chosen.

In case of learning with kernels, [35] introduced how multitask learning can be performed with kernel methods. Jacob and Vert [36] provided the following formulation for sharing of information between tasks using a multitask kernel. For any two samples s_A and s_B from tasks t_A and t_B respectively, $K_{MTL}((s_A, t_A), (s_B, t_B))$ is the multitask kernel providing a measure of similarity between these tuples. Mathematically, $K_{MTL}((s_A, t_A), (s_B, t_B)) = K_S(s_A, s_B) \cdot K_T(t_A, t_B)$ where K_S is the kernel on the samples, and K_T gives the kernel value between two tasks. Jacob and Vert [36] used this formulation for predicting peptide–MHC-I binding. An overview of MTL applications for problems in computational biology is presented by [37].

Pipeline for predicting long-range chromatin interactions

Contact matrix output by any experiment profiling chromatin interactions must be subjected to normalization and extraction of significant contacts. Details of the motivation and various approaches for doing so are reviewed Ay and Noble [38]. Also, these experiments are usually performed for multiple biological replicates to assess the impact of experimental errors and other variations.

Figure 1 illustrates our approach for predicting long-range chromatin interactions. The normalization and peak-calling procedures that we adopted for analyzing the 5C data used in this study are described in "Materials" section. Once a raw contact matrix has been normalized and the significant interactions have been called, we binarize the contact matrix as follows. Genomic loci (along the rows) not called significant interaction partners of a particular locus (along the columns) in either replicate constitute the negative class (see Fig. 1, cells denoted by filled black boxes) and those called significant in all replicates constitute the positive class (see Fig. 1, cells denoted by filled orange boxes). This leaves a lot of uncalled loci (along the rows). These are denoted by unfilled boxes (Fig. 1). Then, we build a classifier corresponding to each locus along the column of the matrix. We call these loci the 'model-defining' loci. For each individual classifier we collect loci along the rows and falling under the relevant column of the contact matrix as loci belonging to the positive and negative class for this classifier or it may not be called at all. This is shown in Fig. 1. Clearly, any locus that belongs to the positive class in one model, may belong to either the positive or negative class in another model. Given a set of sequences belonging to either class, 80% were used for training a classifier while 20% were held-out as test sequences.

The classifiers are based on SVMs with the ODH kernel. The cost parameter for each SVM is varied in the range $10^{\{-3,\ldots,3\}}$. For each model, we perform a 5-fold nested cross-validation to select the best performing SVM cost-value while the ODH feature representation parameters are fixed as described in "Prediction of long-range chromatin interactions is possible from the sequence alone using non-linear SVMs" section. Our pipeline also accounts for class-imbalance by proportionately up-weighting the misclassification cost for the minority class (here, positive class) [39].

Our pipeline, named 'Samarth', is available for download at the supplemental website http://bioinf.mpi-inf.mpg.de/publications/samarth/.

Visualizing the important features for each prediction model

Absolute Max Per Distance (AMPD) visualizations: Recall from "A feature representation based on oligomer distance histograms (ODH) and the ODH kernel" section that the dimensionality of the SVM weight vector for a model with the DNA sequence alphabet, using oligomer length K and distances up to D is $[(|\sum|^K)^2 \times (D + 1)]$ (i.e., of 413,696 and 105,906,176 dimensions for oligomer length 3 and 5 respectively). Due to the oligomer distance histograms-based feature vector representation used in our models, each entry of the SVM weight vector is the coefficient assigned to a K-mer pair separated by a distance $d \in [0, 1, \ldots, D]$. For each of our locus-specific models, the 5-fold outer cross validation resulted in 5 different SVM weight vectors. These five individual weight vectors were averaged to obtain one representative weight vector for a per-locus model. From this averaged weight vector, we noted two K-mer pairs per distance value, one that was assigned the most positive coefficient and the other, most negative. A positive coefficient means the d-separated K-mer pair is an important feature among the positive sequences, while a negative coefficient means it is an important feature to classify the sequence as negative. All such selected K-mers at the various distance values are visualized to provide a distance-centric view of the important features. Such a visualization for an example *region* (*region* 9) for cell line GM12878 is shown in Fig. 3. We call these visualizations 'Absolute Max Per Distance' (AMPD) visualizations. For better readability, the K-mer pairs at even distance values are arranged in the outer column and those at odd distance values in the inner column. Figure 3 and Supplementary Figure S6, S9 and S12 in the Additional file 1 show examples of 'AMPD' visualizations for different regions across the three cell lines GM12878, K562 and HeLa-S3.

Position-Wise Weight Matrix (PWWM)-based 'TopN' visualizations: Independently, the entries of the averaged

Table 3 A dummy PWWM for selected 3-mer pairs at certain distance d. $|w_1|$, $|w_2|$, and $|w_3|$ are magnitudes of the weights for the example 3-mer pairs

	3-mer pairs					
$|w_1|$	AAA				GAA	
$|w_2|$	GAA				AGA	
$|w_3|$	AAG				AAA	
'A'	$\frac{1}{D}(|w_1| + |w_3|)$	$\frac{1}{D}(|w_1| + |w_2| + |w_3|)$	$\frac{1}{D}(|w_1| + |w_2|)$	$\frac{1}{D}(|w_2| + |w_3|)$	$\frac{1}{D}(|w_1| + |w_3|)$	$\frac{1}{D}(|w_1| + |w_2| + |w_3|)$
'C'	0	0	0	0	0	0
'G'	$\frac{1}{D}(|w_2|)$	0	$\frac{1}{D}(|w_3|)$	$\frac{1}{D}(|w_1|)$	$\frac{1}{D}(|w_2|)$	0
'T'	0	0	0	0	0	0
p	1	2	3	4	5	6

'A', 'C', 'G' and 'T' are the rows corresponding to the nucleotides. Position, $p \in \{1,\ldots,6\}$. Each cell is divided by $D = (|w_1| + |w_2| + |w_3|)$

weight vector were sorted in descending order and then thresholded to reveal the top 25 scoring entries. Figure 4 visualizes only those selected top-25 K-mer pairs. Here, the $(D + 1)$ distances are arranged radially. Each spoke gives the magnitude of the highest-scoring K-mer pair at the corresponding distance. If the magnitude crosses the threshold value, that spoke is plotted in either 'blue' (see Fig. 4) or 'red' (see Additional file 1) for positive and negative contribution respectively, while otherwise plotted in gray. We call these visualizations 'Top25', or more generally, 'TopN' visualizations where one can choose a suitable value for 'N'. Note that there can be several entries at the same distance among the top-25 leading to sequence logo-like representations. At any distance d, all motifs that exceeded the threshold are collected along with their weight magnitudes and stacked one over the other to finally represent them with a consensus motif. This is done by constructing a 'Position-Wise Weight Matrix' (PWWM) of dimension ($|\sum| \times 2K$) which represents the nucleotides appearing at each position from 1 to $2K$ along with their relative contribution to the weight vector. A dummy example illustrating this is shown in Table 3. This PWWM is computed as follows. For position $p \in \{1, \ldots, 2K\}$, the matrix cell ('A' / 'C' / 'G' / 'T', p) is populated with the sum of the weight contribution of those motifs in which the given nucleotide is present at position p. The matrix is then normalized for the column entries to sum up to 1. The resulting consensus sequences are represented as sequence logos [40] in the 'Top25' visualizations in Fig. 4. Supplementary Figures S7, S8, S10, S11, S13 and S14 in the Additional file 1 show example 'Top25' visualizations for various *regions* from the cell lines GM12878, K562 and HeLa-S3.

Additional file

Additional file 1: This file provides additional performance plots and visualizations, and more detailed description of the data. Figure S1: Z-scores for various cell lines at different FDRs Figure S2: Lengths of restriction fragments for various regions in different cell lines Figure S3: Box-plots of SVC performances for all regions (numbered 'A0-A9') in GM12878 Figure S4: Box-plots of SVC performances for all regions (numbered 'B0-B9') in K562 Figure S5: Box-plots of SVC performances for all regions (numbered 'C0-C9') in HeLa-S3 Figure S6: 'AMPD' visualization of the informative K-mer pairs from the classifier for region 9 in GM12878 Figure S7: 'Top25' visualization of the informative 3-mer pairs separated by various distances and their magnitudes from the classifier for region 7 and 9 in GM12878 Figure S8: 'Top25' visualization of the informative 3-mer pairs separated by various distances and their magnitudes from the classifier for region 7 and 9 in GM12878 Figure S9: 'AMPD' visualization of the informative K-mer pairs from the classifier for region 7 in K562 Figure S10: 'Top25' visualization of the informative 3-mer pairs separated by various distances and their magnitudes from the classifier for region 7 in K562 Figure S11: 'Top25' visualization of the informative 3-mer pairs separated by various distances and their magnitudes from the classifier for region 7 in K562 Figure S12: 'AMPD' visualization of the informative K-mer pairs from the classifier for region 6 in HeLa Figure S13: 'Top25' visualization of the informative 3-mer pairs separated by various distances and their magnitudes from the classifier for region 6 in HeLa Figure S14: 'Top25'

visualization of the informative 3-mer pairs separated by various distances and their magnitudes from the classifier for region 6 in HeLa Table S1: Details of the genomic regions from each cell line Table S2: Overlap of candidate loci among regions for cell line GM12878 Table S3: Overlap of candidate loci among regions for cell line K562 Table S4: Overlap of candidate loci among regions for cell line HeLa.

Abbreviations
AMPD: Absolute max per distance; AUC: Area Under the receiver operating characteristic (ROC) Curve; auPRC: Area under precision-recall curve; bp: Base pairs; DRMs: Dinucleotide repeat motifs; FDR: False discovery rate; kb: kilobases; LoI: Locus of interest; MTL: Multitask learning; ODH: Oligomer distance histograms; PWWM: Position-wise weight matrix; SVM: Support vector machine; TCR: TSS-containing region; ToI: TCR of interest; TSS: Transcription start site; VNTRs: Variable number tandem repeats

Acknowledgements
The authors wish to acknowledge the anonymous reviewers for their comments and suggestions which helped improve the manuscript. The authors also wish to thank Thomas Lengauer for many helpful discussions during the study.

Funding
Not Applicable.

Authors' contributions
SN designed, implemented and performed the computational experiments, discussed and interpreted the model performances, and drafted the manuscript. NP conceived, designed and supervised the study, discussed and interpreted the model performances, and edited the manuscript. Both authors read and approved the final manuscript.

Competing interests
The authors declare that they have no competing interests.

Author details
[1]Computational Biology & Applied Algorithmics, Max Planck Institute for Informatics, Saarland Informatics Campus, Building E1.4, D-66123 Saarbruecken, Germany. [2]Present address: Department of Computer Science, University of Tübingen, Sand 14, D-72076 Tübingen, Germany.

References
1. Cope N, Fraser P, Eskiw C. The yin and yang of chromatin spatial organization. Genome Biol. 2010;11(3):204. doi:10.1186/gb-2010-11-3-204.
2. Bickmore WA. The spatial organization of the human genome. Annu Rev Genomics Hum Genet. 2013;14(1):67–84. doi:10.1146/annurev-genom-091212-153515. PMID: 23875797. http://dx.doi.org/10.1146/annurev-genom-091212-153515.
3. de Wit E, de Laat W. A decade of 3C technologies: insights into nuclear organization. Gene Dev. 2012;26(1):11–24. doi:10.1101/gad.179804.111.
4. Dekker J, et al. Capturing chromosome conformation. Science. 2002;295(5558):1306–11. doi:10.1126/science.1067799. http://www.sciencemag.org/content/295/5558/1306.full.pdf.
5. Simonis M, et al. Nuclear organization of active and inactive chromatin domains uncovered by chromosome conformation capture-on-chip (4C). Nat Genet. 2006;38(11):1348–54. doi:10.1038/ng1896.
6. Zhao Z, et al. Circular chromosome conformation capture (4C) uncovers extensive networks of epigenetically regulated intra- and interchromosomal interactions. Nat Genet. 2006;38(11):1341–7. doi:10.1038/ng1891.
7. Dostie J, et al. Chromosome conformation capture carbon copy (5C): A massively parallel solution for mapping interactions between genomic elements. Genome Res. 2006;16(10):1299–309. doi:10.1101/gr.5571506. http://genome.cshlp.org/content/16/10/1299.full.pdf+html.
8. Fullwood MJ, et al. An oestrogen-receptor-[agr]-bound human chromatin interactome. Nature. 2009;462(7269):58–64. doi:10.1038/nature08497.

9.   Lieberman-Aiden E, et al. Comprehensive mapping of long-range interactions reveals folding principles of the human genome. Science. 2009;326(5950):289–93. doi:10.1126/science.1181369. http://www.sciencemag.org/content/326/5950/289.full.pdf.

10. Rao SSP, et al. A 3d map of the human genome at kilobase resolution reveals principles of chromatin looping. Cell. 2014;159(7):1665–80. doi:10.1016/j.cell.2014.11.021.

11. Heidari N, et al. Genome-wide map of regulatory interactions in the human genome. Genome Res. 2014;24(12):1905–1917. doi:10.1101/gr.176586.114. http://genome.cshlp.org/content/24/12/1905.full.pdf+html.

12. Zeitz MJ, et al. Genomic interaction profiles in breast cancer reveal altered chromatin architecture. PLoS ONE. 2013;8(9):73974. doi:10.1371/journal.pone.0073974.

13. Dekker J, Marti-Renom MA, Mirny LA. Exploring the three-dimensional organization of genomes: interpreting chromatin interaction data. Nat Rev Genet. 2013;14(6):390–403. Review.

14. Roy S, et al. A predictive modeling approach for cell line-specific long-range regulatory interactions. Nucleic Acids Res. 2015. doi:10.1093/nar/gkv865. http://nar.oxfordjournals.org/content/early/2015/09/03/nar.gkv865.full.pdf+html.

15. Dixon JR, et al. Topological domains in mammalian genomes identified by analysis of chromatin interactions. Nature. 2012;485(7398):376–80. doi:10.1038/nature11082.

16. Varoquaux N, et al. A statistical approach for inferring the 3D structure of the genome. Bioinformatics (Oxford). 2014;30(12):26–33. doi:10.1093/bioinformatics/btu268.

17. Boser BE, Guyon IM, Vapnik VN. A training algorithm for optimal margin classifiers. In: Proceedings of the Fifth Annual Workshop on Computational Learning Theory. COLT '92. New York: ACM; 1992. p. 144–52. doi:10.1145/130385.130401. http://doi.acm.org/10.1145/130385.130401.

18. Sanyal A, et al. The long-range interaction landscape of gene promoters. Nature. 2012;489(7414):109–13. doi:10.1038/nature11279.

19. Lingner T, Meinicke P. Remote homology detection based on oligomer distances. Bioinformatics (Oxford). 2006;22(18):2224–31. doi:10.1093/bioinformatics/btl376. Accessed 24 May 2011.

20. Vogt P. Potential genetic functions of tandem repeated dna sequence blocks in the human genome are based on a highly conserved "chromatin folding code". Hum Genet. 1990;84(4):301–36. doi:10.1007/bf00196228.

21. Hamada H, et al. Characterization of genomic poly(dt-dg).poly(dc-da) sequences: structure, organization, and conformation. Mol Cell Biol. 1984;4(12):2610–21. 6098814.

22. Yáñez-Cuna JO, et al. Dissection of thousands of cell type-specific enhancers identifies dinucleotide repeat motifs as general enhancer features. Genome Res. 2014;24(7):1147–56. doi:10.1101/gr.169243.113. http://genome.cshlp.org/content/24/7/1147.full.pdf+html.

23. Ramamoorthy S, et al. Length and sequence dependent accumulation of simple sequence repeats in vertebrates: Potential role in genome organization and regulation. Gene. 2014;551(2):167–75. doi:10.1016/j.gene.2014.08.052.

24. Malaspina A, et al. A survey of trinucleotide/tandem repeat-containing transcripts (tnrts) isolated from human spinal cord to identify genes containing unstable {DNA} regions as candidates for disorders of motor function. Brain Res Bull. 2001;56(3-4):299–306. doi:10.1016/S0361-9230(01)00597-4. Triplet Repeat Diseases.

25. Brookes KJ. The {VNTR} in complex disorders: The forgotten polymorphisms? a functional way forward?. Genomics. 2013;101(5):273–81. doi:10.1016/j.ygeno.2013.03.003.

26. Knight PA, Ruiz D. A fast algorithm for matrix balancing. IMA J Numer Anal. 2012. doi:10.1093/imanum/drs019. http://imajna.oxfordjournals.org/content/early/2012/10/26/imanum.drs019.full.pdf+html.

27. Hughes JR, et al. Analysis of hundreds of cis-regulatory landscapes at high resolution in a single, high-throughput experiment. Nat Genet. 2014;46(2):205–12. Technical Report.

28. Tanay A. Extensive low-affinity transcriptional interactions in the yeast genome. Genome Res. 2006;16(8):962–72. doi:10.1101/gr.5113606. http://genome.cshlp.org/content/16/8/962.full.pdf+html.

29. Harrow J, et al. Gencode: The reference human genome annotation for the encode project. Genome Res. 2012;22(9):1760–74. doi:10.1101/gr.135350.111. http://genome.cshlp.org/content/22/9/1760.full.pdf+html.

30. Edgar R, Domrachev M, Lash AE. Gene expression omnibus: Ncbi gene expression and hybridization array data repository. Nucleic Acids Res. 2002;30(1):207–10. doi:10.1093/nar/30.1.207. http://nar.oxfordjournals.org/content/30/1/207.full.pdf+html.

31. Meinicke P, et al. Oligo kernels for datamining on biological sequences: a case study on prokaryotic translation initiation sites. BMC Bioinforma. 2004;5(1):169. doi:10.1186/1471-2105-5-169.

32. Rätsch G, Sonnenburg S. Accurate splice site prediction for caenorhabditis elegans. In: Kernel Methods in Computational Biology. MIT Press series on Computational Molecular Biology. Cambridge: MIT Press; 2004. p. 277–98.

33. Rätsch G, et al. Rase: recognition of alternatively spliced exons in c.elegans. Bioinformatics. 2005;21(suppl 1):369–77. doi:10.1093/bioinformatics/bti1053. http://bioinformatics.oxfordjournals.org/content/21/suppl_1/i369.full.pdf+html.

34. Evgeniou T, Pontil M. Regularized multi–task learning. In: Proceedings of the Tenth ACM SIGKDD International Conference on Knowledge Discovery and Data Mining. KDD '04. New York: ACM; 2004. p. 109–17. doi:10.1145/1014052.1014067. http://doi.acm.org/10.1145/1014052.1014067.

35. Evgeniou T, et al. Learning multiple tasks with kernel methods. J Mach Learn Res. 2005;6:615–37.

36. Jacob L, Vert JP. Efficient peptide—mhc-i binding prediction for alleles with few known binders. Bioinformatics. 2008;24(3):358–66. doi:10.1093/bioinformatics/btm611. http://bioinformatics.oxfordjournals.org/content/24/3/358.full.pdf+html.

37. Widmer C, Rätsch G. Multitask learning in computational biology. JMLR W&CP. ICML 2011 Unsupervised and Transfer Learning Workshop. 2012;27:207–16.

38. Ay F, Noble W. Analysis methods for studying the 3d architecture of the genome. Genome Biol. 2015;16(1):183. doi:10.1186/s13059-015-0745-7.

39. Elkan C. The foundations of cost-sensitive learning. In: Proceedings of the 17th International Joint Conference on Artificial Intelligence - Volume 2. IJCAI'01. San Francisco: Morgan Kaufmann Publishers Inc.; 2001. p. 973–8. http://dl.acm.org/citation.cfm?id=1642194.1642224.

40. Schneider TD, Stephens RM. Sequence logos: a new way to display consensus sequences. Nucleic Acids Res. 1990;18:6097–100.

11

NaviGO: interactive tool for visualization and functional similarity and coherence analysis with gene ontology

Qing Wei[1], Ishita K. Khan[1], Ziyun Ding[2], Satwica Yerneni[3] and Daisuke Kihara[2,1]* (iD)

Abstract

Background: The number of genomics and proteomics experiments is growing rapidly, producing an ever-increasing amount of data that are awaiting functional interpretation. A number of function prediction algorithms were developed and improved to enable fast and automatic function annotation. With the well-defined structure and manual curation, Gene Ontology (GO) is the most frequently used vocabulary for representing gene functions. To understand relationship and similarity between GO annotations of genes, it is important to have a convenient pipeline that quantifies and visualizes the GO function analyses in a systematic fashion.

Results: NaviGO is a web-based tool for interactive visualization, retrieval, and computation of functional similarity and associations of GO terms and genes. Similarity of GO terms and gene functions is quantified with six different scores including protein-protein interaction and context based association scores we have developed in our previous works. Interactive navigation of the GO function space provides intuitive and effective real-time visualization of functional groupings of GO terms and genes as well as statistical analysis of enriched functions.

Conclusions: We developed NaviGO, which visualizes and analyses functional similarity and associations of GO terms and genes. The NaviGO webserver is freely available at: http://kiharalab.org/web/navigo.

Keywords: Gene function, Gene ontology, GO, Ontology, GO directed acyclic graph, Function similarity, Gene function prediction, GO annotation, function enrichment analysis, GO parental terms, GO association score

Background

Functional elucidation of genes is one of the central problems in modern biology including bioinformatics. For systematic function annotation, GO is widely used as the vocabulary of gene functions [1]. GO terms are arranged in a hierarchical directed acyclic graph (DAG), where parental relationships between terms are represented. GO is updated periodically by the Gene Ontology Consortium [2], and currently holds over 44,000 terms. The DAG structure is divided into three different GO categories (three disconnected roots), namely, Biological Process (BP), Molecular Function (MF), and Cellular Component (CC). The large volume of the vocabulary and

their parental relationship make it non-trivial to provide an intuitive summary of GO annotations of genes.

AmiGO [3] is an online tool maintained by the Gene Ontology Consortium [2] that is widely used to search and browse the gene ontology database. Apart from this, there are other existing tools [4–6] that can be used for GO visualization and comparison. QuickGO, a tool that is developed under the Gene Ontology Annotation (GOA) project, allows searches of GO terms and genes with a specified GO, and provides static clickable maps [7]. A drawback of the existing works provide a static way of visualizing GO DAG topology either in a static image or in the SVG format. Once the topologies are generated in the frontend of these servers, users are unable to explore different branches of the GO hierarchy interactively in real-time. Simple tasks of GO terms, for example, listing all the parental terms from query GO terms, or mapping GO terms and visualize them on the GO DAG, are not trivial in the existing web-based tools.

* Correspondence: dkihara@purdue.edu
[2]Department of Biological Science, Purdue University, West Lafayette, IN 47907, USA
[1]Department of Computer Science, Purdue University, West Lafayette, IN 47907, USA
Full list of author information is available at the end of the article

Browsing parental GO terms, visualizing GO terms in the DAG topology interactively to find related terms is fundamental in grasping function annotations of genes under studies.

In this work, we present a new interactive web-based tool, NaviGO, for comprehensive analysis of GO terms and gene functions that provide advantages over the existing GO-based web tools in three aspects: first, NaviGO is equipped with an interactive and fast rendering of the GO DAG named GO Visualizer [8], which instantly maps user-input GO terms on the GO DAG. The mapping will color parental terms of user-input GO terms and provide intuitive understanding of the similarity among them (since similarity scores are based on the topology of GO DAGs). On the GO Visualizer, users can interactively expand the hierarchy or change the view, which is advantageous over static pictures offered by AmiGO and other existing GO web tools. Second, we provide an interactive GO relationship analysis and in-depth quantification of GO similarity/divergence by incorporating six scoring schemes. The six scores reflect a variety of relationships of GO terms, ranging from GO topological structure, protein-protein interaction (PPI) association, contextual association, and annotation frequency. Particularly, in NaviGO we have implemented three scoring schemes developed previously in our group for assessing functional coherence of GO terms, namely, Co-occurrence Association Score (CAS), Pubmed Association Score (PAS) [9], and Interaction Association Score (IAS) [10], which are based on statistics of GO term pairs that are observed to co-occur in gene annotations, literature abstracts in PubMed, and physically interacting proteins, respectively. These three scoring schemes enable quantification of GO term distance based on these different contextual associations and provide cross-domain GO term comparison, unlike the three other semantic similarity based scoring schemes (Resnik, Lin and Relevance Similarity [11]; see Methods). Third, leveraging the different GO scoring schemes, in NaviGO we provide quantitative analysis of functional similarities for a group of genes and visualization of functionally similar gene clusters using a scoring scheme of users' choice. NaviGO is also linked from gene function prediction webservers, PFP [12, 13] and ESG [8, 14], so that function prediction can be readily analyzed.

Besides the exisiting GO visualization and comparison tools mentioned above [3–7], there are other GO-based tools that are more focused on particular biological analysis of genes. GSEA is focused on GO enrichment analysis and linked to a gene annotation database [15]. GeneWeaver is a data repository of genes from multiple species, which includes gene annotations, expressions, QTL, GWAS, and other biological data [16]. Tools associated with GeneWeaver can link a gene dataset to stored data, and compare datasets considering homology and gene overlaps [16]. DAVID stores annotations, domain architectures, and pathway information of genes, and users can classify a gene set which considers common GO annotations of genes [17]. VLAD performs GO enrichment analysis of genes and visualizes results on the GO hierarchy [18]. Compared to these tools, NavGO is unique in that it provides multiple different definitions of GO term similarity and associations, and can visualize and tell parental relationships of GO terms, and also linked from state-of-the-art function prediction servers.

Overall, NaviGO provides tools for exploring and understanding GO vocabulary as a basis of GO function annotation and also offers tools for biological analyses of GO annotations of genes. In addition to the web-based tools, the source codes are made available to download for local use of the software. NaviGO is a useful tool for both computational biologists who deal with GO terms and biologists who perform functional analyses of genes.

Implementation
NaviGO is a web-based software for analyzing functional similarity and associations of GO terms and genes. NaviGO is equipped with four types of analyses users can perform, which are accessible from each tab of the NaviGO page. They are, "GO Parents" for retrieving and visualizing parental terms of query GO terms, "GO Set" for computing similarity and associations of query GO terms, "GO Enrichment" for identifying enriched GO annotations in a set of query proteins, and "Protein Set" for performing functional similarity and association analysis for a set of query proteins. Each of them is described in details in the following subsections.

The input page and the logical architecture of NaviGO is provided in Fig. 1. NaviGO can either take a list of GO terms for (Fig. 1a) or a list of annotated genes (Fig. 1b) as input of analysis. NaviGO first queries in its underlying MySQL database and retrieves pre-calculated pairwise GO similarity scores computed with the six different scoring schemes. Then, based on the job type, it either constructs similarity matrices based on the input GO terms and further continues to compute functional similarity among gene products/proteins based on the GO similarity matrices or it moves onto performing an enrichment analysis by calculating p-values for the over-represented GO terms in the input. In either case, the final result of these analysis can be visualized by NaviGO's interactive GO visualizer or in Cytoscape [19].

Results and discussion
Real-time and interactive rendering of GO terms
Retrieving and mapping parental GO terms on the GO hierarchy for query GO terms is implemented as a basic functionality. In the GO Parents page, NaviGO retrieves

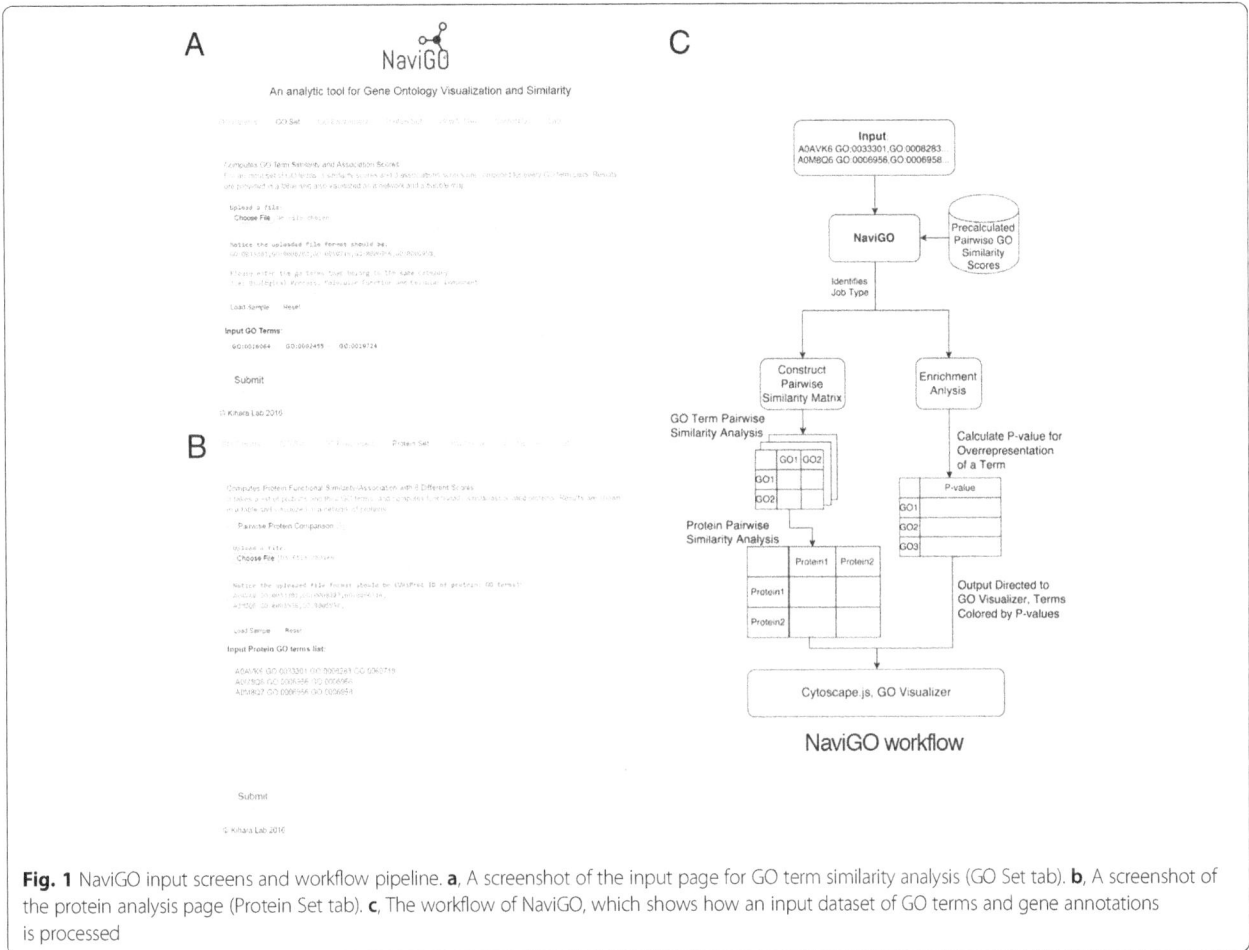

Fig. 1 NaviGO input screens and workflow pipeline. **a**, A screenshot of the input page for GO term similarity analysis (GO Set tab). **b**, A screenshot of the protein analysis page (Protein Set tab). **c**, The workflow of NaviGO, which shows how an input dataset of GO terms and gene annotations is processed

parental terms for a query and visualizes those GO terms in the DAG interactively (Fig. 2). Users can investigate the relationship of multiple GO terms by looking for their common parent and relative position in the hierarchy. This function is useful for better understanding gene annotations. For example, in UniProt, a number of GO terms are listed as function annotation for a gene, but it is often difficult to understand which terms are closely related and which are not. In such case, visualizing GO annotations on NaviGO can provide clear picture of the annotation.

In the visualization, query GO terms will be circled with bold black in the hierarchy and parental terms for the input GO terms will be listed in the text area so that users can copy and paste them for further use or for writing a document. Branches of the GO DAG can be expanded interactively by clicking a node. Hovering any node in the GO visualization will update the node information on the upper right corner of the frame. To boost the rendering speed, an option is provided that avoids newly expanded nodes from checking and updating parental relationship with the whole set of existing nodes.

Quantification of GO term association

The tool from the next tab "GO Set" computes pairwise GO association scores for a list of input GO terms and outputs them in three formats: a similarity graph (Fig. 3), a bubble chart (Fig. 4), and a score table (Fig. 5). From the result page, users can choose the three different output formats. For each type of GO scoring schemes described in Methods, four cutoff thresholds are provided, which were computed from the overall distribution of GO pair scores under each scheme (top 1%, 5%, 10%, and 20%) and shown in the color-scale (red to pink for high to low) in the table format (Fig. 5). The last column in the table contains common parents between GO pairs and a link to the interactive visualization with GO Visualizer (Fig. 2). The result table can be downloaded in a CSV format file.

Here for illustration, we used a set of 48 GO terms as input. From the GO pairs from the 48 GO terms, Table 1 lists six pairs where GO pairs has a high IAS but a low SS score. IAS indicates likelihood that protein pairs with high IAS GO terms have physical interaction, and thus different from functional similarity represented by SS

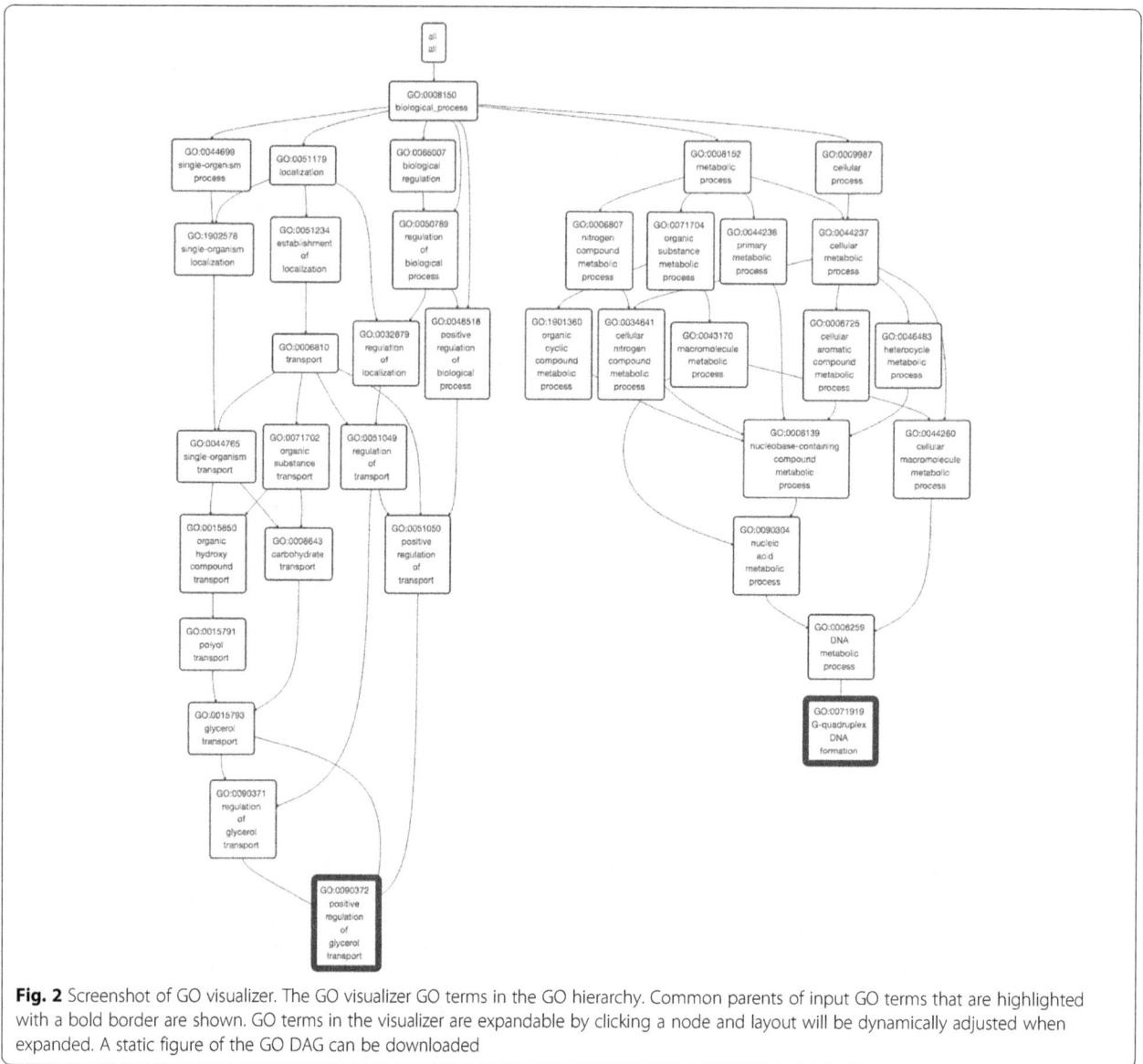

Fig. 2 Screenshot of GO visualizer. The GO visualizer GO terms in the GO hierarchy. Common parents of input GO terms that are highlighted with a bold border are shown. GO terms in the visualizer are expandable by clicking a node and layout will be dynamically adjusted when expanded. A static figure of the GO DAG can be downloaded

scores. These six pairs are highlighted in bold in the network view of the GO terms in Fig. 3 (dashed lines for cross-domain pairs). The first three pairs in Table 1 have high IAS from the same GO category while the last six GO pairs in the table are pairs from different categories. The first example from the first group is GO:0071919 *G-quadruplex DNA formation* and GO:0090372 *positive regulation of glycerol transport*, both in BP. Only one common ancestral GO terms in the hierarchy (Fig. 2), GO:0008150 *biological process* is found for this GO term pair at the depth of 0. Since the lowest common ancestor of this pair is too shallow (i.e. general) in the GO hierarchy, the SS score for this pair is low (0.003). On the other hand, due to the large number of occurrence in interacting proteins in PPI, the IAS score of this GO pair is very high (2961.47).

The last three examples illustrate cases where IAS can identify related GO terms across GO categories. The first example is a pair of BP GO term GO:0000279 *M phase* and CC GO term GO:0071065 *alpha9-beta integrin-vascular cell adhesion molecule-1 complex*. The SS score of these two terms is not calculated since these terms are from different categories, and SS is only defined for pairs from the same GO category. However, the pair has a very high IAS score (5922.95) because of the large amount of protein interactions in PPI network have these terms (Table 1).

Relationships of GO terms can be visualized in two ways, in a network or in a bubble chart, which provides intuitive understanding of similarity and relationship of GO terms. Fig. 3 shows a GO term association graph of IAS for the 48 GO terms. In this graph, GO terms are

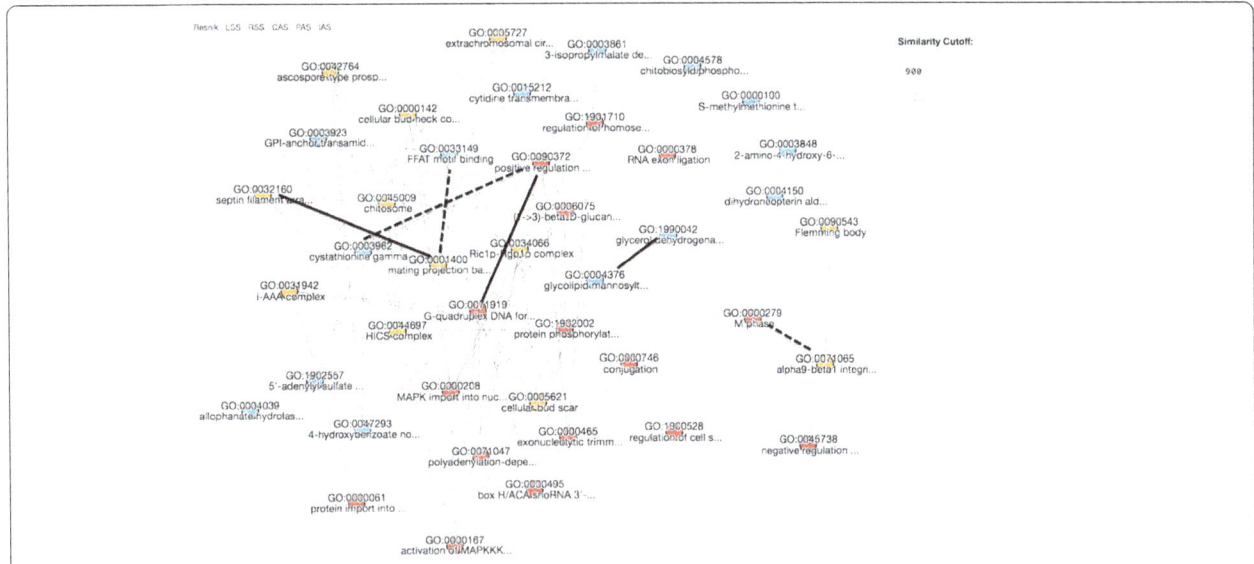

Fig. 3 Network view of GO term association. A screenshot of the network view of GO term association in the resulted page. The score to consider can be switched by clicking the *bottom panel* on the *upper left corner*. GO terms from different GO categories are mapped in different colors (BP: *red*; MF: *blue*; CC: *yellow*). *Upper right corner* panel is used for adjusting a score cut-off threshold. Six GO pairs discussed as examples in Table 1 are highlighted in bold (*dashed lines* for cross-domain pairs)

connected if they have an IAS above the threshold value (900 in this example) that can be controlled by the scale bar on the right upper corner of the panel (Fig. 3). GO terms in different categories are shown in different colors (BP: red; MF: blue; CC: yellow). The edges in bold and dashed lines are those pairs that are discussed in Table 1.

The bubble chart maps GO terms in terms of two scores users' choice on the X- and Y-axis (Fig. 4) by a statistical dimension-reduction method named multidimensional scaling (MDS) [20, 21] implemented in R [22]. In case GO terms have the identical score, the centers of the circles/dots of the terms are shifted by a small

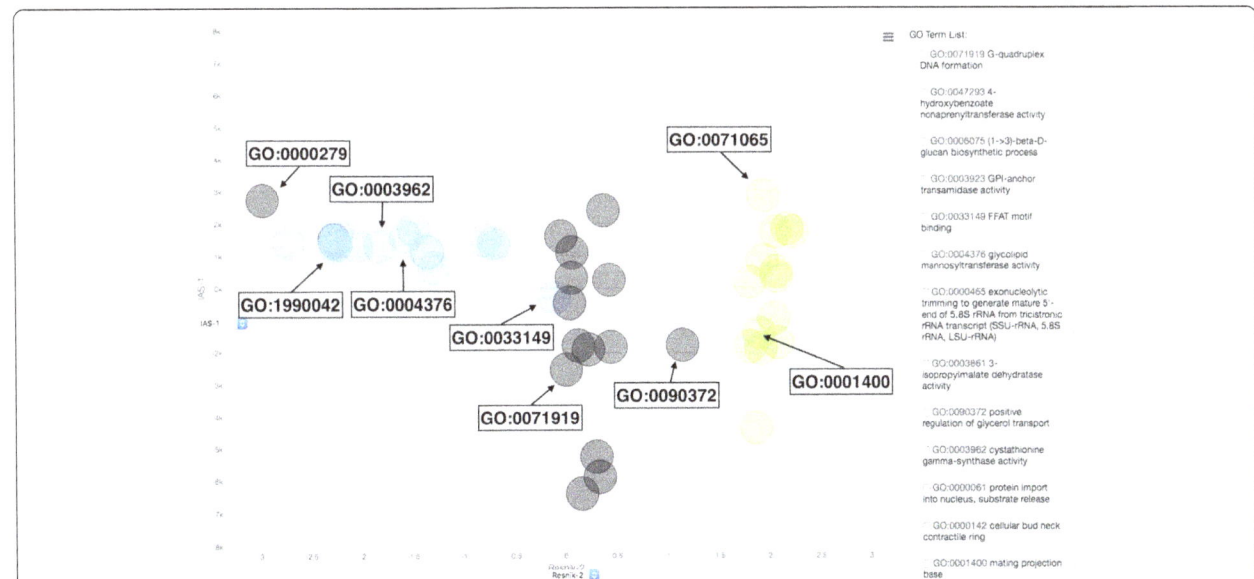

Fig. 4 Bubble chart view of the GO term association. A screenshot of the bubble chart view in the resulted page. A bubble chart maps GO terms in the two dimensional space by considering their similarity in terms of two scores shown on the two axes. Score scheme to be used for each axis can be chosen by the option fields in the middle of the axis. Users can zoom in by holding down the mouse button and then dragging it to the desired area. The *right panel* shows GO terms that are currently visible on the chart. The example of shown here are visualization of the same set of 48 GO terms as used in Fig. 3. In this example plot, the X-axis is the Resnik semantic similarity score and the Y-axis chosen is IAS. For illustration, in this figure GO terms are colored according to their GO category, MF, *blue*; BP: *black*; and CC: *green*. GO terms listed in Table 1 are labelled

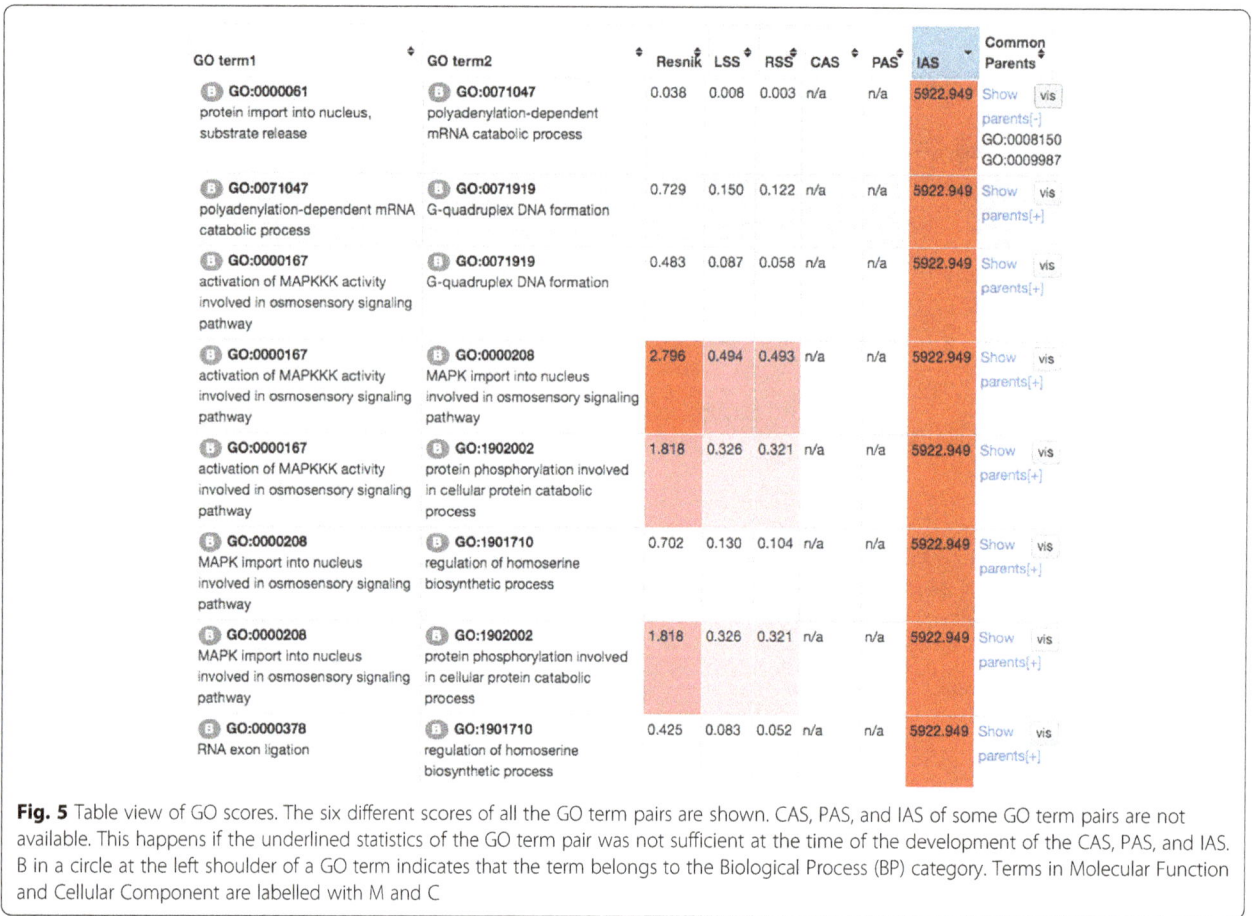

Fig. 5 Table view of GO scores. The six different scores of all the GO term pairs are shown. CAS, PAS, and IAS of some GO term pairs are not available. This happens if the underlined statistics of the GO term pair was not sufficient at the time of the development of the CAS, PAS, and IAS. B in a circle at the left shoulder of a GO term indicates that the term belongs to the Biological Process (BP) category. Terms in Molecular Function and Cellular Component are labelled with M and C

amount to a random direction to avoid complete overlap. The bubble chart is interactive and the coordinate data is exportable. In Fig. 4, interestingly, along the Resnik score (X-axis), the GO terms are clearly separated by their GO category visualized in each different color, because the Resnik score for GO terms of different categories is 0, i.e. not defined, thus very far between them. In contrast, IAS is defined even for terms across GO categories. Thus, some GO pairs across different categories, for example, GO pairs of (GO: 0000279 from BP, GO: 007165 from CC) and (GO: 0033149 from MF, GO: 0001400 from CC) shown in the bottom half of Table 1, are close when mapped on the IAS (Y-axis) but far in Resnik (X-axis).

GO Enrichment analysis

The NaviGO server supports the analysis of GO term enrichment. For an input list of annotated genes, enriched GO terms in the genes relative to the fraction in the entire genome will be identified. This is useful for

Table 1 Examples of IAS that are different from the SS scores

GO ID 1	Description	Domain	GO ID 2	Description	Category	IAS	RSS	LSS	Resnik
GO:0071919	G-quadruplex DNA formation	BP	GO:0090372	Positive regulation of glycerol transport	BP	2961.47	0.003	-0.014	-0.076
GO:0004376	Glycolipid mannosyltransferase activity	MF	GO:1990042	Protein histidine kinase binding	MF	5922.95	0.042	0.074	0.364
GO:0001400	Mating projection base	CC	GO:0032160	Septin filament array	CC	4230.68	0.002	0.008	0.039
GO:0000279	M phase	BP	GO:0071065	Alpha9-beta1 integrin-vascular cell adhesion molecule-1 complex	CC	5922.95	N/A	N/A	N/A
GO:0033149	FFAT motif binding	MF	GO:0001400	Mating projection base	CC	1692.27	N/A	N/A	N/A
GO:0090372	Positive regulation of glycerol transport	BP	GO:0003962	Cystathionine gamma-synthase activity	MF	987.16	N/A	N/A	N/A

finding dominant common functions for a set of genes, which are identified, for example, by gene expression data or protein-protein interaction network. Thus, the analysis can aid to identify the associated proteins involved in certain function within an organism. The server will automatically identify the organism based on the UniProt ID [23] of the first input protein; however, users can specify the organism in the Organism window manually. NaviGO will connect to the UniProt database by using their RESTful service and automatically retrieve the organism information and the background GO annotation information of the organism.

The result page lists GO terms sorted by calculated p-value. The p-value tells how rare (significant) it is to have enrichment of the GO term in the protein set considering the number of proteins in the set, the number of proteins with that GO term in the organism, and the number of proteins in the organism. GO terms of significant p-value (0.00005) (or top 30 GO terms, whichever smaller) will be visualized in the GO hierarchy (Fig. 6). The number of GO terms to visualize can be controlled manually by users. The enriched GO terms are color-mapped according to the p-value of enrichment on the GO DAG visualizer. In the example in Fig. 6, a GO enrichment analysis is shown for 20 annotated proteins that are involved in the MAPK signalling pathway, which were found in a SNP-targeted GWAS studies as set of proteins involved in the Rheumatoid Arthritis disease [24]. Enrichment analysis helps to identify the GO terms that are prominent among these disease associated proteins. GO terms, such as GO:0051403 *stress-activated MAPK cascade* with p-value of 4.04E-11

and GO:0000186 *activation of MAPKK activity* with p-value of 3.13E-9, are in the top enriched results. Due to the fact that the activation of the TLRs signalling pathway can trigger the activation of the MAPK pathways [24], GO terms like GO:0034166 *toll-like receptor 10 signaling pathway* also has a low p-value (5.06E-13) and ranked the top in the list (indicated with red circle). In Fig. 6, GO terms which has e-value of less than 1.0E-10 are circled in orange and their function descriptions are shown.

Quantifying functional association of proteins

This functionality available from the "Protein Set" tab in the NaviGO website, takes a list of annotated proteins as input and computes functional relevance between each protein pair. This function help identifying functional groups in a set of proteins. The server can take a function annotation file in the CAFA format. The result will be provided in a table as well as an interactive clustering view. An example result shown in Fig. 7 are for the 33 protein pairs that have high IAS mong all the protein pairs from the human genome. Since IAS are defined for GO term pairs taken from physically interacting proteins, protein pairs with a high IAS are highly likely to be interacting with each other. In the result table (Fig. 7a), significance of similarity predictions is classified into five levels shown in a color-scale (red to pink for high to low). The scale indicates that the score is within top 1, 5, 10 and 20%, which are computed relative to the score distribution of all the protein pairs of the organism chosen by the user at the pull-down menu. The median option in the pull-down menu shows significance based on the average values of 5th and 6th genomes (i.e. median) when the 10

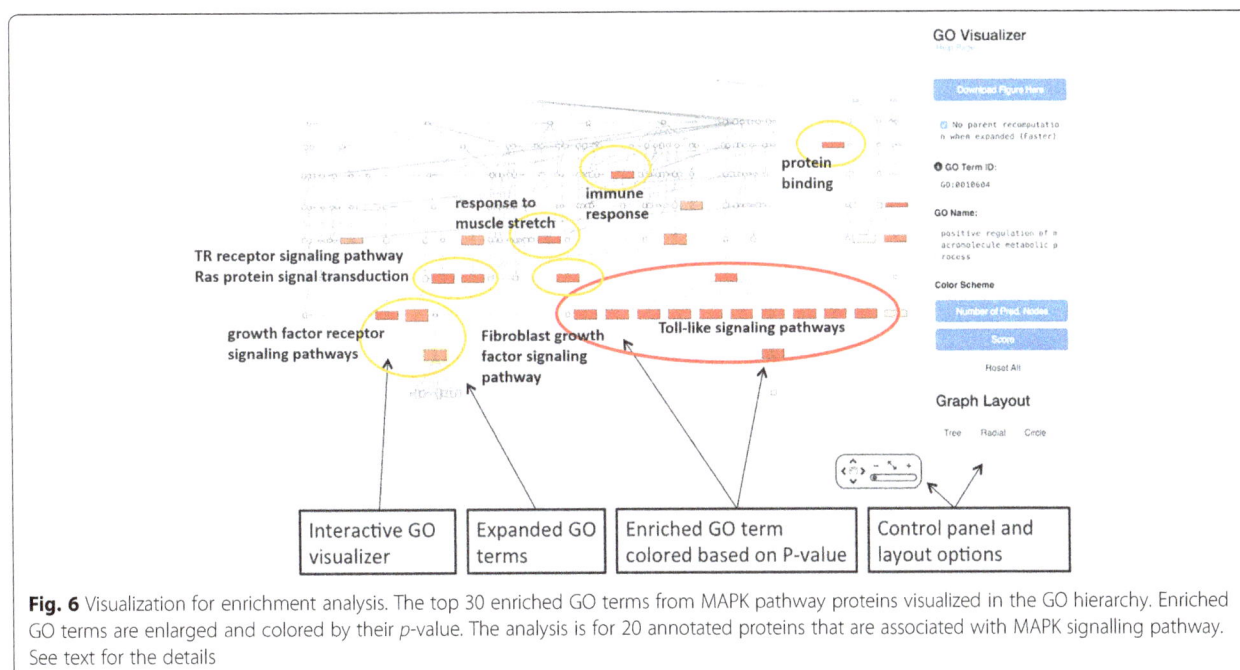

Fig. 6 Visualization for enrichment analysis. The top 30 enriched GO terms from MAPK pathway proteins visualized in the GO hierarchy. Enriched GO terms are enlarged and colored by their p-value. The analysis is for 20 annotated proteins that are associated with MAPK signalling pathway. See text for the details

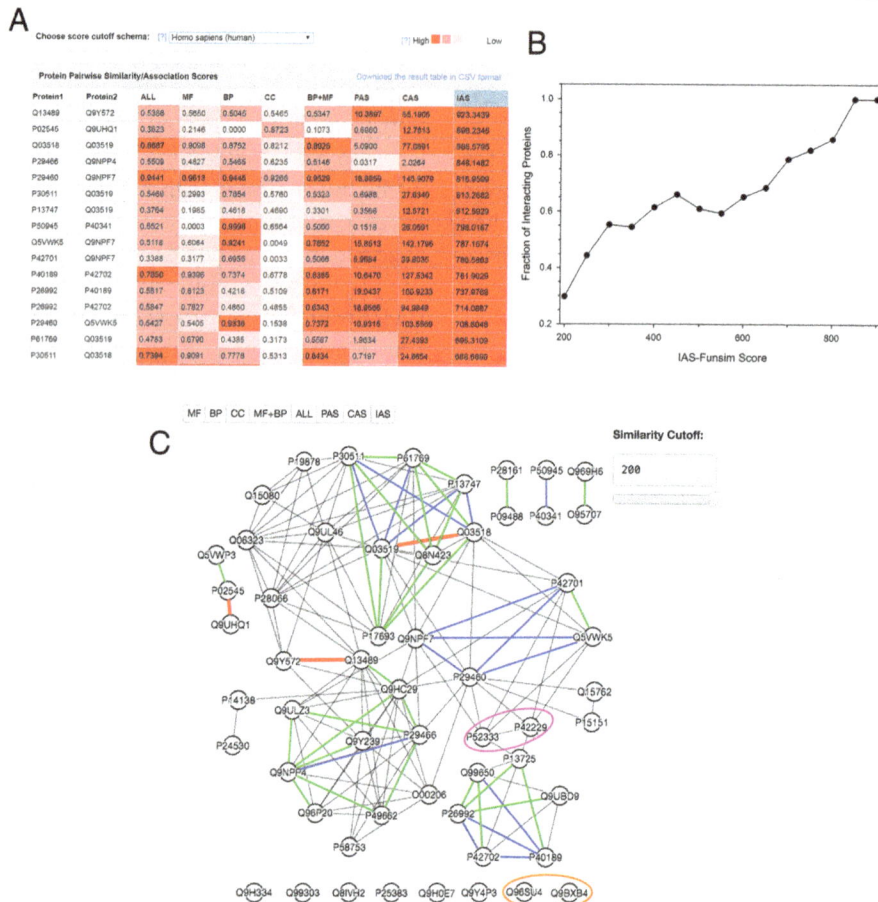

Fig. 7 Example of the protein set analysis. Pairwise IAS scores among all protein pairs in the human genome were computed. **a**, A snap shot of the table of protein pairs in the result page sorted by the IAS score. The color level shows the significance of the scores. **b**, The fraction of the protein pairs that are actually physically interacting among those above IAS score cutoffs (x-axis). Pairs that have a score of 200 or higher are considered. Physically interacting protein pairs were checked with the BIOGRID database. For example, 100% of pairs that have a score of 850 or higher and 78.5% of pairs with a score of 700 or higher actually interact with each other. **c**, A network view of the 56 unique proteins from the top high-scoring 33 protein pairs. Protein pairs that have an IAS score of 200 or higher are connected by edges. Protein pairs that have a high IAS score of over 850, 650, and 450, are connected with thick color lines in red, blue, and green. There are three, 19, and 47 such pairs, respectively. Two protein pairs that are discussed in Table 2 are circled. The *magenta circle* shows an example of physically interacting pairs and the *circle in orange* shows a functionally similar protein pair that do not physically interact with each other, which is correctly identified with a low IAS. See text for more details

genomes in the list are sorted in the descending order of their corresponding cut-off values. The full result table is available to download in the CSV format.

In Fig. 7a, pairs are sorted by IAS. Notice that pairs with high IAS do not always have significant scores in terms of the other scores, e.g. ALL (functional similarity of proteins using Relevance Similarity scores (Eq. 6), indicating that IAS captures a unique feature of GO annotations of proteins, which are different from functional similarity. Figure 7b shows the fraction of physically interacting protein pairs among pairs with IAS above cutoffs. The graph shows that IAS has a substantial correlation to the fraction of interacting proteins, indicating IAS indeed detects physically interacting protein pairs. For example, all three protein pairs with a score above

850 actually interact with each other, and 68.4% (13 pairs) among 19 pairs that have an IAS over 650 have physical interactions between each other. Figure 7c provides the network view of the 33 protein pairs, where nodes are proteins and edges indicate pairs with IAS above a custom similarity cut-off value (200 is used). Pairs with significant scores, 850, 650, and 450, are highlighted in, red, green, and blue, respectively.

In Fig. 7c, two protein pairs are circled for discussion. These two pairs are listed in Table 2, a pair of P52333 and P42229 and another one, Q96SU4 and Q9BXB4. In the first pair, P52333 is a kinase that phosphorylates and the second one is STAT protein (P42229), which are involved in signal transduction and activation of transcription. These two proteins are not similar in function but

Table 2 Comparison of Protein-pair Scores

Organism	Protein1	Function	Protein2	Function	Funsim-IAS score	Funsim-RSS score
Homo sapiens	P52333	Tyrosine-protein kinase JAK3	P42229	STAT 5A	395.6000	0.2879
Homo sapiens	Q96SU4	Oxysterol-binding protein-related protein 9	Q9BXB4	Oxysterol-binding protein-related protein 11	11.7716	0.9033

physically interact with each other according to the BIOGRID database [25]. In contrast, the second pair, Q96SU4 and Q9BXB4 (orange circle in Fig. 7c), have high functional similarity (0.9033, right column), since they are both oxysterol-binding proteins (OSBP)-related proteins. However, as suggested with a low *funsim* IAS score (11.7716), they do not interact. These two pairs illustrate that *funsim* of IAS, which indicates possibe protein interactions, are different from conventional functional similarity. As shown here, by changing the underlying score for computing *funsim* protein pair score, we can see how proteins are similar or distinct in different aspects of GO term relationships.

Analysing function prediction results with NaviGO

NaviGO is linked from two function prediction servers, PFP [12, 13] and ESG [14], so that predicted GO terms for a gene by the servers can be easily further analysed. These two servers were ranked among top in gene function prediction contests held in recent years, the function prediction category in the Critical Assessment of techniques in protein Structure Prediction (CASP7) [26] and in the two rounds of the Critical Assessment of Function Annotation (CAFA) [27, 28]. PFP and ESG are available at http://kiharalab.org/pfp.php and http://kiharalab.org/esg.php [8]. Function prediction have become increasingly important because a substantial fraction of genes in a genome are unannotated [13].

In the output page of PFP and ESG, where predicted GO terms are listed with confidence scores, the GO terms will be sent to NaviGO by clicking a link "Analyze with NaviGO" (Fig. 8a). Here we show a case that the analysis of predicted function by NaviGO revealed that a query protein has two distinct functions. The function prediction was performed by the ESG server for human aconitase (UniProt ID: Q99798). This protein is known as a moonlighting protein, which has two distinct independent functions [29, 30]. The primary function of this protein is an enzyme as aconitase while it is also known to be involved in iron homeostasis [29, 30]. Visualization of the predicted GO terms by NaviGO shows that the predicted GO terms in the MF category (Fig. 8b) are indeed separated in two branches, one on the left with the enzymatic activity (including lyase activity, aconitase hydrolase) and another branch on the right, iron-sulfer clustering binding, which is related to iron homeostasis (Fig. 8c), showing that the prediction correctly captured

two distinct functions of this protein. The network view of the predicted GO terms also clarified that the protein has two distinct MF functions (shown in blue nodes), indicated by the two clusters.

Conclusions

A web-based tool for analysing GO terms and gene annotation was developed. Results are visualized by a user-friendly interactive panel, which provides intuitive understanding of gene function. A strength of NaviGO is that similarity or association of GO terms can be quantified in six different scores and it is equipped with real-time rendering of GO terms in the GO hierarchy. The unique feature of NaviGO should provide great convenience in functional analysis with GO for both bioinformatics researchers and biologists.

Methods

NaviGO is as a web-based tool at http://kiharalab.org/web/navigo. The source codes are made available at Github, https://github.com/kiharalab/NaviGO and https://github.com/kiharalab/GOVisualizer.

GO similarity/association scores

In NaviGO, six scores can be used to quantify similarity or association relationship of GO terms. Three scores are for quantifying semantic similarity of GO terms: Resnik's, Lin's, and the relevant semantic similarity score. The other three scores, CAS, PAS, and IAS are for quantifying GO associations. Detailed explanation of the scores is provided in separate sections below.

To quantify the functional similarity of two genes, the *funsim* score [4, 5] is used. Funsim of two sets of terms, i.e. GO annotations of two genes, is calculated from an all-by-all similarity matrix, where each entry of the matrix is a similarity score of users' choice between a GO pair.

CAS and PAS

We previously developed two function association scores, Co-occurrence Association Score (CAS) and PubMed Association Score (PAS) [3]. CAS quantifies frequency of co-occurring GO terms within the gene annotations in the GOA database while PAS takes consideration of co-occurrence of GO terms in PubMed abstracts. A characteristic differentiating the two methods from other methods is that the two scores can be defined cross-domain associations between GO terms, i.e. terms from Molecular

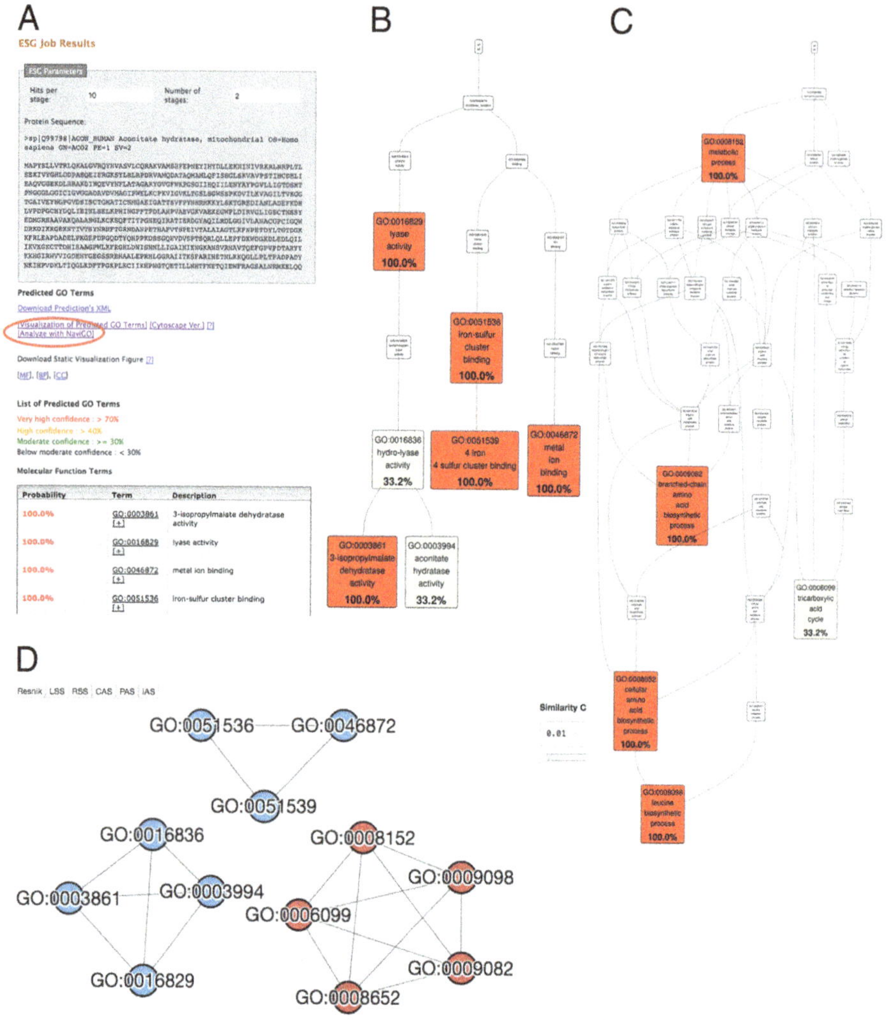

Fig. 8 Analysing function prediction results with NaviGO. NaviGO is linked from the PFP and ESG function prediction webservers, which predict GO terms for input protein sequence. This example shows function prediction for human aconitate hydrolase (UniProt ID: Q99798). This protein is a moonlighting protein, which has two distinct function, aconitase and involvement of iron homeostatis. **a**, An output page of ESG. The output page has a link to NaviGO, which is indicated by a red circle in the figure. Clicking this link will send predicted GO terms of the query protein listed below in the table, which has the medium confidence or higher, to NaviGO's GO Set input page, so that users can further analyse the predicted GO terms. **b**, Predicted GO terms in the MF category visualized in NaviGO. Color codes shows the confidence of prediction. **c**, Predicted GO terms in the BP category visualized in NaviGO. **d**, The network view of the predicted GO terms in NaviGO using the RSS, showing functionally similar GO terms in clusters. GO terms in MF and BP are colored in *blue* and *red*, respectively. We see two clusters for MF, indicating that this protein has two distinct functions

Function (MF) and Biological Process (BP), those from MF and Cellular Component (CC), and those from BP and CC.

$$CAS(i,j) = \frac{\dfrac{c(i,j)}{\sum_{ij} c(i,j)}}{\left(\dfrac{c(i)}{\sum_k c(k)}\right)\left(\dfrac{c(j)}{\sum_k c(k)}\right)} \qquad (1)$$

where $C(i,j)$ is the number of sequences in the database that contain both the GO terms i and j. Similarly, $C(i)$ is the total number of sequences annotated with the GO

term i, and so is the $C(j)$. The numerator of Eq. 1, $\dfrac{c(i,j)}{\sum_{ij} c(i,j)}$ is essentially the fraction of sequences that are annotated with two particular GO terms, i and j, among all the sequences in the database. The denominator multiplies the fraction (probability) of sequences in the database that are annotated with GO term i and the fraction of sequences in the database that are annotated with GO term j. Thus, it is the expected fraction of sequences in the database with the two GO annotations, i and j, if i and j are randomly assigned to sequences.

Using the numerator and the denominator, altogether CAS quantifies how often two GO terms i and j co-annotate sequences relative to the random chance. CAS = 1 means that the observation of co-annotation of i and j is the same as expected by the random chance, and a larger value indicates that i and j are correlated in gene annotation.

Similarly, PAS is defined as:

$$PAS(i,j) = \frac{\frac{Pub(i,j)}{\sum_{i,j} Pub(i,j)}}{\left(\frac{Pub(i)}{\sum_k Pub(k)}\right)\left(\frac{Pub(j)}{\sum_k Pub(k)}\right)} \qquad (2)$$

$$= \frac{Pub(i,j)}{Pub(i)Pub(j)} \cdot \frac{\left(\sum_k Pub(k)\right)^2}{\sum_{k,l} Pub(k,l)}$$

Here, $Pub(i,j)$ is the number of PubMed abstracts which contain both the GO terms i and j. Similarly, $Pub(i)$ is the number of abstracts that contain GO term i and the same is applicable for $Pub(j)$. The numerator of Eq. 2, $\frac{Pub(i,j)}{\sum_{ij} Pub(i,j)}$, is the fraction of abstracts in PubMed that mention two particular GO terms, i and j, among all the abstracts in the PubMed database. The denominator multiplies the fraction (probability) of abstracts in PubMed that mention GO term i and the fraction of abstracts that mention GO term j. Thus, it is the expected fraction of abstracts in the database with the two GO annotations, i and j, if i and j randomly show up in abstracts. Altogether, PAS quantifies how often two GO terms i and j are co-mentioned in PubMed abstracts relative to the random chance. PAS = 1 means that GO term i and j are not related, and a larger value indicates that i and j are related and frequently co-mentioned in biological contexts. Importantly, it is possible that GO terms that do not have a high functional similar scores (Resnik, Lin's, and Relevance Similarity scores) have a high CAS or PAS. High PAS and CAS implies that proteins with the GO term annotation are functionally related and play roles in the same biological context, e.g. pathways.

IAS

The Interaction Association Score (IAS) [7] captures the propensity of GO term pairs to occur in interacting proteins by counting the number of GO term pair that occur in interacting proteins normalized by random chance. Thus, high IAS between a protein pair indicates a high possibility that the protein pairs interact with each other. The GO_IAS for each GO term pair was computed as follows:

$$GO\_IAS(GOx, GOy) = \frac{\frac{N(GOx,GOy)}{\#T.Edges}}{\left(\frac{N(GOx)}{\#T.Nodes}\right)\left(\frac{N(GOy)}{\#T.Nodes}\right)} \qquad (3)$$

where $N(GOx-GOy)$ is the number of times GO term pair GOx and GOy interact in PPI networks, $\#T.Edges$ is the total number of interactions (edges) in PPI networks, $N(GOx)$ and $N(GOy)$ are the number of times GOx and GOy independently occur in proteins the networks, and $\#T.Nodes$ is the total number of proteins in the PPI networks. Figure 9 shows an example of a small PPI network. This network has 5 edges between 5 proteins; 3 proteins are annotated with GO:1, and 2 proteins with GO:2. There are 2 edges that connects between GO:1 and GO:2 (P1 to P2 and P2 to P4). From this network, GO_IAS for GO:1 and GO:2 is computed as $(2/5)/((3/5)(2/5)) = 1.67$. Similar to PAS and CAS, IAS quantifies how often two GO terms i and j are observed in physically interacting proteins in a protein-protein interaction network relative to the expected number of observations by the random chance. If two proteins are annotated with GO terms that have high IAS, it suggests that the proteins may physically interact with each other.

Significant difference between CAS, PAS, and IAS from conventional GO functional similarity scores described in the next section is that the former three scores quantifies functional relevance of GO term pairs in biological contexts, co-annotation to genes (CAS), co-mention in PubMed abstracts (PAS), and interacting protein pairs (IAS). Due to the design, these scores are capable of identifying proteins in the same pathways (CAS, PAS) [3] and physical interacting proteins (IAS) [7]. Correlation of CAS/PAS/IAS to regular functional similarity scores (below) is not very high [3, 7], because proteins in the same pathway and physically interacting proteins are not necessarily have similar function.

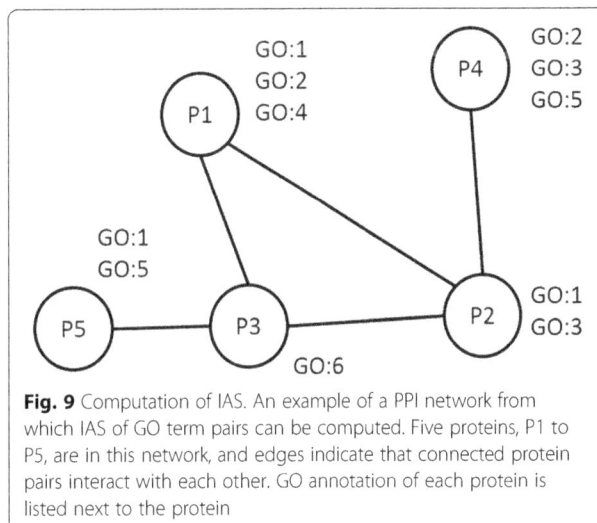

Fig. 9 Computation of IAS. An example of a PPI network from which IAS of GO term pairs can be computed. Five proteins, P1 to P5, are in this network, and edges indicate that connected protein pairs interact with each other. GO annotation of each protein is listed next to the protein

Resnik, Lin's, and relevance similarity scores

For quantifying GO term similarity, NaviGO provides three score options. The Resnik's [5] similarity score measures the semantic similarity of a GO term pair according to the lowest common ancestor (LCA) of the GO term pair, while the Lin's similarity is based on the information content of LCA and the GO term pair queried [3].

$$sim_{Resnik}(c_1, c_2) = \max_{c \in S(c_1, c_2)}(-\log p(c)) \tag{4}$$

$$sim_{Lin}(c_1, c_2) = \max_{c \in S(c_1, c_2)} \left(\frac{2 \cdot \log p(c)}{\log p(c_1) + \log p(c_2)} \right) \tag{5}$$

Here $p(c)$ is the probability of a GO term c, which is defined as the fraction of the occurrence of c in the GO Database. $s(c1, c2)$ is the set of common ancestors of the GO terms c1 and c2. The root of the ontology has a probability of 1.0.

The relevance semantic similarity score (sim_{Rel}) [4] for computing functional similarity of a pair of GO terms, c1 and c2:

$$sim_{Rel}(c_1, c_2) = \max_{c \in S(c_1, c_2)} \left(\frac{2 \cdot \log p(c)}{\log p(c_1) + \log p(c_2)} \cdot (1 - p(c)) \right) \tag{6}$$

The first term considers the relative depth of the common ancestor c to the average depth of the two terms c1 and c2 while the second term takes into account how rare it is to identify the common ancestor c by chance.

Functional similarity score of gene pairs

To quantify the functional similarity of two annotated genes, we used the *funsim* score [4, 5]. The *funsim* score of two sets of terms, GO^A and GO^B for gene A and B, of a respective size of N and M, is calculated from an all-by-all similarity matrix s_{ij}.

$$Sij = sim(GO_i^A, GO_j^B)_{\forall i \in \{1..N\}, \forall j \in \{1..M\}} \tag{7}$$

For $sim(GO_i^A, GO_j^B)$, the relevance similarity score is usually used but other scores can be used, too. Since the relevance similarity score is defined only for GO pairs of the same category, a matrix is computed separately for the three categories, BP, CC, and MF:

$$GO_{score} = \max \left(\frac{1}{N} \sum_{i=1}^{N} \max_{1 \leq i \leq M} s_{ij}, \frac{1}{M} \sum_{i=1}^{M} \max_{1 \leq i \leq N} s_{ij} \right) \tag{8}$$

GOscore will be any of the three category scores (MFscore, BPscore, CCscore). Finally, the *funsim* score is computed as

$$funsim = \frac{1}{3} \left[\left(\frac{MFscore}{\max(MFscore)} \right)^2 + \left(\frac{BPscore}{\max(BPscore)} \right)^2 + \left(\frac{CCscore}{\max(CCscore)} \right)^2 \right] \tag{9}$$

where $max(GOscore) = 1$ (maximum possible GOscore) and the range of the *funSim* score is [0, 1].

Gene ontology enrichment analysis

The probability of a GO term X being annotated to a protein in the cluster is computed by:

$$f(k; N, m, n) = \frac{\binom{m}{k}\binom{N-m}{n-k}}{\binom{N}{n}} \tag{10}$$

where k is the number of proteins in the cluster annotated with X, N is the number of annotated proteins in the organism, m is the number of proteins in the organism annotated with X, and n is the number of annotated proteins in the cluster. To calculate a p-value for over-representation of a term, we use the following equation:

$$P_{hg}(X) = \sum_{i=k}^{n} f(i; N, m, k) \tag{11}$$

Abbreviations

BP: Biological process; CAS: Co-occurrence association score; CC: Cellular component; DAG: Directed acyclic graph; GO: Gene ontology; IAS: Interaction association score; MDS: Multi-dimensional scaling; MF: Molecular function; PAS: PubMed association score; PPI: Protein-protein interaction; RSS: Relevance semantic similarity; SS: Semantic similarity

Acknowledgements

We thank Steve Wilson for technical support.

Funding

This work was supported partly by the National Institutes of Health (R01GM097528), the National Science Foundation (IIS1319551, DBI1262189, IOS1127027, DMS1614777). The funding bodies had no role in the design or conclusion of the current study.

Authors' contributions

DK and QW developed the original idea. QW developed the software. IKK and SY helped develop and update the databases. ZD performed data analysis for Figs. 7 and 8. QW, IKK, and DK wrote the manuscript. All authors read and approved the final manuscript.

Competing interests

The authors declare that they have no competing interests.

Author details

[1]Department of Computer Science, Purdue University, West Lafayette, IN 47907, USA. [2]Department of Biological Science, Purdue University, West Lafayette, IN 47907, USA. [3]Division of Biomedical Statistics and Informatics, Mayo Clinic, Rochester, MN 55905, USA.

vertexes simulated. Since the run-time for the algorithm at every iteration is in order of $O(N^2)$, decreasing the number of vertexes would theoretically have a tremendous effect on reducing the run-time. However, the mapping of the structure into vertexes and edges must be done in a way that still produces visually pleasing layouts and RNA diagrams. Inspired by VARNA, we have employed a representation which maps each RNA loop, as well as every stem base pair, to a vertex, and connects those via edges. (Fig. 2).

Furthermore, we have implemented the system in such a way that repulsion only occurs between loops, and not base pairs. Since the repulsion step is the main time consuming step of each iteration (having run time in order of $O(N^2)$), decreasing the number of participating vertexes in the repulsion interaction should greatly reduce the run time of the algorithm. Constructing the system in such a way that only loops experience repulsion ensures that loops will be pushed away from each other, thus not intersecting each other. In theory, this should aid the structure in adopting a final layout that has minimal or no intersection of any structural elements.

At the initial step of the simulation, the RNA graphs are placed in a naive initial layout which is inspired from a circular representation of the RNA molecule (Fig. 3). Then, an iterative process begins in which the structure is slowly brought to a stable position by Newtonian inspired spring and repulsion forces.

jViz.RNA and the Newtonian model

Originally jViz.RNA mapped the RNA structure into a detailed graph, $G = \{V, E\}$. In the detailed graph representation, each nucleotide is a vertex $v \in V$, and each chemical bond corresponds to an edge $e \in E$. The entire structure is initially laid out in a circle, and an iterative process designed to move the structure into a stable layout begins. In this paper, we employ the ***compressed graph*** representation, where each structural element (loops and base pairs) is a vertex $v \in V$, and the edges are graph elements which connect the different structure elements (Fig. 2b).

For the purposes of the following sections, the notation $\vec{P}$ will be used to denote the positions of the different vertexes which represent the RNA structures, and $\vec{P}^i_{n,k}$ will be used to denote the position of vertex i (since there are many vertexes), at time-step n (since the simulation operates in discrete time-steps), during the k-th Newton iteration (since a portion of the experiments in this paper use Newton's method for converging on the position of vertex i at time-step n).

The most basic computation done at each iteration is the unit vector function, $\vec{U}(\vec{P}^i, \vec{P}^j)$. This function calculates the unit vector pointing from point $\vec{P}^j$ (the position

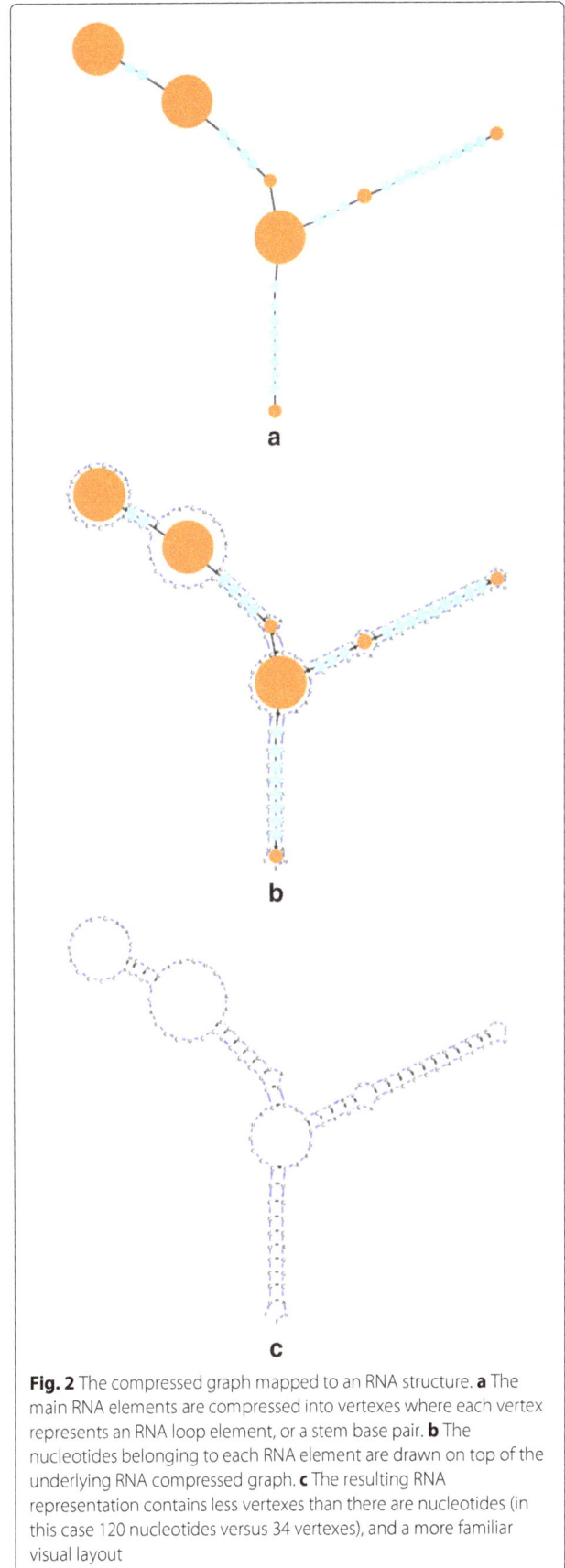

Fig. 2 The compressed graph mapped to an RNA structure. **a** The main RNA elements are compressed into vertexes where each vertex represents an RNA loop element, or a stem base pair. **b** The nucleotides belonging to each RNA element are drawn on top of the underlying RNA compressed graph. **c** The resulting RNA representation contains less vertexes than there are nucleotides (in this case 120 nucleotides versus 34 vertexes), and a more familiar visual layout

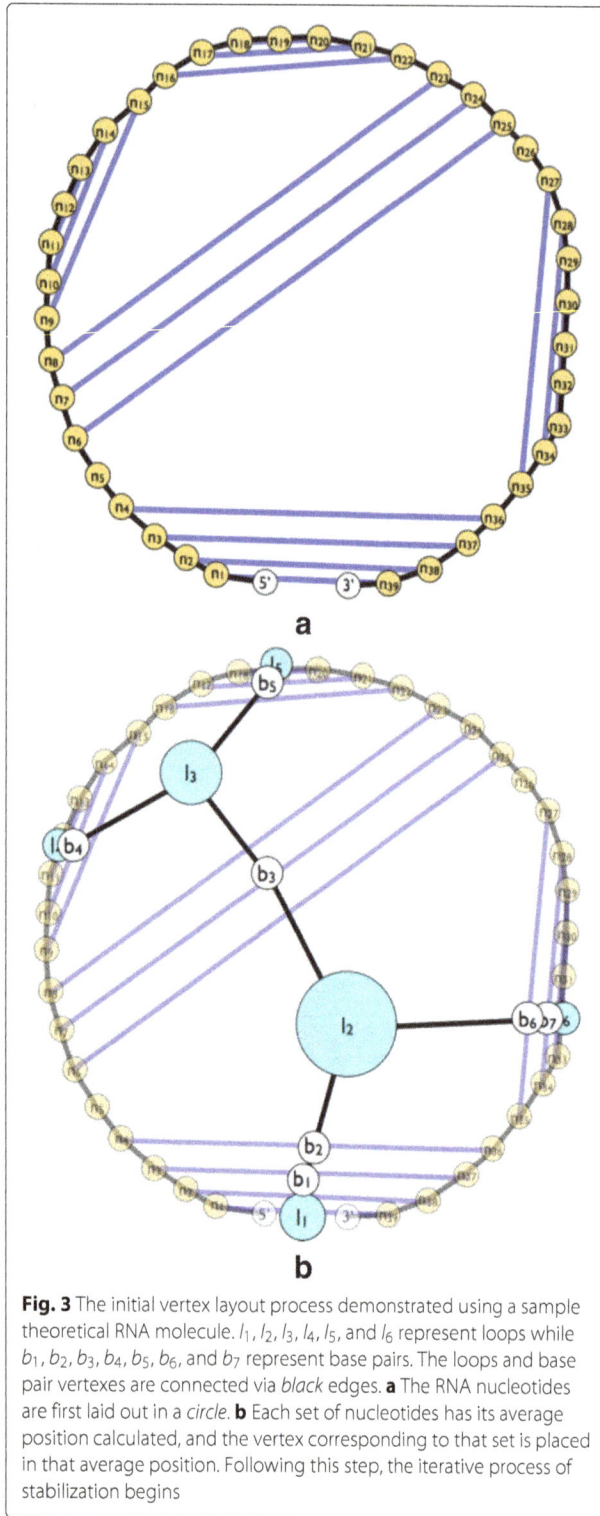

Fig. 3 The initial vertex layout process demonstrated using a sample theoretical RNA molecule. l_1, l_2, l_3, l_4, l_5, and l_6 represent loops while b_1, b_2, b_3, b_4, b_5, b_6, and b_7 represent base pairs. The loops and base pair vertexes are connected via *black* edges. **a** The RNA nucleotides are first laid out in a *circle*. **b** Each set of nucleotides has its average position calculated, and the vertex corresponding to that set is placed in that average position. Following this step, the iterative process of stabilization begins

In each iteration, each vertex moves based on two forces: repulsion and attraction. The repulsion forces each vertex v^i experiences from vertex v^j can be described as:

$$\vec{R}\left(\vec{P}^i, \vec{P}^j\right) = \frac{G}{\left|\vec{P}^i - \vec{P}^j\right|^2} \times \vec{U}\left(\vec{P}^i, \vec{P}^j\right) \qquad (2)$$

where $\vec{U}(\vec{P}^i, \vec{P}^j)$ is the unit vector function showing the direction from vertex v^j to vertex v^i, and G is a coefficient to control the size of the force experienced. The attraction forces each vertex v^i experiences from vertex v^j can be described as:

$$\vec{A}\left(\vec{P}^i, \vec{P}^j\right) = K \times \left[\vec{P}^j - \vec{P}^i + \left(r_{ideal} \times \vec{U}\left(\vec{P}^i, \vec{P}^j\right)\right)\right] \qquad (3)$$

where $\vec{U}(\vec{P}^i, \vec{P}^j)$ is again the unit vector function, K is an attraction coefficient to control for the size of the force, and r_{ideal} is the ideal desired distance between the vertexes v^i and v^j.

The iterative process stops when the forces for all vertexes have reached equilibrium, or when for all vertexes $\{v^i | 1 \leq i \leq N\}$ the following holds:

$$\forall v^i, \left|\sum_{v^j \in L, j \neq i} \vec{R}\left(\vec{P}^i, \vec{P}^j\right) + \sum_{v^j \in C^i} \vec{A}\left(\vec{P}^i, \vec{P}^j\right)\right| \leq \epsilon \qquad (4)$$

Where C^i is the set of all vertexes connected to vertex v^i (in other words, $v^j \in C^i$ iff there is an edge between v^j and v^i), and L is the set of all loops (that is, $v^j \in L$ iff v^j is a vertex representing a loop).

That is to say, the iterative process stops when the sum of the forces acting on the vertexes is smaller than ϵ. Setting $\epsilon = 0$ will force the simulation to continue to calculate until the forces are perfectly at odds, but setting a small value for ϵ allows the layout algorithm to stop sooner when achieving a stable structure. ϵ as such, controls the degree of stability required before simulation of the structure's movement stops. In this work, we chose to explore two methods of implementing the physics based RNA model: The Forward Euler method, and the Backward Euler method. The two methods make it possible to evaluate the movement of the vertexes. However, the latter is more numerically stable than the former, and allows for greater time steps and faster visualizations, as well as a more stable user interaction experience.

The forward Euler method

The Euler method is a first-order integration method which belongs to a larger class called the Runge-Kutta methods (most famous being the fourth-order method [31]). The simplest version is called the Forward Euler Method [32]. This method of calculating each time step can be expressed in the following manner:

of vertex j, v^j) to point $\vec{P}^i$ (the position of vertex i, v^i) as follows:

$$\vec{U}\left(\vec{P}^i, \vec{P}^j\right) = \frac{\vec{P}^i - \vec{P}^j}{\left|\vec{P}^i - \vec{P}^j\right|} \qquad (1)$$

$$\begin{aligned} t_{n+1} &= t_n + \Delta t \\ \vec{P}^i_{n+1} &= \vec{P}^i_n + \Delta t f(t_n, \vec{P}^i_n) \end{aligned} \qquad (5)$$

which means the time t_n is advanced by the time-step Δt and then the position of the vertex v^i is updated based on the size of the time-step, and the current behaviour of the particle, f (which is usually a function which depends on the particle's current state and/or the time).

When applied to the movement of the RNA vertexes, the Forward Euler method can be written as:

$$
\begin{aligned}
t_{n+1} &= t_n + \Delta t \\
f(t_n, \vec{P}_n^i) &= \sum_{v^j \in L, j \neq i} \vec{R}\left(\vec{P}_n^i, \vec{P}_n^j\right) + \sum_{v^j \in C^i} \vec{A}\left(\vec{P}_n^i, \vec{P}_n^j\right) \\
\vec{P}_{n+1}^i &= \vec{P}_n^i + \Delta t f\left(t_n, \vec{P}_n^i\right)
\end{aligned}
\tag{6}
$$

where in this case, the behaviour of the particle with regard to its position is the sum of repulsion ($\vec{R}$) forces (over all loops $v^j \in L$, where L is the set of all loops) and attraction ($\vec{A}$) forces acting on the particle.

Since base pair vertexes do no participate in the repulsion step, the expression for a base pair vertex's Forward Euler implementation will be:

$$
\begin{aligned}
t_{n+1} &= t_n + \Delta t \\
f\left(t_n, \vec{P}_n^i\right) &= \sum_{v^j \in C^i} \vec{A}\left(\vec{P}_n^i, \vec{P}_n^j\right) \\
\vec{P}_{n+1}^i &= \vec{P}_n^i + \Delta t f\left(t_n, \vec{P}_n^i\right)
\end{aligned}
\tag{7}
$$

However, since the implementation of this expression is trivially similar to the expression in (6), the remainder of this text will focus on that expression, with the implications for the expression in (7) being omitted for brevity.

The main drawback presented by the Forward Euler method is its numerical instability. Simply put, when the time-step Δt is too long, or the coefficients which control the simulation become too large, the simulation does not stabilize into an equilibrium. In fact, it can become increasingly unstable. The solution to this drawback lies in the implementation of the Backward Euler method, which takes this instability into account.

The backward Euler method
Much like the Forward Euler method is described as explicit, there is an implicit Euler method; the Backward Euler method. Generally, it is defined as:

$$
\begin{aligned}
t_{n+1} &= t_n + \Delta t \\
\vec{P}_{n+1}^i &= \vec{P}_n^i + \Delta t f\left(t_n, \vec{P}_{n+1}^i\right)
\end{aligned}
\tag{8}
$$

where $f(t_n, \vec{P}_{n+1}^i)$ is again a function that describes the movement of the object. Notice it is very similar to the explicit method, but the term $\vec{P}_{n+1}^i$ appears on both sides of the equation. As a result, finding $\vec{P}_{n+1}^i$ is no longer a simple issue of updating the timestep, but it is that of solving for it algebraically.

In the case of the current simulation, the Backward Euler method would yield the following expression:

$$
\begin{aligned}
t_{n+1} &= t_n + \Delta t \\
f\left(t_n, \vec{P}_{n+1}^i\right) &= \sum_{v^j \in L, j \neq i} \vec{R}\left(\vec{P}_{n+1}^i, \vec{P}_n^j\right) + \sum_{v^j \in C^i} \vec{A}\left(\vec{P}_{n+1}^i, \vec{P}_n^j\right) \\
\vec{P}_{n+1}^i &= \vec{P}_n^i + \Delta t f\left(t_n, \vec{P}_{n+1}^i\right)
\end{aligned}
\tag{9}
$$

which becomes a fairly difficult equation to solve for $\vec{P}_{n+1}^i$ directly. Instead, an approximation is used to solve for $\vec{P}_{n+1}^i$.

Applying Newton's method to solve the Backward Euler expression
The expression in (9) can be rearranged to produce the following equation:

$$
\vec{P}_{n+1}^i = \vec{P}_n^i + \Delta t \left[\sum_{v^j \in L, j \neq i} \vec{R}\left(\vec{P}_{n+1}^i, \vec{P}_n^j\right) + \sum_{v^j \in C^i} \vec{A}\left(\vec{P}_{n+1}^i, \vec{P}_n^j\right) \right]
\tag{10}
$$

which can be rewritten as:

$$
\begin{aligned}
\vec{F}\left(\vec{P}_{n+1}^i\right) = \vec{0} &= -\vec{P}_{n+1}^i + \vec{P}_n^i \\
&+ \Delta t \left[\sum_{v^j \in L, j \neq i} \vec{R}\left(\vec{P}_{n+1}^i, \vec{P}_n^j\right) + \sum_{v^j \in C^i} \vec{A}\left(\vec{P}_{n+1}^i, \vec{P}_n^j\right) \right]
\end{aligned}
\tag{11}
$$

meaning the solution for $\vec{P}_{n+1}^i$ is the root of the function $\vec{F}(\vec{P}_{n+1}^i)$. While it may be difficult to solve for the root directly, Newton's method offers an approach for approximating the root of the vector function $\vec{F}\left(\vec{P}_{n+1}^i\right)$ [33].

Defining the vector function's components
As outlined in [33], it is necessary to define each of the components in $\vec{F}$ individually so that their derivatives can then be found with respect to each of the variables. In the case of the RNA simulation, the function $\vec{F}$ contains only two components; f_x and f_y which are each defined as:

$$
\begin{aligned}
f_x\left(\vec{P}_{n+1}^i\right) &= -x_{n+1}^i + x_n^i + \Delta t \left[\sum_{v^j \in L, j \neq i} R_x\left(\vec{P}_{n+1}^i, \vec{P}_n^j\right) \right. \\
&\left. + \sum_{v^j \in C^i} A_x\left(\vec{P}_{n+1}^i, \vec{P}_n^j\right) \right] \\
f_y\left(\vec{P}_{n+1}^i\right) &= -y_{n+1}^i + y_n^i + \Delta t \left[\sum_{v^j \in L, j \neq i} R_y\left(\vec{P}_{n+1}^i, \vec{P}_n^j\right) \right. \\
&\left. + \sum_{v^j \in C^i} A_y\left(\vec{P}_{n+1}^i, \vec{P}_n^j\right) \right]
\end{aligned}
\tag{12}
$$

This definition requires both $\vec{R}$ and $\vec{A}$ (as well as $\vec{U}$) to be defined in terms of their x and y components as:

$$R_x\left(\vec{P}_{n+1}^i, \vec{P}_n^j\right) = \frac{G}{\left(x_{n+1}^i - x_n^j\right)^2 + \left(y_{n+1}^i - y_n^j\right)^2} \times U_x\left(\vec{P}_{n+1}^i, \vec{P}_n^j\right)$$

$$R_y\left(\vec{P}_{n+1}^i, \vec{P}_n^j\right) = \frac{G}{\left(x_{n+1}^i - x_n^j\right)^2 + \left(y_{n+1}^i - y_n^j\right)^2} \times U_y\left(\vec{P}_{n+1}^i, \vec{P}_n^j\right)$$

$$A_x\left(\vec{P}_{n+1}^i, \vec{P}_n^j\right) = K \times \left[x_n^j - x_{n+1}^i + \left(r_{ideal} \times U_x\left(\vec{P}_{n+1}^i, \vec{P}_n^j\right)\right)\right]$$

$$A_y\left(\vec{P}_{n+1}^i, \vec{P}_n^j\right) = K \times \left[y_n^j - y_{n+1}^i + \left(r_{ideal} \times U_y\left(\vec{P}_{n+1}^i, \vec{P}_n^j\right)\right)\right]$$

$$U_x\left(\vec{P}_{n+1}^i, \vec{P}_n^j\right) = \frac{x_{n+1}^i - x_n^j}{\sqrt{\left(x_{n+1}^i - x_n^j\right)^2 + \left(y_{n+1}^i - y_n^j\right)^2}}$$

$$U_y\left(\vec{P}_{n+1}^i, \vec{P}_n^j\right) = \frac{y_{n+1}^i - y_n^j}{\sqrt{\left(x_{n+1}^i - x_n^j\right)^2 + \left(y_{n+1}^i - y_n^j\right)^2}}$$

$$(13)$$

Finding the components' derivatives

In order to apply Newton's method to the RNA model, the Jacobian matrix D of the vector function $\vec{F}$ needs to be defined. In order to do so, expressions for all partial derivatives of the components in Eqs. (12) - (13) need to be defined, where each component has two partial derivatives; with respect to x_{n+1}^i and with respect to y_{n+1}^i. The derivation of each component's partial derivatives is quite long and is not the main focus of this article. Therefore, for brevity purposes, the individual derivatives are outlined in the set of Eqs. (14)-(17):

$$\frac{\delta f_x}{\delta x_{n+1}^i}\left(\vec{P}_{n+1}^i\right) = -1 + \Delta t\left[\sum_{v^j \in L, j \neq i} \frac{\delta R_x}{\delta x_{n+1}^i}\left(\vec{P}_{n+1}^i, \vec{P}_n^j\right)\right.$$

$$\left. + \sum_{v^j \in C^i} \frac{\delta A_x}{\delta x_{n+1}^i}\left(\vec{P}_{n+1}^i, \vec{P}_n^j\right)\right]$$

$$\frac{\delta f_x}{\delta y_{n+1}^i}\left(\vec{P}_{n+1}^i\right) = \Delta t\left[\sum_{v^j \in L, j \neq i} \frac{\delta R_x}{\delta y_{n+1}^i}\left(\vec{P}_{n+1}^i, \vec{P}_n^j\right)\right.$$

$$\left. + \sum_{v^j \in C^i} \frac{\delta A_x}{\delta y_{n+1}^i}\left(\vec{P}_{n+1}^i, \vec{P}_n^j\right)\right]$$

$$\frac{\delta f_y}{\delta x_{n+1}^i}\left(\vec{P}_{n+1}^i\right) = \Delta t\left[\sum_{v^j \in L, j \neq i} \frac{\delta R_y}{\delta x_{n+1}^i}\left(\vec{P}_{n+1}^i, \vec{P}_n^j\right)\right.$$

$$\left. + \sum_{v^j \in C^i} \frac{\delta A_y}{\delta x_{n+1}^i}\left(\vec{P}_{n+1}^i, \vec{P}_n^j\right)\right]$$

$$\frac{\delta f_y}{\delta y_{n+1}^i}\left(\vec{P}_{n+1}^i\right) = -1 + \Delta t\left[\sum_{v^j \in L, j \neq i} \frac{\delta R_y}{\delta y_{n+1}^i}\left(\vec{P}_{n+1}^i, \vec{P}_n^j\right)\right.$$

$$\left. + \sum_{v^j \in C^i} \frac{\delta A_y}{\delta y_{n+1}^i}\left(\vec{P}_{n+1}^i, \vec{P}_n^j\right)\right]$$

$$(14)$$

$$r = \left(x_{n+1}^i - x_n^j\right)^2 + \left(y_{n+1}^i - y_n^j\right)^2$$

$$\frac{\delta R_x}{\delta x_{n+1}^i}\left(\vec{P}_{n+1}^i, \vec{P}_n^j\right) = \left\{\left[\frac{\delta U_x}{\delta x_{n+1}^i}\left(\vec{P}_{n+1}^i, \vec{P}_n^j\right) \times \frac{G}{r}\right]\right.$$
$$\left. - \left[\frac{2G\left(x_{n+1}^i - x_n^j\right)}{r^2} \times U_x\left(\vec{P}_{n+1}^i, \vec{P}_n^j\right)\right]\right\}$$

$$\frac{\delta R_x}{\delta y_{n+1}^i}\left(\vec{P}_{n+1}^i, \vec{P}_n^j\right) = \left\{\left[\frac{\delta U_x}{\delta y_{n+1}^i}\left(\vec{P}_{n+1}^i, \vec{P}_n^j\right) \times \frac{G}{r}\right]\right.$$
$$\left. - \left[\frac{2G\left(y_{n+1}^i - y_n^j\right)}{r^2} \times U_x\left(\vec{P}_{n+1}^i, \vec{P}_n^j\right)\right]\right\}$$

$$\frac{\delta R_y}{\delta x_{n+1}^i}\left(\vec{P}_{n+1}^i, \vec{P}_n^j\right) = \left\{\left[\frac{\delta U_y}{\delta x_{n+1}^i}\left(\vec{P}_{n+1}^i, \vec{P}_n^j\right) \times \frac{G}{r}\right]\right.$$
$$\left. - \left[\frac{2G\left(x_{n+1}^i - x_n^j\right)}{r^2} \times U_y\left(\vec{P}_{n+1}^i, \vec{P}_n^j\right)\right]\right\}$$

$$\frac{\delta R_y}{\delta y_{n+1}^i}\left(\vec{P}_{n+1}^i, \vec{P}_n^j\right) = \left\{\left[\frac{\delta U_y}{\delta y_{n+1}^i}\left(\vec{P}_{n+1}^i, \vec{P}_n^j\right) \times \frac{G}{r}\right]\right.$$
$$\left. - \left[\frac{2G\left(y_{n+1}^i - y_n^j\right)}{r^2} \times U_y\left(\vec{P}_{n+1}^i, \vec{P}_n^j\right)\right]\right\}$$

$$(15)$$

$$\frac{\delta A_x}{\delta x_{n+1}^i}\left(\vec{P}_{n+1}^i, \vec{P}_n^j\right) = K \times \left[-1 + \left(r_{ideal} \times \frac{\delta U_x}{\delta x_{n+1}^i}\left(\vec{P}_{n+1}^i, \vec{P}_n^j\right)\right)\right]$$

$$\frac{\delta A_x}{\delta y_{n+1}^i}\left(\vec{P}_{n+1}^i, \vec{P}_n^j\right) = K \times \left(r_{ideal} \times \frac{\delta U_x}{\delta y_{n+1}^i}\left(\vec{P}_{n+1}^i, \vec{P}_n^j\right)\right)$$

$$\frac{\delta A_y}{\delta x_{n+1}^i}\left(\vec{P}_{n+1}^i, \vec{P}_n^j\right) = K \times \left(r_{ideal} \times \frac{\delta U_y}{\delta x_{n+1}^i}\left(\vec{P}_{n+1}^i, \vec{P}_n^j\right)\right)$$

$$\frac{\delta A_y}{\delta y_{n+1}^i}\left(\vec{P}_{n+1}^i, \vec{P}_n^j\right) = K \times \left[-1 + \left(r_{ideal} \times \frac{\delta U_y}{\delta y_{n+1}^i}\left(\vec{P}_{n+1}^i, \vec{P}_n^j\right)\right)\right]$$

$$(16)$$

$$r = \sqrt{\left(x_{n+1}^i - x_n^j\right)^2 + \left(y_{n+1}^i - y_n^j\right)^2}$$

$$\frac{\delta U_x}{\delta x_{n+1}^i}\left(\vec{P}_{n+1}^i, \vec{P}_n^j\right) = \frac{\left(y_{n+1}^i - y_n^j\right)^2}{r^3}$$

$$\frac{\delta U_x}{\delta y_{n+1}^i}\left(\vec{P}_{n+1}^i, \vec{P}_n^j\right) = \frac{-\left(y_{n+1}^i - y_n^j\right)\left(x_{n+1}^i - x_n^j\right)}{r^3}$$

$$\frac{\delta U_y}{\delta x_{n+1}^i}\left(\vec{P}_{n+1}^i, \vec{P}_n^j\right) = \frac{-\left(y_{n+1}^i - y_n^j\right)\left(x_{n+1}^i - x_n^j\right)}{r^3}$$

$$\frac{\delta U_y}{\delta y_{n+1}^i}\left(\vec{P}_{n+1}^i, \vec{P}_n^j\right) = \frac{\left(x_{n+1}^i - x_n^j\right)^2}{r^3}$$

$$(17)$$

and the matrix D is defined as:

$$D\left(\vec{P}_{n+1}^i\right) = \begin{bmatrix} \frac{\delta f_x}{\delta x_{n+1}^i}\left(\vec{P}_{n+1}^i\right) & \frac{\delta f_x}{\delta y_{n+1}^i}\left(\vec{P}_{n+1}^i\right) \\ \frac{\delta f_y}{\delta x_{n+1}^i}\left(\vec{P}_{n+1}^i\right) & \frac{\delta f_y}{\delta y_{n+1}^i}\left(\vec{P}_{n+1}^i\right) \end{bmatrix}.$$

Constructing the Newton step

Given the function $\vec{F}$ and the matrix D, progressively better estimates for the value of $\vec{P}^i_{n+1}$ can be found by applying the following Newton step:

$$\vec{P}^i_{n+1,k+1} = \vec{P}^i_{n+1,k} - \vec{F}\left(\vec{P}^i_{n+1,k}\right) \times D^{-1}\left(\vec{P}^i_{n+1,k}\right) \quad (18)$$

where $D^{-1}\left(\vec{P}^i_{n+1,k}\right)$ is the inverse matrix of $D\left(\vec{P}^i_{n+1,k}\right)$. That is, at every Newton step $k+1$, the value of both the function $\vec{F}$ and its components' derivatives, encapsulated in the matrix D^{-1}, are evaluated at the point $\vec{P}^i_{n+1,k}$, that is, the point $\vec{P}^i_{n+1}$ from the previous Newton step. The initial estimate, $\vec{P}^i_{n+1,0}$ can be obtained by applying the Forward Euler. As more Newton steps are repeated, a better and better estimate for $\vec{P}^i_{n+1}$ emerges. However, each Newton step increases the run-time of each iteration of the algorithm. In general, each additional Newton step increases the run time of the physics based simulation by $O(L^2)$ where L is the number of loops in the simulation.

Experimental parameters and test-bed structures

For the purposes of these experiments, 17 RNA molecules were chosen from the RNA STRAND v2.0 database [34], and were run under two different configurations. The configurations and their parameters can be found in Table 1, while the structure details can be found in Table 2[1]. The structure lengths are given in "nt," which stands for "nucleotides."

Different time-steps were chosen for the different configurations (Table 1). Configuration 1 was assigned the highest time-steps it can support without losing stability. Configuration 2 can handle larger time steps, but the choice of time-step influences the choice for the number of Newton iterations (such that larger time steps required more Newton iterations to reach convergence). Therefore, a value of 3.0 was chosen to support satisfactory convergence within 5 Newton iterations.

Each structure was run 20 times and the CPU time of the run was measured until the structure stabilized (that is, until the large movement of any of its components was less than ϵ). The average run-time was calculated and plotted. If a structure's stabilization process took more than 30 mins (1800 s) it was terminated and its stabilization time was taken as 1800 s.

Table 1 The parameters for the two experimental configurations

Configuration #	1	2
Movement update	Forward Euler	Backward Euler
K	10.0	10.0
G	0.01	0.01
Time-step (Δt)	0.01	3.0
Minimal stablization movement (ϵ)	0.0001	0.3
Newton iterations	N/A	5

Table 2 The RNA structures chosen for comparison between the forward and backward Euler methods

#	RNA STRAND ID	Original ID	Length (nt)	Reference
1	NDB_00051	PDB: 1VTQ	75	[40]
2	PDB_01255	PDB: 2R8S	159	[41]
3	PDB_01076	PDB: 2GO5	217	[42]
4	PDB_00985	PDB: 2CZJ	248	[43]
5	PDB_00528	PDB: 1KOG	304	[44]
6	PDB_00398	PDB:1FCW	380	[45]
7	PDB_01144	PDB: 2J37	408	[46]
8	SRP_00288	SRPDB: Sacc.cere._M28116	522	[47]
9	RFA_00829	Rfam: RF00551	551	[48]
10	CRW_00736	CRW: a.I2.c.N.tabacum.B.ND2	696	[35]
11	CRW_00731	CRW: a.I2.c.N.tabacum.A.trnI.i1	772	[35]
12	CRW_00757	CRW: a.I2.m.Z.mays.A.OX2.i1	912	[35]
13	CRW_00533	CRW: d.233.m.C.elegans	953	[35]
14	CRW_00540	CRW: d.233.m.L.terrestris	1279	[35]
15	CRW_00539	CRW: d.233.m.L.bleekeri	1333	[35]
16	CRW_00742	CRW: a.I2.m.A.aegerita.B.LSU.2059	1857	[35]
17	CRW_00534	CRW: d.233.m.C.eugametos	1915	[35]

Improving the attraction force calculations

The system of forces described in the previous section allowed the RNA structure simulation to stabilize and present the RNA structural elements much better than the former jViz.RNA implementation (Fig. 4a). However, the resulting stable layouts were not satisfactory due to the overlap artefacts created (Fig. 4b–c). Stems would often overlap loops and would not stabilize into their correct position based on their connectivity to the loops. While a user could, in theory, address such a problem manually, we felt there is room for further improvements. In order to correct the overlap artefacts, a slight modification to the attraction force calculation was implemented.

Originally, the attraction forces would apply attraction between the centres of two vertexes (Fig. 5a–b). However, with a slight modification, each vertex can store the ideal positions for each stem protruding from it (Fig. 5c). Using these ideal positions in the equation for $\vec{A}(P^i, P^j)$ to move each vertex to its ideal position and orientation (Fig. 5d). The resulting layouts prove to be much more visually appealing and containing much less overlap, especially for smaller RNA structures (Figs. 8, 9, 10, 11, 12 and 13).

Results

Comparison of jViz.RNA's performance employing the forward and backward Euler methods

Figure 6 shows the run times of jViz.RNA when employing the Forward and Backward Euler method. As expected,

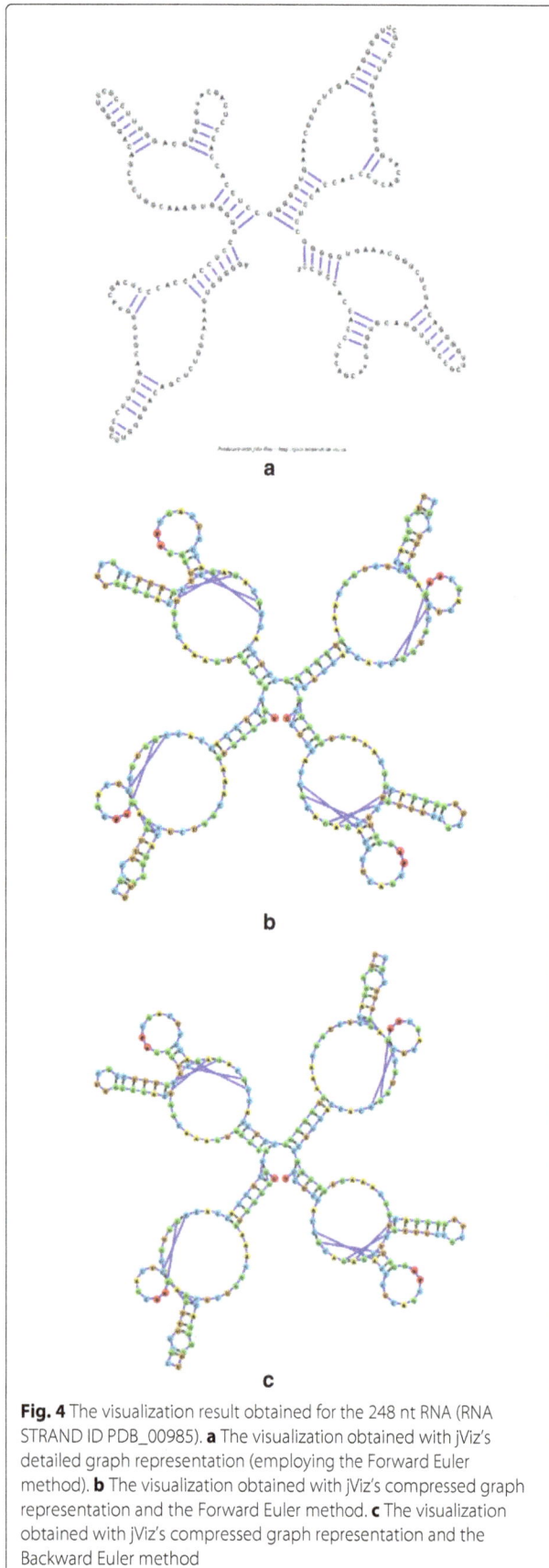

Fig. 4 The visualization result obtained for the 248 nt RNA (RNA STRAND ID PDB_00985). **a** The visualization obtained with jViz's detailed graph representation (employing the Forward Euler method). **b** The visualization obtained with jViz's compressed graph representation and the Forward Euler method. **c** The visualization obtained with jViz's compressed graph representation and the Backward Euler method

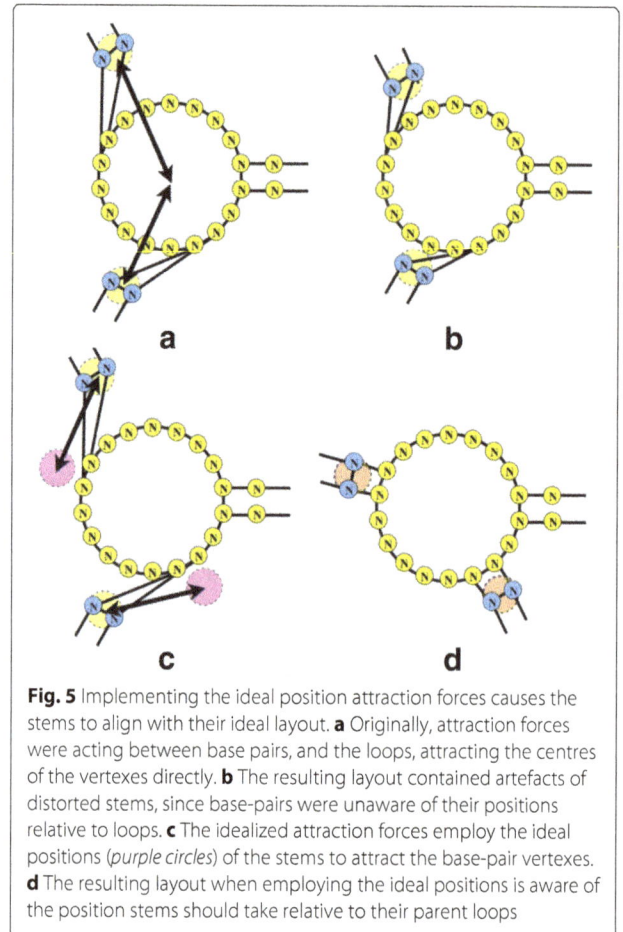

Fig. 5 Implementing the ideal position attraction forces causes the stems to align with their ideal layout. **a** Originally, attraction forces were acting between base pairs, and the loops, attracting the centres of the vertexes directly. **b** The resulting layout contained artefacts of distorted stems, since base-pairs were unaware of their positions relative to loops. **c** The idealized attraction forces employ the ideal positions (*purple circles*) of the stems to attract the base-pair vertexes. **d** The resulting layout when employing the ideal positions is aware of the position stems should take relative to their parent loops

since the Backward Euler method takes a much larger time step, the structures subject to the Backward Euler simulation converge to a stable layout much more rapidly than when subject to the Forward Euler. In fact, to truly appreciate the difference, the log_{10} of the run times was taken and plotted in Fig. 7. As can be seen, the run times of the Forward Euler method are often $\approx$ 100 times longer than the Backward Euler run times. Considering the fact that no structure was allowed longer than 1800 seconds to stabilize, it is fair to assume that under the current parameters of K and G, the difference in run time could have been even greater for some structures.

One would expect that the run-times would increase in a quadratic order to the number of nucleotides. However, while there is a general increase in run time with structure size, some small structures take longer than larger ones to stabilize. This observation points to the fact that the connectivity of the structure plays a very important role in its stabilization time. Overall, a structure X composed of 3 times as many nucleotides as structure Y would take longer to stabilize, but it may not be straightforward to deduce exactly how much longer. Even the number of vertexes in a given structure does not provide a good heuristic

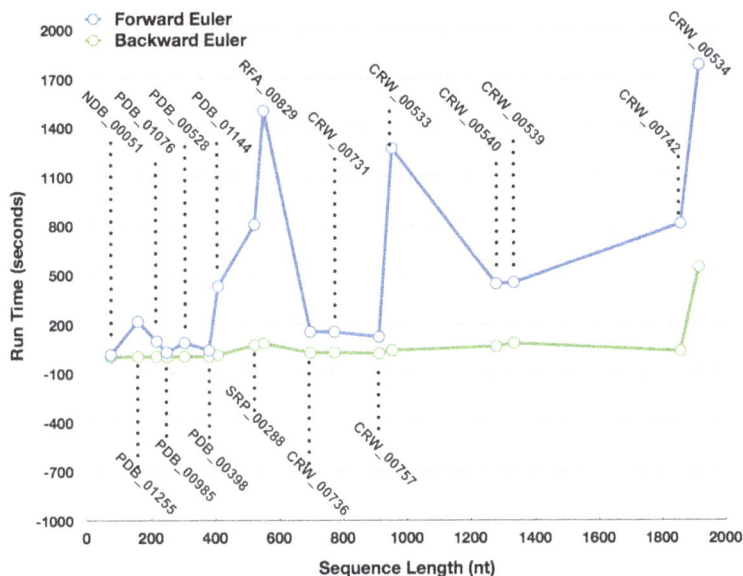

Fig. 6 The Run-times (expressed in seconds) of jViz.RNA's compressed graph representations employing both the Forward and Backward Euler methods

to calculating the difference in stabilization time for both the Forward and Backward Euler.

Despite the relative uncertainty in the relationship between a given structure's run time and its size, there is a great deal of certainty that the Backward Euler proved superior when compared to the Forward Euler. First, it can produce stable layouts employing a time step 300 times larger than the Forward Euler method without losing stability. Second, it exhibits much faster run-time

performance. As demonstrated in this work, some large structures may pose a challenge to a system which takes smaller time step since the topology of the structure itself dictates how long it will take to stabilize

Visual comparison of the different algorithms

In order to get a full appreciation of the advantages of the different methods explored in this work, as well as potential future improvements, it is necessary to look at

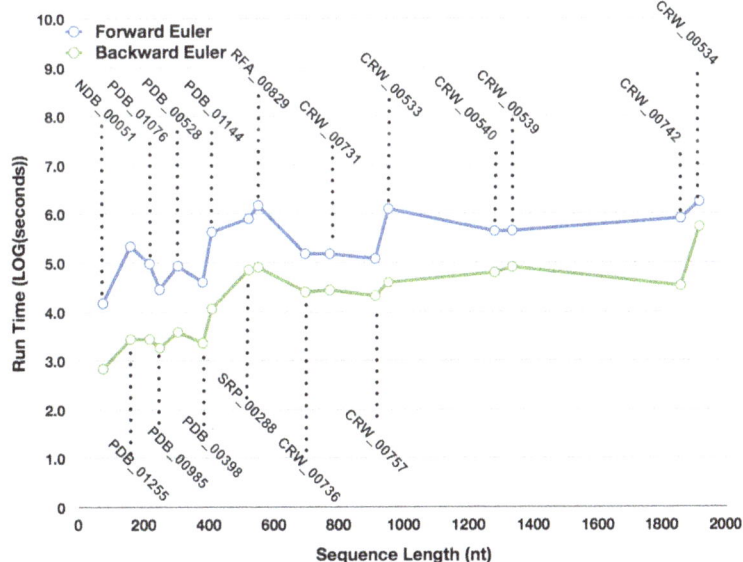

Fig. 7 The Run-times (expressed in log_{10}(seconds)) of jViz.RNA's compressed graph representations employing both the Forward and Backward Euler methods

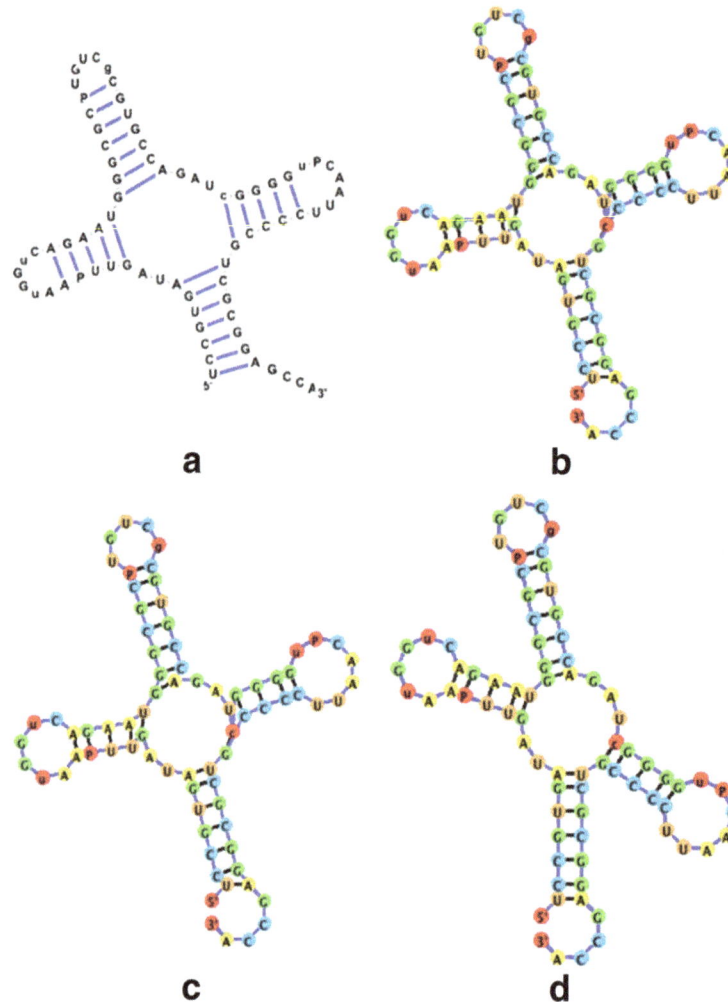

Fig. 8 The visualization result obtained for the 75 nt RNA (RNA STRAND ID NDB_00051) utilizing: **a** The detailed graph representation (employing the Forward Euler method). **b** The compressed graph representation and the Forward Euler method. **c** The compressed graph representation and the Backward Euler method. **d** The compressed graph representation and the Backward Euler method while employing ideal positions attraction forces

both the run-times reported in Figs. 6 and 7, as well as at the resulting visualizations each method produced. Figures 8, 9, 10, 11, 12 and 13 demonstrate the visualizations produced when employing the four different methods explored in this paper (all Euler implementation of the compressed graph as well as their detailed graph counter-part).

Comparing the images produced by the compressed and detailed graphs reveals additional differences between the methods. It is immediately evident that the figures produced by employing the compressed graph adhere more strictly to RNA visualization conventions; namely, the circular loops, and the constant distance between base pairs. However, at the same time, the detailed graph representation demonstrates some advantages over the compressed graph images. First, there are no cases of

stems that intersect, which contributes to less user intervention being required to "untangle" the structure. The compressed graph representations, on the other hand, occasionally have stems that intersect and would require the user to explore the structure to resolve such conflicts. Though this drawback is addressed to a large extent by substituting the ideal positions as the attraction points for the vertexes, the need to manually untangle the structure may persist in certain cases.

Figure 14 demonstrates the layout algorithm as it is applied to a few instances of related RNAs. All three RNA molecules shown in Fig. 14 are tRNA molecules for different amino acids. The utilization of the ideal layout algorithm allows the related RNA molecules to be laid out in a similar conformation. Though different users may wish to align RNA structures differently, related RNA

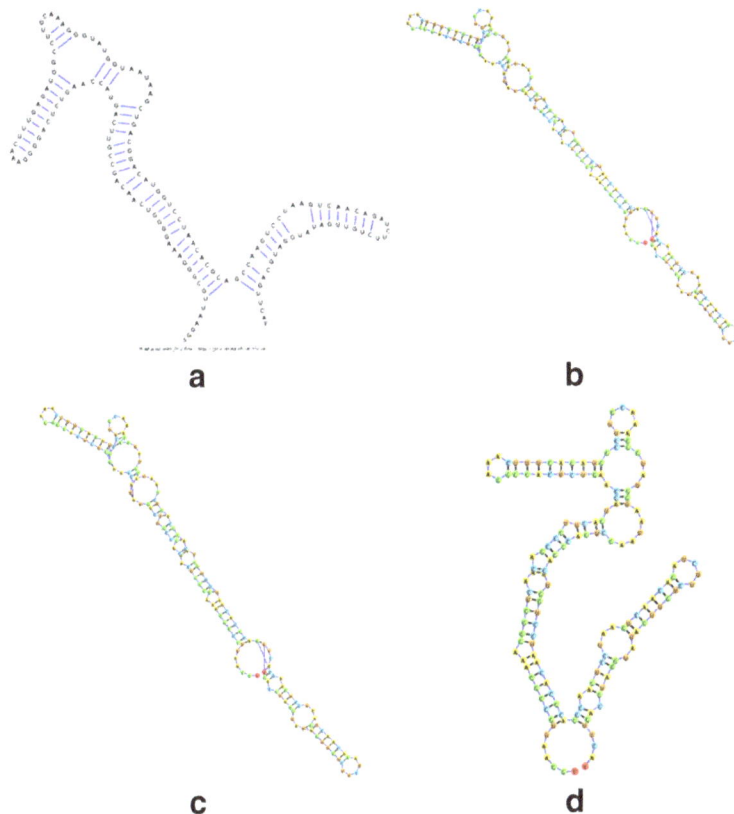

Fig. 9 The visualization result obtained for the 159 nt RNA (RNA STRAND ID PDB_01255) utilizing: **a** The detailed graph representation (employing the Forward Euler method). **b** The compressed graph representation and the Forward Euler method. **c** The compressed graph representation and the Backward Euler method. **d** The compressed graph representation and the Backward Euler method while employing ideal positions attraction forces

molecules should be drawn in a similar fashion to easily highlight homologous structural regions, which may share functional roles (such as the anticodons located on the tRNA middle stem, and the binding site for the amino acid located at the 3' end).

Discussion

Two major objectives have been set out for this work: The first was to improve jViz.RNA's visualization through a new representation, the second was to design an enhanced automatic layout algorithm in light of this new representation and to improve its run-time performance. Both objectives have been achieved through our employment of the compressed graph representation and the Backward Euler method.

Improving jViz.RNA's visualization

Comparing the detailed graphs and compressed graph visualization demonstrated that employing the compressed graphs produces visualizations more consistent with current RNA visualization methods, and does so at a fraction of the time. However, for large molecules that gain in time may be offset by the time required by

the user to examine and untangle the RNA structure in the case of intersecting stems. This can be addressed by the modification of the attraction force to act as both an attraction and rotation force. However, the existence of some overlap may still be present for certain structures (such as those seen in Fig. 11d). Future work to address this limitation would focus on configuring the system to find an equilibrium between the correct positioning of stems and repulsion between the various RNA components to prevent them from overlaying each other.

Extending the automatic layout algorithm

Comparing the two methods of calculating movement for the RNA components of the compressed graph revealed that employing the Backward Euler method produces more stable simulations of the vertexes' movement, and does so in less time than the Forward Euler method (Figs. 6 and 7). Though it is evident the connectivity of the structure determines the degree of run-time improvement, it is evident that all structures we have tested stabilize faster using the Backward Euler method. Given that the Backward Euler implementation is much

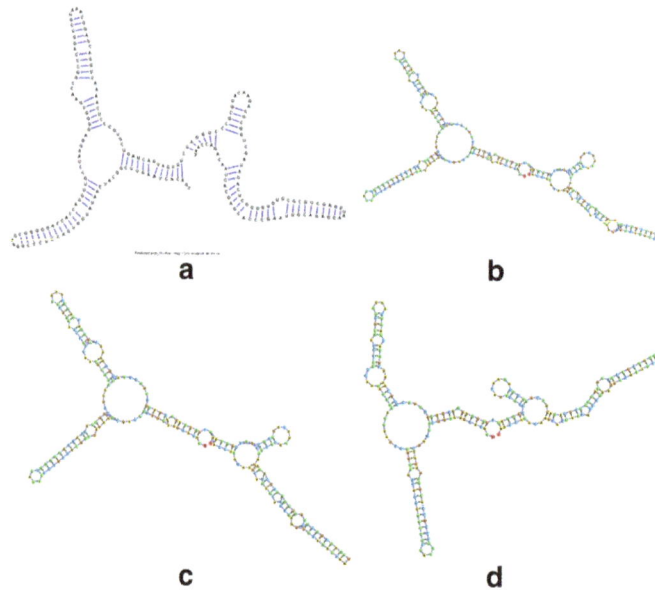

Fig. 10 The visualization result obtained for the 217 nt RNA (RNA STRAND ID PDB_01076) utilizing: **a** The detailed graph representation (employing the Forward Euler method. **b** The compressed graph representation and the Forward Euler method. **c** The compressed graph representation and the Backward Euler method. **d** The compressed graph representation and the Backward Euler method while employing ideal positions attraction forces

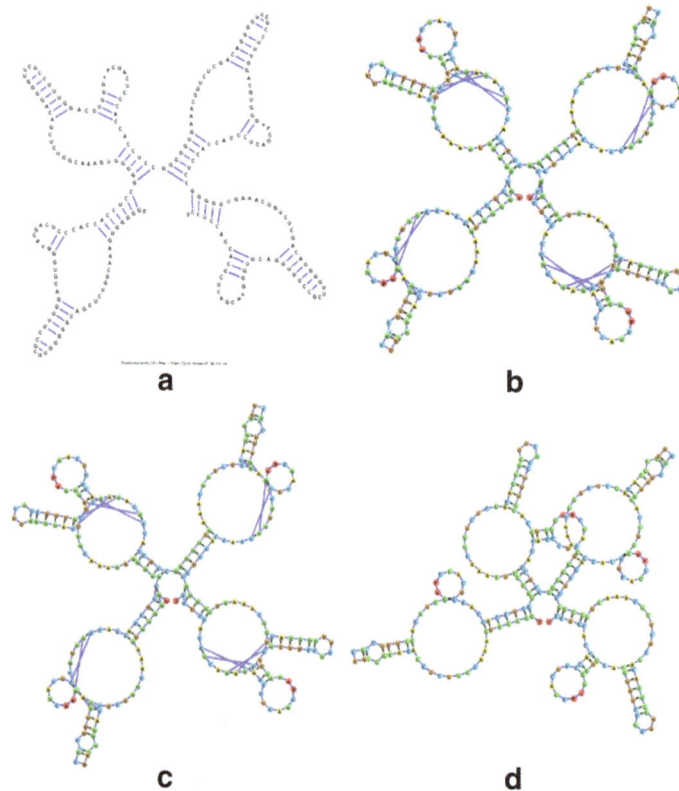

Fig. 11 The visualization result obtained for the 248 nt RNA (RNA STRAND ID PDB_00985) utilizing: **a** The detailed graph representation (employing the Forward Euler method). **b** The compressed graph representation and the Forward Euler method. **c** The compressed graph representation and the Backward Euler method. **d** The compressed graph representation and the Backward Euler method while employing ideal positions attraction forces

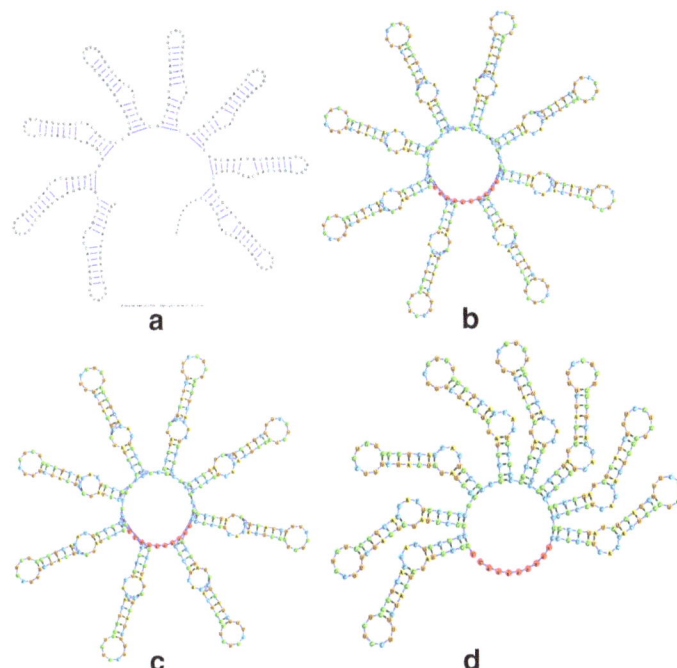

Fig. 12 The visualization result obtained for the 304 nt RNA (RNA STRAND ID PDB_00528) utilizing: **a** The detailed graph representation (employing the Forward Euler method). **b** The compressed graph representation and the Forward Euler method. **c** The compressed graph representation and the Backward Euler method. **d** The compressed graph representation and the Backward Euler method while employing ideal positions attraction forces

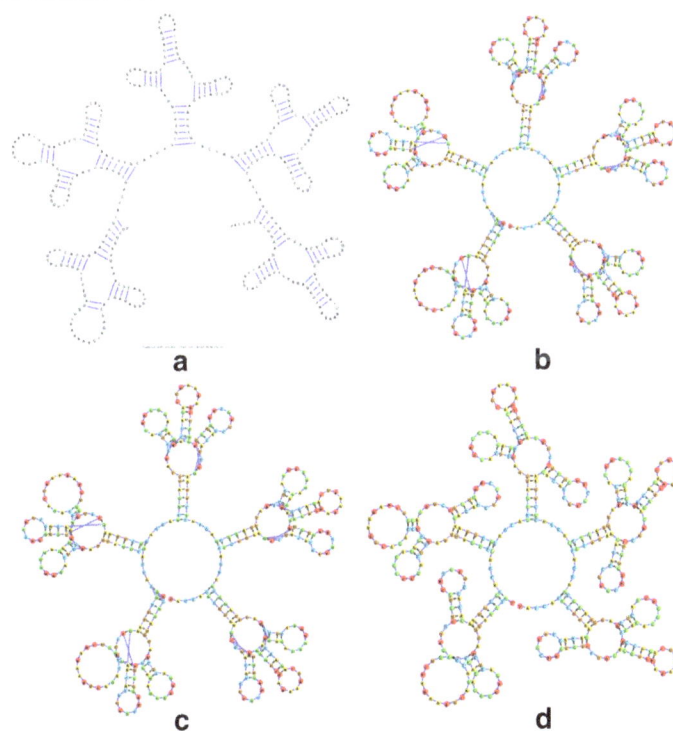

Fig. 13 The visualization result obtained for the 380 nt RNA (RNA STRAND ID PDB_00398) utilizing: **a** The detailed graph representation (employing the Forward Euler method). **b** The compressed graph representation and the Forward Euler method. **c** The compressed graph representation and the Backward Euler method. **d** The compressed graph representation and the Backward Euler method while employing ideal positions attraction forces

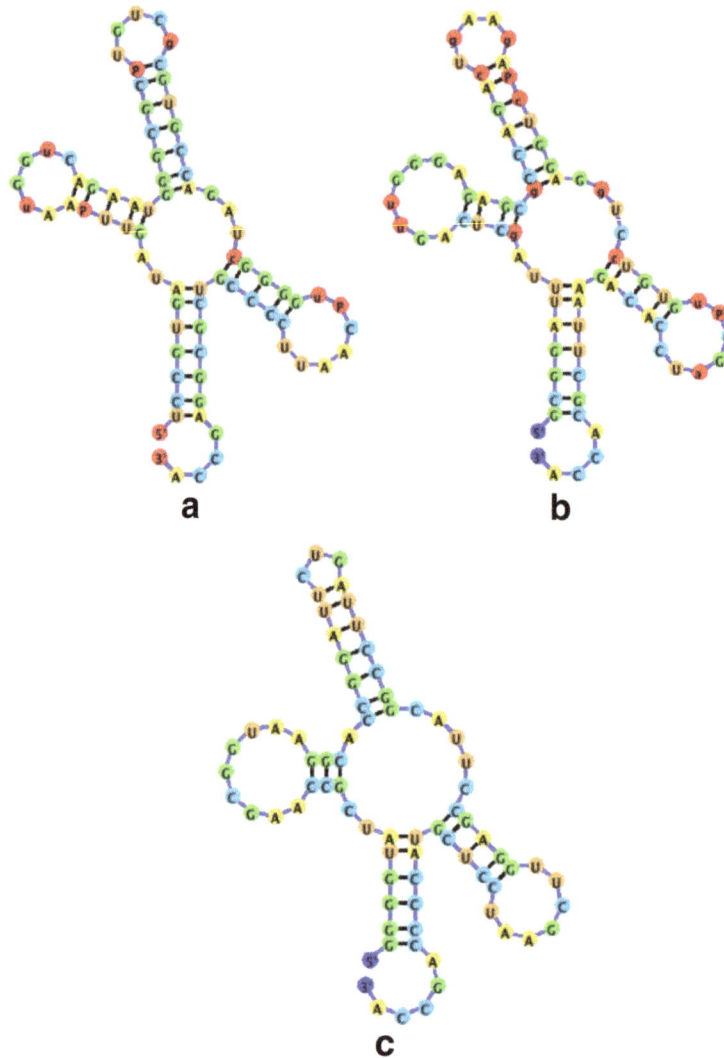

Fig. 14 A visualization comparison of tRNA molecules employing the Backward Euler method and the ideal positions attraction forces. **a** The visualization for a Yeast tRNA utilizing jViz.RNA (RNA STRAND ID: NDB_00051, PDB ID: 1VTQ). **b** The visualization for a Yeast tRNA utilizing jViz.RNA (RNA STRAND ID: PDB_00045, PDB ID: 1EHZ). **c** The visualization for an E.coli tRNA utilizing jViz.RNA (RNA STRAND ID: PDB_00426, NDB ID: 1GTS)

more stable than its Forward counter-part, it is the Backward Euler method that becomes the more desirable choice.

Conclusion

In this article we described advances made in the representation and layout algorithms for dynamic RNA secondary structure visualization. We reviewed existing tools and algorithms and showed that only few allow for dynamic visualization. One such tool is jViz.RNA. We discussed its shortcomings in terms of layout, stability and run-time performance and proposed several improvements based on a compressed graph representation and advanced numerical integration methods. We presented two graph based representations for RNA visualization, as

well as two methods to create dynamic RNA structures that lay themselves out automatically and respond to user interaction.

The utilization of compressed graphs as a model of the RNA structure, the experiments shown here profiling their performance, and examining the underlying physics, demonstrate a substantial improvement to the original representation and layout algorithms employed in jViz.RNA. The work on layout conventions was greatly influenced by feedback from life science collaborators. The new algorithms have increased stability for automatic layout, reduced overlap of structural elements that decreases the need for user intervention, and increased run time performance to allow for the handling of larger RNA structures. Having examined the basic properties of compressed graph behaviour, we can verify it is a better tool to produce dynamic and responsive RNA models than its detailed counter-part. We have also discovered a few important areas of improvement such as stem orientation relative to loops, and prevention of structural element intersection, and provided improvements to the layout algorithm to handle these issues. By comparing the Forward and Backward Euler methods we demonstrated the superiority of the Backward Euler method in its ability to support larger time steps without losing stability, and consequently, allowing for faster simulations of the RNA compressed graph.

Our future work will focus on incorporating pseudoknot representation into the compressed graph model, to allow for the visualization of pseudoknotted structures. Additionally, constructing and modifying RNA structures will also be explored.

Overall, this manuscript has explored the use of compressed graphs to improve the layout of RNA secondary structures, as well as the best physics based simulation method for such an implementation. The results presented demonstrate that both the compressed graph representation, as well as the Backward Euler integrator, greatly enhance the run-time performance and usability.

We anticipate these findings to benefit other researchers in RNA structure visualization, or more generally, biological structure visualization, as the underlying ideas are transferrable. We also provide a tool that is platform independent, easy to use, and can quickly render publication quality structure images. We would anticipate that many researchers in RNA structure will find this tool useful and that it will find wide acceptance. We believe the technical details discussed in this manuscript will impact how other visualization researchers think about dynamic structure visualization and we see the impact of this work in both providing a useful software tool as well as presenting a methodology for other visualization tools to adopt.

Abbreviations
CRW: Comparative RNA Website; DNA: Deoxyribo-nucleic Acid; nt: nucleotide; PDB: Protein Data Bank; Rfam: RNA Families Database; RNA: Ribo-nucleic Acid; rRNA: Ribosomal RNA; SRPDB: Signal Recognition Particle Database; tRNA: Transfer RNA; UTR: Untranslated Region

Acknowledgements
The authors would like to acknowledge feedback on the visualization aspects of the software received from Peter Unrau, Professor of Molecular Biology and Biochemistry, Simon Fraser University, his lab members, and the many international users of earlier versions of the software who have provided their feedback.

Funding
We would like to thank the Natural Sciences and Engineering Research Council of Canada (NSERC) for their support under grant number 611362. The first author would also like to thank NSERC for providing funding in the form of a CGS-D scholarship.

Authors' contributions
Both authors have contributed equally to the writing of this manuscript. Both authors read and approved the final manuscript.

About the Authors
Boris Shabash received a Bachelor of Health Sciences (BHSc), specializing in bioinformatics, at the University of Calgary (2009), a Master of Science (MSc), specializing in Computing Science, at Simon Fraser University (2011), and is currently working on his PhD at Simon Fraser University. His areas of expertise lie in Evolutionary Computation techniques for solving bioinformatics based problems, as well as bioinformatics related visualization. He has also extensive experience working on synthetic biology based applications, both in sillico and in vitro.

Kay C. Wiese is an Associate Professor in the School of Computing Science at Simon Fraser University, BC, Canada, where he is the Director of the Bioinformatics Research Lab. His research interests are in computational intelligence and bioinformatics, particularly algorithms for RNA secondary structure prediction and visualization. Dr. Wiese has served as a member of the organizing committee of Research in Computational Molecular Biology (RECOMB) and the IEEE Symposium on Computational Intelligence in Bioinformatics and Computational Biology (CIBCB) in various roles including Program Chair. Dr. Wiese was the Chair of the IEEE Computational Intelligence Society Bioinformatics and Bioengineering Technical Committee for 2006-2007 and served as the Vice President for Technical Activities of the IEEE Computational Intelligence Society during 2008-2009. He also serves on the steering committee of 3 IEEE journals and on the editorial board of 2 international journals in the area of Computational Intelligence and Bioinformatics.

Competing interests
The authors declare that they have no competing interests.

References
1. Rietveld K, Poelgeest RV, Pleij A, Boom JHV, Bosch L. The tRNA-like structure at the 3' terminus of turnip yellow mosaic virus RNA. Differences and similarities with canonical tRNA. Nucleic Acids Res. 1982;10(6): 1929–46.
2. Joshi RL, Joshi S, Chapeville F, Haenni AL. tRNA-like structures of plant viral RNAs: conformational requirements for adenylation and aminoacylation. EMBO J. 1983;2(7):1123–27.
3. Rietveld K, Pleij CWA, Bosch L. Three-dimensional models of the tRNA-like 3' termini of some plant viral RNAs. EMBO J. 1983;2(7):1079–85.
4. Rao ALN, Dreher TW, Marsh LE, Hall TC. Telomeric function of the tRNA-like structure of brome mosaic virus RNA. In: Proceedings of the National Academy of Science of the United States of America, vol. 86. Washington: National Academy of Sciences; 1989. p. 5335–39.
5. Matsuda D, Dreher TW. The tRNA-like structure of Turnip yellow mosaic virus RNA is a 3'-translational enhancer. Virology. 2003;321:36–46.
6. Powers T, Noller HF. A functional pseudoknot in 16S ribosomal RNA. EMBO J. 1991;10(8):2203–14.
7. Finken M, Kirschner P, Meier A, Wrede A, Bottger EC. Molecular basis of streptomycin resistance in Mycobacterium tuberculosis: alterations of the ribosomal protein S12 gene and point mutations within a functional 16S ribosomal RNA pseudoknot. Mol Microbiol. 1993;9(6):1239–46.
8. Chaloin L, Lehmann MJ, Sczakiel G, Restle T. Endogenous expression of a high-affinity pseudoknot rna aptamer suppresses replication of hiv-1. Nucleic Acids Res. 2002;30(18):4001–008. doi:10.1093/nar/gkf522. http://nar.oxfordjournals.org/content/30/18/4001.full.pdf+html.
9. Shabash B, Wiese KC. RNA Visualization: Relevance and the Current State-of-the-art Focusing on Pseudoknots. IEEE/ACM Trans Comput Biol Bioinforma. 2016;PP(99):1–1. doi:10.1109/TCBB.2016.2522421.
10. Ponty Y, Leclerc F. Drawing and editing the secondary structure(s) of RNA In: Picardi E, editor. RNA Bioinformatics. New York: Springer; 2015. p. 63–100.
11. Darty K, Denise A, Ponty Y. VARNA: Interactive drawing and editing of the RNA secondary structure. Bioinformatics. 2009;25(15):1974–5.
12. Wiese KC, Glen E, Vasudevan A. jViz.RNA - A Java Tool for RNA Secondary Structure Visualization. IEEE Trans Nanobioscience. 2005;4(3): 212–8.

13. Wiese KC, Glen E. jViz.RNA - An Interactive Graphical Tool for Visualizing RNA Secondary Structures Including Pseudoknots. In: The 19th Symposium on Computer-Based Medical Systems. New York: Institute of Electrical and Electronics Engineers; 2006.

14. Glen E. JVIZ.RNA - A TOOL FOR VISUAL COMPARISON AND ANALYSIS OF RNA SECONDARY STRUCTURES Master's thesis, Simon Fraser University. 2007.

15. Shabash B. Improving the Portability and Performance of jViz.RNA - A Dynamic RNA Visualization Software. Master's thesis, Simon Fraser University. 2009.

16. Shabash B, Wiese KC, Glen E. Improving the Portability and Performance of jViz.RNA - A Dynamic RNA Visualization Software In: Shibuya T, Kashima H, Sese J, Ahmad S, editors. Pattern Recognition in Bioinformatics. Lecture Notes in Computer Science. Cham: Springer; 2012. p. 82–93.

17. Kerpedjiev P, Hammer S, Hofacker IL. Forna (force-directed RNA): Simple and effective online RNA secondary structure diagrams. Bioinformatics. 2015;31(20):3377. doi:10.1093/bioinformatics/btv372.

18. Kim W, Lee Y, Han K. Visualization of RNA Pseudoknot Structure In: Palma JLM, Sousa AA, Dongarra J, Hernández V, editors. High Performance Computing for Computational Science — VECPAR 2002. Lecture Notes in Computer Science. Berlin: Springer; 2003. p. 181–94.

19. Han K, Byun Y. PseudoViewer2: visualization of RNA pseudoknots of any type. Nucleic Acids Res. 2003;31(13):3432–40.

20. Byun Y, Han K. PseudoViewer3: generating planar drawings of large scale RNA structures with pseudoknots. Bioinformatics. 2009;25(11):1435–7.

21. Byun YA, Han KS. Visualization method of RNA Pseudoknot structures. Google Patents. EP Patent App. EP20,040,012,477. 2004. http://www.google.com/patents/EP1482439A2?cl=en. Accessed 9 Feb 2017.

22. Byun Y, Han K. PseudoViewer: web application and web service for visualizing RNA pseudoknots and secondary structure. Nucleic Acids Res. 2006;34:416–22.

23. Seibel PN, Müller T, Dandekar T, Schultz J, Wolf M. 4SALE - A tool for synchronous RNA sequences and secondary structure alignment and editing. BMC Bioinforma. 2006;7(498):498–504.

24. Seibel PN, Müller T, Dandekar T, Wolf M. Synchronous visual analysis and editing of RNA sequences and secondary structure alignments using 4SALE. BMC Res Notes. 2008;1(91):91–7.

25. Jossinet F, Westhof E. Sequence to Structure (S2S): display, manipulate and interconnect RNA data from sequence to structure. Bioinformatics. 2005;21(15):3320–21.

26. Jossinet F, Westhof E. S2S-Assemble2: a Semi-Automatic Bioinformatics Framework to Study and Model RNA 3D Architectures. Hoboken: Wiley-VCH Verlog GmbH & Co. KGaA; 2014, pp. 667–86.

27. Xu W, Wongsa A, ad Lei Shang JL, Cannone JJ, Gutell RR. RNA2DMap: A Visual Exploration Tool of the Information in RNA's Higher-Order Structure. In: Proceedings of the 2011 IEEE International Conference on Bioinformatics and Biomedicine. New York: Institute of Electrical and Electronics Engineers; 2011. p. 613–7.

28. Weinberg Z, Breaker RR. R2R - software to speed the depiction of aesthetic consensus RNA secondary structures. BMC Bioinforma. 2011;12(1):3–11.

29. Lai D, Proctor JR, Zhu JYA, Meyer IM. R-chie: a web server and R package for visualizing RNA secondary structures. Nucleic Acids Res. 2012; 40(12):95.

30. Ponty Y. VARNA: Visualization Applet for RNA. http://varna.lri.fr/index.php?lang=en&page=demo&css=varna. Accessed 9 Feb 2017.

31. Kutta W. Beitrag zur näherungweisen integration totaler differentialgleichungen. 1901.

32. Weisstein EW. Euler Forward Method. http://mathworld.wolfram.com/EulerForwardMethod.html. Accessed 9 Feb 2017.

33. University O. Lecture 13 Nonlinear Systems - Newton's Method. https://www.math.ohiou.edu/courses/math3600/lecture13.pdf. Accessed 9 Feb 2017.

34. Andronescu M, Bereg V, Hoos H, Condon A. RNA STRAND: The RNA Secondary Structure and Statistical Analysis Database. BMC Bioinforma. 2008;9(1):340. doi:10.1186/1471-2105-9-340.

35. Cannone JJ, Subramanian S, Schnare MN, Collett JR, D'Souza LM, Du Y, Feng B, Lin N, Madabusi LV, Müller KM, Pande N, Shang Z, Yu N, Gutell RR. The Comparative RNA Web (CRW) Site: an online database of comparative sequence and structure information for ribosomal, intron, and other RNAs. BMC Bioinforma. 2002;3(1):1–31. doi:10.1186/1471-2105-3-2.

36. Berman HM, Westbrook J, Feng Z, Gilliland G, Bhat TN, Weissig H, Shindyalov IM, Bourne PE. The protein data bank. Nucleic Acids Res. 2000;28:235–42.

37. Zweib C. SRPDB (Signal Recognition Particle Database). 2017. http://rth.dk/resources/rnp/SRPDB/SRPDB.html. Accessed 9 Feb 2017.

38. Nawrocki EP, Burge SW, Bateman A, Daub J, Eberhardt RY, Eddy SR, Floden EW, Gardner PP, Jones TA, Tate J, Finn RD. Rfam 12.0: updates to the RNA families database. Nucleic Acids Res. 2015;43(D1):130. doi:10.1093/nar/gku1063.

39. Silva PAGC, Pereira CF, Dalebout TJ, Spaan WJM, Bredenbeek PJ. An RNA Pseudoknot Is Required for Production of Yellow Fever Virus Subgenomic RNA by the Host Nuclease XRN1. J Virol. 2010;84(21):11395–406.

40. Comarmond MB, Giege R, Thierry JC, Morast D. Three-dimensional structure of yeast tRNA^{Asp}.I,structure determination. Acta Crystallogr. 1986;B(42):272–80.

41. Ye JD, Tereshko V, Frederiksen JK, Koide A, Fellouse FA, Sidhu SS, Koide S, Kossiakoff AA, Piccirilli JA. Synthetic antibodies for specific recognition and crystallization of structured RNA. Proc Natl Acad Sci. 2008;105(1):82–7.

42. Halic M, Gartmann M, Schlenker O, Mielke T, Pool MR, Sinning I, Beckmann R. Signal recognition particle receptor exposes the ribosomal translocon binding site. Science. 2006;312(5774):745–7. doi:10.1126/science.1124864. http://science.sciencemag.org/content/312/5774/745.full.pdf.

43. Bessho Y, Shibata R, Sekine S-i, Murayama K, Higashijima K, Hori-Takemoto C, Shirouzu M, Kuramitsu S, Yokoyama S. Structural basis for functional mimicry of long-variable-arm tRNA by transfer-messenger RNA. Proc Natl Acad Sci. 2007;104(20):8293–98. doi:10.1073/pnas.0700402104. http://www.pnas.org/content/104/20/8293.full.pdf.

44. Torres-Larios A, Anne-Catherine, Dock-Bregeon, Romby P, Rees B, Sankaranarayanan R, Caillet J, Springer M, Ehresmann C, Ehresmann B, Moras D. Structural basis of translational control by Escherichia coli threonyl tRNA synthetase. Nat Struct Biol. 2002;9:343–7.

45. Agrawal RK, Spahn CMT, Penczek P, Grassucci RA, Nierhaus KH, Frank J. Visualization of tRNA movements on the escherichia coli 70s ribosome during the elongation cycle. J Cell Biol. 2000;150(3):447–60.

46. Halic M, Blau M, Becker T, Mielke T, Pool MR, Wild K, Sinning I, Beckmann R. Following the signal sequence from ribosomal tunnel exit to signal recognition particle. Nature. 2006;444(7118):507–11.

47. Dallas A, Moore PB. The loop E–loop D region of Escherichia coli 5S rRNA: the solution structure reveals an unusual loop that may be important for binding ribosomal proteins. Structure. 1997;5(12):1639–53. doi:10.1016/S0969-2126(97)00311-0.

48. Griffiths-Jones S, Moxon S, Marshall M, Khanna A, Eddy SR, Bateman A. Rfam: annotating non-coding RNAs in complete genomes. Nucleic Acids Res. 2005;33(suppl 1):121–4. doi:10.1093/nar/gki081. http://nar.oxfordjournals.org/content/33/suppl_1/D121.full.pdf+html.

13

Homology to peptide pattern for annotation of carbohydrate-active enzymes and prediction of function

P. K. Busk[*] (ID), B. Pilgaard, M. J. Lezyk, A. S. Meyer and L. Lange

Abstract

Background: Carbohydrate-active enzymes are found in all organisms and participate in key biological processes. These enzymes are classified in 274 families in the CAZy database but the sequence diversity within each family makes it a major task to identify new family members and to provide basis for prediction of enzyme function. A fast and reliable method for *de novo* annotation of genes encoding carbohydrate-active enzymes is to identify conserved peptides in the curated enzyme families followed by matching of the conserved peptides to the sequence of interest as demonstrated for the glycosyl hydrolase and the lytic polysaccharide monooxygenase families. This approach not only assigns the enzymes to families but also provides functional prediction of the enzymes with high accuracy.

Results: We identified conserved peptides for all enzyme families in the CAZy database with Peptide Pattern Recognition. The conserved peptides were matched to protein sequence for *de novo* annotation and functional prediction of carbohydrate-active enzymes with the Hotpep method. Annotation of protein sequences from 12 bacterial and 16 fungal genomes to families with Hotpep had an accuracy of 0.84 (measured as F1-score) compared to semiautomatic annotation by the CAZy database whereas the dbCAN HMM-based method had an accuracy of 0.77 with optimized parameters. Furthermore, Hotpep provided a functional prediction with 86% accuracy for the annotated genes. Hotpep is available as a stand-alone application for MS Windows.

Conclusions: Hotpep is a state-of-the-art method for automatic annotation and functional prediction of carbohydrate-active enzymes.

Keywords: Carbohydrate-active enzymes, Genomics, Annotation, Software

Background

Carbohydrate-active enzymes are produced by all organisms to accomplish enzymatic modification of carbohydrate-containing compound both intra- and extracellularly. Hence, this enzyme group is relevant for understanding central biological processes such as sugar metabolism, protein glycosylation and, on an ecological level, for global biomass synthesis and degradation. It is not surprising that carbohydrate-active enzymes are used in medical and industrial biotechnology. The CAZy database (http://www.cazy.org/) was founded in 1991 and contains a unique classification of carbohydrate-active enzymes including carefully curated information about enzyme sequence, structure and function [1]. Currently, the publicly available information in the CAZy database consists of almost 400.000 unique protein sequences classified in more than 300 families.

Despite the abundant information in the CAZy database, *de novo* annotation of carbohydrate-active enzymes is not a trivial task. State-of-the-art methods involve automatic identification by matching the sequences of interest to protein models generated directly from sequences in the CAZy database or indirectly from protein domain models from other databases or by BLAST search followed by manual curation of the data [1–4].

Entirely automatic annotation methods have been developed based on hidden Markov model (HMM) recognition of all or a subset of the enzymes in the CAZy database and are available as web-based services [5–7]. E.g., the dbCAN method was made by refining

* Correspondence: pbus@kt.dtu.dk
Department of Chemical and Biochemical Engineering, Technical University of Denmark, Søltofts Plads, Building 229, 2800 Kgs. Lyngby, Denmark

HMM models from the Conserved Domain Database to fit the families in the CAZy database and supplementing the database with new HMM models for the families in the CAZy database that are not modelled in the Conserved Domain Database [7].

Even when it is possible to annotate a protein to a specific family this does not necessarily allow an exact prediction of its enzymatic activity. This is due to that the classification of the carbohydrate-active enzymes in the CAZy database is based on protein sequence and structure similarity [1]. Thus, in many cases the classification does not reflect enzymatic activity [1]. Hence, proteins with identical enzymatic activity are classified in different families and most of the families contain proteins with different enzymatic activities.

Identification of short, conserved motifs can be used to group related protein sequences and will often pinpoint proteins with the same enzymatic activity [8, 9]. Furthermore, the method Homology to Peptide Pattern (Hotpep) matches the short, conserved motifs to undescribed protein sequences to obtain a fast, sensitive and precise annotation of carbohydrate-active enzymes to families [10]. Moreover, when experimental data on enzymatic activity is available Hotpep allows prediction of the enzymatic activity of the proteins. In practice, the experimental data on enzyme activity collected in the CAZy database can be used to predict the enzymatic activity of approximately 75% of the carbohydrate-active enzymes in a genome with 80% accuracy [9, 10].

We used the method Peptide Pattern Recognition (PPR) to identify short, conserved sequence motifs for all enzyme families in the CAZy database. The peptide patterns were combined with Hotpep to obtain a stand-alone software for automatic annotation and functional prediction of carbohydrate-active enzymes. As an example, to illustrate the workability of the approach, annotation of protein sequences from 12 bacterial and 16 fungal genomes was addressed. Hotpep had an F1 score of 0.86 (sensitivity = 0.88, precision = 0.84) for predicting carbohydrate-active enzymes in 12 bacterial genomes and an F1 score of 0.82 (sensitivity = 0.77, precision = 0.88) for predicting carbohydrate-active enzymes in 16 fungal genomes compared to semiautomatic annotation by the CAZy database tools for carbohydrate-active enzyme annotation [1, 4]. Moreover, Hotpep correctly predicted the activity of 86% of the characterized carbohydrate-active enzymes in the CAZy database.

The carbohydrate binding modules (CBM) are not defined as carbohydrate-active enzymes *per se* but are carbohydrate binding domains within multidomain carbohydrate-active enzymes [11]. Using short, conserved peptides for the CBM families in the CAZy database Hotpep annotates the CBMs with an F1 score of 0.87.

The Hotpep stand-alone application is available for download from Sourceforge for use on desktop computers with the MS Windows operative system.

Implementation

Development and testing of Hotpep for carbohydrate-active enzymes followed a number of steps as outlined (Fig. 1).

Protein sequences

The first step was to download sequences for all members of each carbohydrate-active enzyme family in the CAZy database (www.cazy.org [1]) from Genbank (https://www.ncbi.nlm.nih.gov/ [12]) in August, 2016. The CBM families were downloaded in February, 2017. Sequences that were 100% redundant or 100% identical to a part of another sequence were removed.

Identification of short, conserved peptides

PPR was used for identification of short, conserved peptides in each family of carbohydrate-active enzymes as previously described [9, 10, 13]. Briefly, for each family

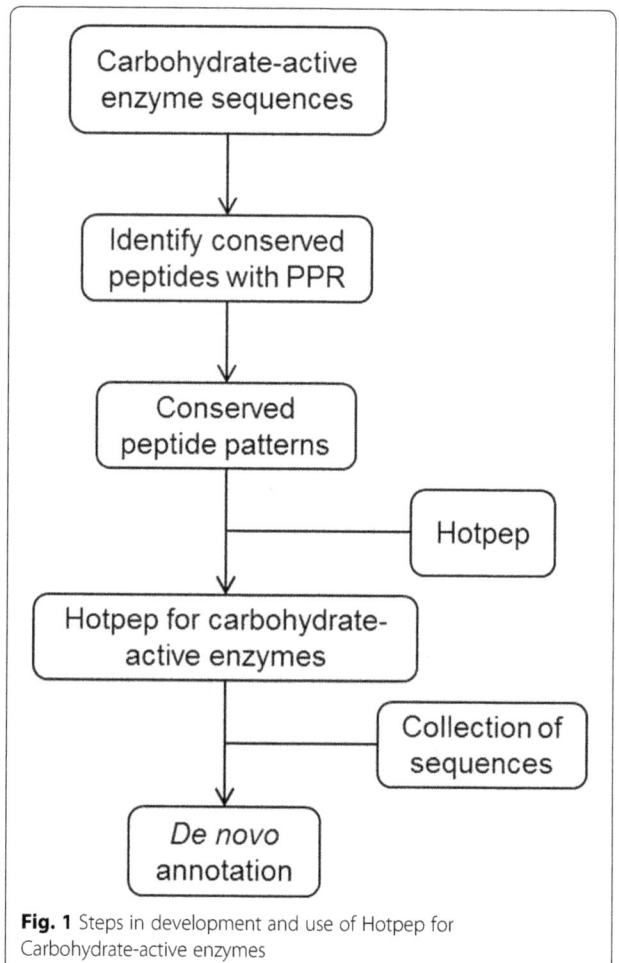

Fig. 1 Steps in development and use of Hotpep for Carbohydrate-active enzymes

PPR found the largest group of proteins that contained at least 10 of 70 conserved hexamer peptides. The length of the conserved peptides (hexamers), the number of conserved peptides per protein (10) and the total number of conserved peptides per group (70) were chosen as they were the conditions that gave the best rate of prediction of protein function in empirical testing of peptide lengths from trimers to decamers, 5 – 40 conserved peptides per protein and 30 – 200 conserved peptides per group [9]. Moreover, the minimum frequency of each conserved peptide in a group was 0.20 as this threshold gives the best rate of prediction of protein function [9]. For CBM domains the parameters 30 conserved hexapeptides per PPR group and 3 conserved peptides per protein were used for PPR analysis.

The first group of proteins identified by this method was named group 1. Next, PPR found the second largest group of proteins, not including any proteins from group 1. This group of proteins was named group 2 and so on. The analysis was stopped when less than five proteins were grouped together.

In this way a number of groups consisting of a list of protein sequences and a list of conserved peptides were generated for each family in the CAZy database. Groups including proteins with a described enzyme activity as reported in the CAZy database were assigned the same function as the enzymes as previously described [9].

For AA families 9, 10 and 11 the conserved peptide lists of the previously described expanded families were used [13].

Sequence collections

Genome-annotated protein products ("*_protein.faa.gz" files) were downloaded from Genbank for 12 bacterial (Table 1) and 16 fungal species (Table 2). For comparison of annotation from genomes and from predicted proteins the files *_genomic.fna.gz (genome assembly) and *_protein.faa.gz (protein products annotated on the genome assembly) for the following fungi *Thermothelomyces thermophile* (Accession: GCF_000226095.1), *Talaromyces stipitatus* (Accession: GCA_000003125.1), *Botryobasidium botryosum* (Accession: GCA_000697 705.1), *Coprinopsis cinerea* (Accession: GCA_0001828 95.1), *Serendipita indica* (Accession: GCA_000313 545.1), *Mucor circinelloides* (Accession: GCA_000401 635.1) and *Rhizopus delemar* (Accession: GCA_00014 9305.1) were downloaded from Genbank.

Annotation with Hotpep

Genomic fragments were annotated as previously described [10]. Annotation of protein products from genome assemblies was performed on full-length predicted protein sequences essentially as described [9]. Briefly, protein sequence was given a score for each group-specific peptide lists for each family by:

1. Finding all the conserved peptides from the list that were present in the sequence.
2. Sum the frequency of these peptides to obtain the group-specific frequency score.

A hit was considered significant if the protein sequence:

1. Included three or more conserved peptides from a group.
2. The frequency score for the peptides was higher than 1.0
3. The conserved peptides represented at least ten amino acids of the protein sequence.

If a protein satisfied all three conditions it was assigned to the family and to the PPR group with the

Table 1 Bacterial strains and accession numbers

Name	Phylum	Isolated from	Accession numbers
Bacteroides cellulosilyticus WH2	Bacteroidetes	Gut and stomach	GCA_000463315.1
Caldicellulosiruptor saccharolyticus DSM8903	Firmicutes	Wood Thermophilic anaerobe	GCA_000016545.1
Deinococcus peraridilitoris DSM19664	Deinococcus-Thermus	Coastal desert	GCA_000317835.1
Desulfotomaculum gibsoniae DSM7213	Firmicutes	Freshwater ditch	GCA_000233715.3
Enterobacter lignolyticus SCF1	Proteobacteria	Tropical forest soil	GCA_000164865.1
Melioribacter roseus P3M-2	Ignavibacteriae	Wooden surface of a chute	GCA_000279145.1
Prevotella ruminicola 23	Bacteroidetes	Gut	GCA_000025925.1
Rhodococcus jostii RHA1	Actinobacteria	Hexachlorocyclohexane-contaminated soil	GCA_000014565.1
Ruminiclostridium thermocellum ATCC27405	Firmicutes	Soil/manure	GCA_000015865.1
Teredinibacter turnerae T7901	Proteobacteria	Intracellular in shipworm	GCA_000023025.1
Thermacetogenium phaeum DSM12270	Firmicutes	thermophilic anaerobic methanogenic reactor	GCA_000305935.1
Thermoanaerobacterium thermosaccharolyticum DSM571	Firmicutes	Soil	GCA_000145615.1

Table 2 Fungal strains (basidiomycotae) and accession numbers

Name	Order	Life style	Accession numbers
Postia placenta	*Polyporales*	Brown rot	GCA_000006255.1
Fomitopsis pinicola	*Polyporales*	Brown rot	GCA_000344655.2
Gloeophyllum trabeum	*Gloeophyllales*	Brown rot	GCA_000344685.1
Coniophora puteana	*Boletales*	Brown rot	GCA_000271625.1
Dacryopinax sp.	*Dacrymycetales*	Brown rot	GCA_000292625.1
Tremella mesenterica	*Tremellales*	Mycoparasite	GCA_000271645.1
Dichomitus squalens	*Polyporales*	White rot	GCA_000275845.1
Trametes versicolor	*Polyporales*	White rot	GCA_000271585.1
Fomitiporia mediterranea	*Hymenochaetales*	White rot	GCA_000271605.1
Auricularia delicata	*Auriculariales*	White rot	GCA_000265015.1
Punctularia strigosozonata	*Corticiales*	White rot	GCA_000264995.1
Heterobasidion annosum	*Russulales*	White rot	GCA_000320585.2
Stereum hirsutum	*Russulales*	White rot	GCA_000264905.1
Phanerochaete_carnosa	*Polyporales*	White rot	GCA_000300595.1
Ceriporiopsis subvermispora	*Polyporales*	White rot	GCA_000320605.2
Phlebiopsis gigantea	*Polyporales*	White rot	GCA_000832265.1

highest group-specific frequency score. Moreover, if this group had been assigned a function by the PPR analysis, the same function was predicted for the protein [9].

Hotpep including the conserved peptide patterns described here is available for download as an application for the MS Office operative system from Sourceforge.

Annotation with dbCAN

The protein products from each genome were annotated *de novo* with the dbCAN web service for protein annotation with standard parameters and with optimized parameters (E-value < 10^{-18}; coverage > 0.35 for bacteria and E-value < 10^{-17}; coverage > 0.45 for fungi) by downloading scripts and HMMs as described (http://csbl.bmb.uga.edu/dbCAN/annotate.php, [7]).

Statistical analysis

The following values were calculated for pairwise comparison of two annotation methods:

True positives = Number of hits found by both screening methods. False positives = Number of proteins found by the screening method being tested but not by the reference method. False negatives = Number of proteins found by the reference method but not by the screening method being tested.

Sensitivity was calculated as True positives/(True positives + False negatives); Precision (positive prediction value) was calculated as True positives/(True positives + False positives) and F1 score (the harmonic mean of precision and sensitivity) was calculated as (2 × True positives)/(2 × True positives + False positives + False negatives).

Results and discussion

Short, conserved peptides identified in the carbohydrate-active enzyme from the glycoside hydrolase families in the CAZy database can be used for fast, efficient and reliable approach for annotation by the Hotpep method [10]. Moreover, groups of carbohydrate-active proteins sharing the same short, conserved peptides do often have the same enzymatic activity [9]. Thus, by comparing the rich information on experimentally characterized enzymes in the CAZy database with the PPR grouping of the enzymes it is possible to predict the enzymatic activity of the uncharacterized members of the groups with 80% accuracy. In this way, a functional prediction was obtained for 72% of the annotated glycoside hydrolases in 39 fungal genomes [10].

To accomplish automatic annotation of all carbohydrate-active enzymes with Hotpep we downloaded all sequences in the families of the five enzyme classes: Carbohydrate esterases (CE), Glycoside hydrolases (GH), Auxiliary activities (AA), Polysaccharide lyases (PL) and Glycosyl transferases (GT). A total of 594,121 accession numbers were found in the CAZy database and reduced to 380,269 non-redundant protein sequences before each family was sorted into groups of proteins sharing up to 70 short, conserved hexapeptides and assignment of function to each group containing more than two functionally characterized members (Additional file 1). In total 36% of the 5590 PPR groups for all enzyme families included functionally characterized proteins. These groups with associated functions contained 65% of the PPR-grouped proteins. For the glycoside hydrolases, 41% of the groups included functionally characterized proteins and a total of 74% of all proteins, in agreement with the previous report of a functional prediction of 72% of the glycoside hydrolases [10].

For the CBM class of carbohydrate-binding modules we found 71,253 accession numbers in the CAZy database resulting in 45,048 non-redundant protein sequences. Due to the short length of most CBM domains [7, 11] it was uncertain whether the standard parameters of 70 conserved peptides per PPR group and 10 conserved peptides per protein were optimal for annotation of CBMs. Therefore, different parameters for PPR were tested for classification of the isolated CBM domains followed by Hotpep annotation of the full-length proteins and comparison to the annotation in the CAZy database. There was little variation in the F1 score (0.83 - 0.87) within the range of tested parameters (Additional file 2) in agreement with the notion that PPR groups are fairly stable within a large range of parameters [9]. The parameters 30 conserved peptides per PPR group and 3 conserved peptides per protein gave the highest F1 score of 0.87 and were chosen for annotation of CBMs.

Hotpep annotates proteins by matching the lists of conserved peptides of a group to the protein sequences of interest [10, 13, 14]. Any sequence that fulfills a number of criteria (see Implementation) of which the most important is that the sequence should include at least three of the conserved peptides, will be annotated to the protein group. We combined Hotpep with the lists of conserved peptides for all enzyme families in the CAZy database to an application that can identify members of all carbohydrate-active enzyme families and CBMs. The AA9, AA10 and AA11 conserved peptides were substituted with the AA9exp, AA10exp and AA11exp conserved peptides that represent a more complete description of the sequence variation in these families [13]. The complete lists of peptides and frequencies are available for download at Sourceforge together with the accession numbers of the sequences for each group and the library of EC functional scores for each group.

The input for annotation with Hotpep is a text file with predicted protein sequences in fasta format. The algorithm is started by double-clicking the Hotpep icon. This will open a DOS prompt, where the user writes the name of the input file containing the fasta-formatted protein sequences (Fig. 2).

Hotpep screens the input sequences for members of all families in the CAZy database. This will take 5 – 20 min for all predicted genes in a bacterial or fungal genome. Several genomes can be annotated in parallel by running Hotpep several times. The results files are saved in six directories, one for each class of carbohydrate-active enzymes, one for the CBMs and two summary files: One with the number of hits for each family and one with the accession number of each hit and the families annotated for this hit (Fig. 3a). The latter file gives an overview of the number and families for multidomain enzymes.

The results for each enzyme class is a number of text files (Fig. 3b) prepared for import into MS Excel, LibreOffice or similar spreadsheet applications (Fig. 4). The columns in the spread sheet designates the group where the sequence is annotated, the name of the sequence, the sum of the frequencies of the conserved peptides [10, 14], the number of conserved peptides, the protein sequence, length of the sequence and the sequences of the conserved peptides. In addition, the directories contain a subdirectory with files including prediction of the activities of the enzymes arranged according to EC class (Fig. 3c). As the CBMs are binding modules associated to enzyme domains the predicted function is often the predicted function of the associated enzyme domain as described in the CAZy database. The files with functional prediction contain a column with the prediction of the enzymatic function according to EC class (Fig. 5). The information in this column consists of one or more EC numbers each followed by a colon and a number designating the sum of the number of conserved peptides in each characterized protein in the group. The higher this number, the more proteins in the group have the enzymatic activity represented by the EC number. E.g.; in family GH43 group 71 there are 48 conserved peptide matches to enzymes characterized as endoxylanases (EC 3.2.1.8) (Fig. 5). For family GH8 group 3 there are 65 conserved peptide matches to enzymes characterized as endoxylanases (EC 3.2.1.8) but also 41 conserved peptide matches to enzymes characterized as exo-oligoxylanase (EC 3.2.1.156) in addition to matches to enzymes with other activities (Fig. 5). Hence, expression and enzymatic characterization of the sequence with the accession number WP_029428720.1 annotated to this group is necessary to decide whether it is an endoxylanase or an exo-oligoxylanase as the scores for these two activities are similar.

This method correctly predicts 80 – 95% of enzyme activities [9, 10]. To test this further, we used Hotpep to predict the function of 8812 experimentally characterized carbohydrate-active enzymes (Additional file 3). Hotpep correctly predicted the function of 86% of the enzymes. This result supports the previous finding that proteins sharing conserved peptides often but not always have the same activity [9]. Hence, enzymatic activities for individual sequences predicted by Hotpep should be used as a guideline for functional characterization. In an analysis of annotation of glycosyl hydrolases from ORFs in genome fragments with Hotpep it was found that the glycosyl hydrolases that were overlooked by Hotpep could be detected when the full-length amino acid sequence of the enzymes were used for annotation [10]. This finding suggests that more true positive hits are

```
What is the name of the file you wish to screen?
Fungus fungus
Screening Fungus fungus for

CE proteins:. 2 hits
2 functionally annotated

GH proteins:..... 5 hits
5 functionally annotated

AA proteins:. 2 hits
2 functionally annotated

PL proteins:. 2 hits
2 functionally annotated

GT proteins:.. 2 hits
2 functionally annotated

CBM proteins:... 3 hits
2 functionally annotated

Screened Fungus fungus for carbohydrate-active enzymes
The results can be found in Fungus fungus
Please, press "enter" to finish
```

Fig. 2 Hotpep user interface. Double-clicking on the Hotpep icon opens a DOS promt where the name of the sequence directory (e.g., "Fungus fungus") is entered

obtained by examining full-length coding regions rather than ORFs containing single exons. To test this notion we compared the annotation of all carbohydrate-active enzymes in seven fungal genomes to annotation of predicted proteins from the same genomes. The fungi were selected to include genome assemblies and predicted proteins from different research groups to avoid methodical bias. The results showed that 31% more carbohydrate-active enzymes were found by annotation of the predicted proteins from the genomes compared to annotation of ORFs in fragments of the genomes (Additional file 4) in agreement with the previous report [10].

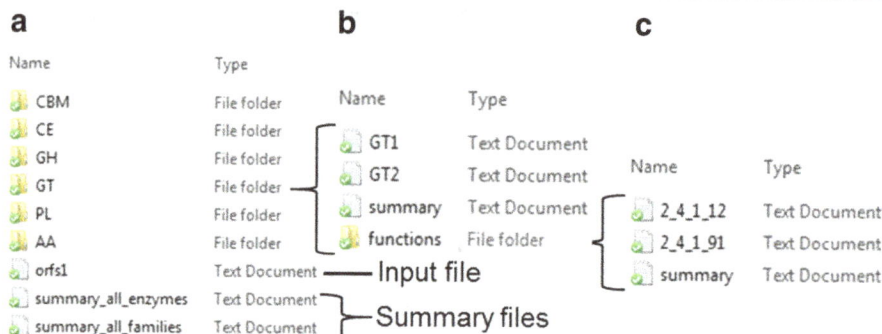

Fig. 3 Organization of the Hotpep output. **a.** The output is delivered in the sequence directory with one directory for each enzyme class in the CAZy database, a file containing a summary of the results and a file with all the families found for each accession number. **b.** Each of the class directories contains files with the hits for each family, a summary and a directory with functional predictions. **c.** The folder with functional predictions contains files for each EC number found and a summary

GH3	seq_name	Frequency	hits	sequence	length	peptides
1	>WP_029426049.1 glyc	11,3	16	MKKVFRKTSLLLATALMGVTGMQAQKAPQDMDRFIDTLMKK	773	PFGYGL,RDPRWG,DVLFGD,FG
1	>WP_029426038.1 glyc	11,3	16	MKKILKRTSLLLISALMSIAAAQAQKSPQDMDRFIDALMKKMT	784	DVLFGD,FGYGLS,RDPRWG,NP
1	>WP_029428699.1 beta	6,65	9	MKTLKIFSACLFLLPFVSCTQVANKGSDAATEKKVESLLSKMTL	750	PFGYGL,RDPRWG,GEDTYL,FG\
1	>WP_029426308.1 beta	5,33	7	MKDLVLKGLRITVVSLALSGCNTPDNLYLDPAQPIEARVDNLM	769	DPRWGR,FGYGLS,GYGLSY,SRL
2	>WP_029426618.1 glyc	4,76	15	MMSFYRLDEKYKSVILKINIMKSGMKCITTGLFKIFILTLSWGSL(	793	NFEYYS,VVSDWG,VLLKNE,FE\
4	>WP_029426465.1 glyc	9,32	28	MKRKLQLLTGIGCLCLCFLSCSQPPYKNPALSPEERANDLVGRL	863	EALHGV,TVFPQA,INIFRD,PRV
4	>WP_029427148.1 glyc	8,32	24	MKRWILIGMVVASGMTIQAQNKLPEKFPYQDTSLTAEERADD	864	FRDPRW,GRGQET,FGHGLS,RD
4	>WP_033160524.1 hypc	7,11	21	MINKKKIIIVFTLLLGGALMKTFAQSFKYPFQNPKLDVEERVKD	881	NIFRDP,RWGRGQ,ETYGED,IFR
4	>WP_029427455.1 glyc	6,9	20	MKKRWIILWLSAMTLNVTAQNEPYKNPELSPSERAWDLLKRN	864	FPQAIG,TYGEDP,GHGLSY,HGL
4	>WP_029429014.1 beta	4,38	15	MKKIIATMAIGACLCSCGGSQKEVYKDSTAPVKDRVEDLLKRN	766	MTLEEK,TYGEDP,KFRLGL,HGLS
10	>WP_029429017.1 beta	10,8	27	MKRTILLLLAILLWGGTSLLHAQAEAPFLPSAADARCKQWVDS	1000	PGHGDT,DLASLT,RRGLGF,TGF
10	>WP_029429016.1 beta	10,3	26	MKKLFLFLLILFSCCARFDAQTHRLPAPQSPVTPVEPILMRPFT!	1046	PGHGDT,GLGFDK,DGEWGL,GI
19	>WP_029426110.1 glyc	23,5	51	MNRKLILSAALSGLMLAATAQTTVAPAIPRDGKIEKKVEALLKK	777	EKIGQM,INAGID,MSRIDD,GP\

Fig. 4 Hotpep output. An output files with hits for the GH3 family opened in MS Excel. The columns (from left to right) contain the group where the sequence is annotated, the name of the sequence, the sum of the frequency of the conserved peptides, the number of conserved peptides, the protein sequence, length of the sequence and the sequences of the conserved peptides

Hence, although exon-intron structure of eukaryotic genes makes them difficult to predict [15] a higher sensitivity in prediction of carbohydrate-active enzymes is obtained by annotating from predicted proteins rather than from ORFs in genome fragments.

Annotation with Hotpep of predicted proteins from 12 bacterial genomes was compared to state-of-the-art semi-automatic annotation reported in the CAZy database [1]. The selected genomes were from bacteria with different lifestyles including bacteria known to degrade extracellular carbohydrates.

The CAZy database reported slightly less carbohydrate-active enzymes than Hotpep for the 12 bacterial genomes (Table 3). We have previously found that Hotpep annotation of fungal genomes are largely in agreement with the results reported in the CAZy database and that the differences between the annotations may be due to genes overlooked by either Hotpep or in the CAZy database [10]. This is a natural effect of the fact that the families in the CAZy database are growing as new members are discovered and some of the families are redefined [1]. E.g.; the lytic polysaccharide monooxygenases (LPMOs) originally classified in the GH61 and CBM33 families [9, 16, 17] were later reclassified to the AA9, AA10 and AA11 families [18, 19]. In view of this plasticity of the CAZy

database it is difficult to precisely determine the correct annotation of carbohydrate-active enzymes in a given dataset [7]. However, if the annotation reported in the CAZy database is defined as correct, then it means that the Hotpep annotation has a sensitivity of 0.88 and a precision of 0.84 (Table 3). This gives an F1 score of 0.86, which means that the methods on average agree on 86% of the number of predicted carbohydrate-active enzymes.

It was reported that automatic identification with the HMM signatures in dbCAN is a highly precise and sensitive method for annotation of carbohydrate-active enzymes [7]. Annotation of the 12 bacterial genomes with the dbCAN web service (http://csbl.bmb.uga.edu/dbCAN/annotate.php) gave a higher number of hits than the annotation in the CAZy database resulting in a sensitivity similar to Hotpep but with lower precision and F1 score (Table 3). However, annotation of the 12 bacterial genomes with the downloaded dbCAN HMMs and optimized parameters [7] gave a lower number of hits than the annotation in the CAZy database resulting in slightly higher sensitivity, precision and F1 score than Hotpep (Table 3). Thus, although the downloadable dbCAN is more difficult to use than the web service as the user has to both download the dbCAN HMMs and install the HMMER 3.0 package [7] the extra effort pays

fam	group	Functions	seq_name	sequence	length	peptides	Frequency	hits
GH5	82	3.2.1.8:148,3.2.1.4:2	>WP_029428723.1	MKRIIKVFFFVGFILLSSCSDNTIC	655	VNLHGF,TQTYSP,FSET	44,5	67
GH30	12	3.2.1.8:5	>WP_029426201.1	MNLKLPLLLGCCLFAAITACAETS	491	DGFGAA,RISIGC,SDFS	35,2	53
GH5	82	3.2.1.8:148,3.2.1.4:2	>WP_029426354.1	MKNITNVFYEFLIALCCLMSSSAI	1205	VVMRPP,MFELAN,DP	19	24
GH43	71	3.2.1.8:48	>WP_029426373.1	MNKRITLFLITLLTVCGVQSQNN	894	TDMVNW,IDDDGQ,A`	15,4	28
GH10	21	3.2.1.8:158	>WP_029428724.1	MKHTKKILGTMLLTAAAVVATS(	721	AWDVVN,FYWQDY,V	11	25
GH30	12	3.2.1.8:5	>WP_029426189.1	MKNKRLFTYVFLFGLLVNGSAC(	516	EYTCCD,TYFVKW,HNY	9,95	16
GH10	46	3.2.1.8:22	>WP_029428714.1	MKLKYLALSVCAAALMSCNSDK	897	RIKGWD,YYNDYG,YPL	8,81	10
GH10	46	3.2.1.8:22	>WP_026367873.1	MKNKIFLLALLIGFSLYSCGSNST`	371	LYYNDY,YYNDYG,GAN	8,64	10
GH8	16	3.2.1.8:70,3.2.1.156:7	>WP_029426464.1	MKNLFYLLLCLIAGASCSQADPTI	419	PSYHVP,SYHVPA,YHV	7	7
GH10	16	3.2.1.8:151	>WP_029429021.1	MKKVKSTLSVGKRIILLSLCMTMI	919	IVNRYK,DVVYAW,VN	6,84	19
GH10	10	3.2.1.8:311,3.2.1.55:17,3.2.1.4:11	>WP_029427770.1	MKHLFKFSLCALALTMGANTGF.	716	YAWDVV,AWDVVN,V	6,22	11
GH8	3	3.2.1.8:65,3.2.1.156:41,3.2.1.132:11,3.2.1.x:10,3.2.1.4:7	>WP_029428720.1	MKLTTLFAVSVSLILSGFCSLGVC	417	EGMSYG,GMSYGM,D\	4,83	8
GH10	21	3.2.1.8:158	>WP_029428711.1	MKLNRIILPLMACALTWSSCDD(	772	VDGIGT,DGIGTQ,NDY	4,68	8

Fig. 5 Hotpep output for functional prediction. Same as Fig. 4 with the addition of a column labelled "Functions" with information on the putative functions of the annotated sequence

Table 3 Annotation of 12 bacterial genomes

Method	CAZy[a]	Hotpep	dbCAN web	dbCAN download
Annotated proteins	1768	1839	2300	1749
True positives	-	1546	1701	1571
False positives	-	296	599	178
False negatives	-	220	67	197
Sensitivity	-	0.88	0.87	0.89
Precision	-	0.84	0.71	0.90
F1 score	-	0.86	0.84	0.89

[a]www.cazy.org

of in the form of a more accurate annotation. In summary, the comparison of the annotation methods showed that the CAZy database, Hotpep and downloaded dbCAN were most in agreement whereas the dbCAN web service annotates a higher number of genes as encoding carbohydrate-active enzymes.

To assess the performance of Hotpep for identification of eukaryotic genes, 16 fungal genomes that have been sequenced and annotated by The Joint Genome Institute and the CAZy database tools by Hori et al. [4] were selected for annotation. Testing on these genomes has the benefit that many of the carbohydrate-active enzymes from these fungi are not part of the CAZy database and has thus not been part of the dataset used to make the conserved peptide patterns used by Hotpep.

In case of the fungal genomes, Hori et al. [4] found slightly more carbohydrate-active enzymes than Hotpep (Table 4). However, Hotpep had an F1 score of 0.82 relative to the annotation by Hori et al., whereas annotation with dbCAN web service and downloaded dbCAN with optimized parameters only had F1 scores of 0.68 and 0.72, respectively (Table 4). Hence, for annotation of the fungal genes Hotpep and Hori et al. gave the most similar result whereas the dbCAN web service and the downloaded dbCAN predicted a higher number of carbohydrate-active enzymes. Summarizing the results for prediction of bacterial and fungal genes Hotpep had a combined F1 score of 0.84, dbCAN web service had an

F1 score of 0.75 and downloaded dbCAN with optimized parameters had an F1 score of 0.77.

The F1 score (0.82) for the comparison of Hotpep with Hori et al. [4] for the 16 fungal genomes is a little lower than the F1 score (0.86) for the annotation of the 12 bacterial genomes. However, the fungal genomes were all from basidiomycetes that are less represented in the CAZy database than carbohydrate-active enzymes from ascomycetes and thus may be more difficult to annotate. To assess this possibility we used previously published data [10] to calculate the F1 score for comparison of annotation of six ascomycete genomes by Hotpep and the CAZy database tools for annotation. The few disagreements between the methods were attributed mainly to differences in gene prediction rather than to differences in annotation [10]. In line with this notion, the F1 score for this dataset of ascomycete genes was 0.92 compared to only 0.82 for the annotation of basidiomycete genes in the present study. This finding suggests that the publicly available CAZy database may not yet account for the complete sequence variation in the carbohydrate-active enzyme families. E.g., the basidiomycete sequences may be underrepresented. This is in agreement with the ongoing addition of new sequences to the CAZy database [1]. A simple expansion of the LPMO enzyme families in the CAZy database by including previously unannotated, publicly available sequences led to the identification of the AA11 enzymes [9] and was shown to give a better representation of the sequence variation of the families, hereby making it possible to identify 31% more LPMOs in 39 fungal genomes [13]. The current version of Hotpep for annotation of carbohydrate-active enzymes include the expanded conserved peptide signatures for the AA9, AA10 and AA11 families. As expanded signatures become available for other families, they will be added to Hotpep.

Hotpep could principally be used for annotation of other enzymes than carbohydrate-active enzymes provided that sufficiently well curated sequence data bases are available.

Table 4 Annotation of 16 fungal genomes

Method	JGI/CAZy[a]	Hotpep	dbCAN web	dbCAN download
Annotated proteins	3985	3534	6238	4490
True positives	-	3084	3463	3057
False positives	-	450	2775	1433
False negatives	-	901	522	928
Sensitivity	-	0.77	0.87	0.77
Precision	-	0.88	0.56	0.68
F1 score	-	0.82	0.68	0.72

[a]Hori et al. [4]

Conclusion

Hotpep is an easy to use tool that performs automatic annotation of carbohydrate-active enzymes with high success rate. The result of annotation with Hotpep is comparable to state-of-the-art semiautomatic annotation by experts [1, 4] and automatic annotation with HMMs [7]. Furthermore, Hotpep also provides a functional prediction of function directly from amino acid sequence.

A downloadable version of Hotpep is available as a stand-alone application that runs on the MS Windows operative system.

Additional files

> **Additional file 1:** Conserved Peptide Patterns for all Carbohydrate-Active Enzyme Families and CBMs. This file includes all conserved peptide patterns for all PPR groups and functional data for the enzymes in each group.
>
> **Additional file 2:** Hotpep annotation of CBMs based on conserved peptides identified by PPR analysis. This file includes the results of Hotpep annotation of CBMs based on conserved peptides identified by PPR analysis with different parameters as indicated.
>
> **Additional file 3:** Hotpep functional prediction of 8812 experimentally characterized enzymes. This file includes experimental activity data from the CAZy database compared to Hotpep predictions for 8812 carbohydrate-active enzymes.
>
> **Additional file 4:** Comparison of Hotpep annotation from genomes and from predicted proteins. This file includes the results of Hotpep annotation of carbohydrate-active enzymes in seven fungal genomes and in the predicted proteins from the genomes.

Abbreviations

AA: Auxiliary activities; CBM: Carbohydrate binding module; CE: Carbohydrate esterases; GH: Glycoside hydrolases; GT: Glycosyl transferases; HMM: Hidden Markov Models; Hotpep: Homology to Peptide Pattern; LPMO: lytic polysaccharide monooxygenase; PL: Polysaccharide lyases; PPR: Peptide Pattern Recognition

Acknowledgements

We thank Kristian Barrett for fruitful discussions on enzyme annotation and on the performance of Hotpep.

Funding

This work was supported by project no.: Mar 14319 from Nordic Innovation; SYNFERON – from the Danish Innovation Fund and by The Villum Foundation. The funding bodies did not play any role in the design of the study, in the collection, analysis, and interpretation of data or in writing the manuscript.

Authors' contributions

PKB wrote the software, downloaded the sequences, made the analysis necessary for to develop Hotpep and performed the comparison of annotations. BP tested the Hotpep algorithm and participated in data requisition and analysis. MJL performed the DBCan annotation and result analysis. ASM discussed the final results and interpretation of the data. LL initiated the study and discussed the final results and interpretation of the data. The manuscript was written by the authors from a draft by PKB. All authors read and approved the final manuscript.

Competing interests

The authors declare that they have no competing interests.

References

1. Lombard V, Golaconda Ramulu H, Drula E, Coutinho PM, Henrissat B. The carbohydrate-active enzymes database (CAZy) in 2013. Nucleic Acids Res. 2014;42:D490–495.
2. Floudas D, Binder M, Riley R, Barry K, Blanchette RA, Henrissat B, et al. The Paleozoic origin of enzymatic lignin decomposition reconstructed from 31 fungal genomes. Science. 2012;336:1715–9.
3. Grigoriev IV, Martinez DA, Salamov AA. 5 - Fungal Genomic Annotation. In: Dilip K. Arora RMB and GBS, editor. Applied Mycology and Biotechnology [Internet]. Elsevier; 2006 [cited 2016 Dec 1]. p. 123–42. Available from: http://www.sciencedirect.com/science/article/pii/S1874533406800080
4. Hori C, Ishida T, Igarashi K, Samejima M, Suzuki H, Master E, et al. Analysis of the Phlebiopsis gigantea Genome, Transcriptome and Secretome Provides Insight into Its Pioneer Colonization Strategies of Wood. PLoS Genet. 2014;10:e1004759.
5. Ekstrom A, Taujale R, McGinn N, Yin Y. PlantCAZyme: a database for plant carbohydrate-active enzymes. Database (Oxford). 2014;2014:bau079.
6. Park BH, Karpinets TV, Syed MH, Leuze MR, Uberbacher EC. CAZymes Analysis Toolkit (CAT): Web service for searching and analyzing carbohydrate-active enzymes in a newly sequenced organism using CAZy database. Glycobiology. 2010;20:1574–84.
7. Yin Y, Mao X, Yang J, Chen X, Mao F, Xu Y. dbCAN: a web resource for automated carbohydrate-active enzyme annotation. Nucleic Acids Res. 2012;40:W445–51.
8. Busk PK, Lange L. A Novel Method of Providing a Library of N-Mers or Biopolymers. WO/2012/101151. [Internet]. 2012 [cited 2012 Dec 11]. Available from: http://www.freepatentsonline.com/WO2012101151A1.html
9. Busk PK, Lange L. Function-based classification of carbohydrate-active enzymes by recognition of short, conserved peptide motifs. Appl Environ Microbiol. 2013;79:3380–91.
10. Busk PK, Lange M, Pilgaard B, Lange L. Several genes encoding enzymes with the same activity are necessary for aerobic fungal degradation of cellulose in nature. PLoS One. 2014;9:e114138.
11. Boraston AB, Bolam DN, Gilbert HJ, Davies GJ. Carbohydrate-binding modules: fine-tuning polysaccharide recognition. Biochem J. 2004;382:769–81.
12. Benson DA, Cavanaugh M, Clark K, Karsch-Mizrachi I, Lipman DJ, Ostell J, et al. GenBank. Nucleic Acids Res. 2013;41:D36–42.
13. Busk PK, Lange L. Classification of fungal and bacterial lytic polysaccharide monooxygenases. BMC Genomics. 2015;16:368.
14. Bech L, Busk PK, Lange L. Cell Wall Degrading Enzymes in Trichoderma asperellum Grown on Wheat Bran. Fungal Genom Biol. 2015;4:116.
15. Brent MR. How does eukaryotic gene prediction work? Nat Biotechnol. 2007;25:883–5.
16. Karlsson J, Saloheimo M, Siika-Aho M, Tenkanen M, Penttilä M, Tjerneld F. Homologous expression and characterization of Cel61A (EG IV) of Trichoderma reesei. Eur J Biochem. 2001;268:6498–507.
17. Watanabe T, Kimura K, Sumiya T, Nikaidou N, Suzuki K, Suzuki M, et al. Genetic analysis of the chitinase system of Serratia marcescens 2170. J Bacteriol. 1997;179:7111–7.
18. Hemsworth GR, Henrissat B, Davies GJ, Walton PH. Discovery and characterization of a new family of lytic polysaccharide monooxygenases. Nat Chem Biol. 2014;10:122–6.
19. Levasseur A, Drula E, Lombard V, Coutinho PM, Henrissat B. Expansion of the enzymatic repertoire of the CAZy database to integrate auxiliary redox enzymes. Biotechnol Biofuels. 2013;6:41.

14

HALC: High throughput algorithm for long read error correction

Ergude Bao[1,2*] and Lingxiao Lan[1]

Abstract

Background: The third generation PacBio SMRT long reads can effectively address the read length issue of the second generation sequencing technology, but contain approximately 15% sequencing errors. Several error correction algorithms have been designed to efficiently reduce the error rate to 1%, but they discard large amounts of uncorrected bases and thus lead to low throughput. This loss of bases could limit the completeness of downstream assemblies and the accuracy of analysis.

Results: Here, we introduce HALC, a high throughput algorithm for long read error correction. HALC aligns the long reads to short read contigs from the same species with a relatively low identity requirement so that a long read region can be aligned to at least one contig region, including its true genome region's repeats in the contigs sufficiently similar to it (similar repeat based alignment approach). It then constructs a contig graph and, for each long read, references the other long reads' alignments to find the most accurate alignment and correct it with the aligned contig regions (long read support based validation approach). Even though some long read regions without the true genome regions in the contigs are corrected with their repeats, this approach makes it possible to further refine these long read regions with the initial insufficient short reads and correct the uncorrected regions in between. In our performance tests on *E. coli*, *A. thaliana* and *Maylandia zebra* data sets, HALC was able to obtain 6.7-41.1% higher throughput than the existing algorithms while maintaining comparable accuracy. The HALC corrected long reads can thus result in 11.4-60.7% longer assembled contigs than the existing algorithms.

Conclusions: The HALC software can be downloaded for free from this site: https://github.com/lanl001/halc.

Keywords: PacBio long reads, Error correction, Throughput

Background

The Illumina sequencing technology, as a representative of second generation sequencing technology, can produce reads of several hundred bases long (called short reads) with an error rate < 1% (dominated by base substitutions) and a cost of approximately $0.03–0.04 per million bases [1]. The low cost of short reads has greatly facilitated the process of sequencing and analyzing new species; however, the limited read length can prohibit sequencing completeness and analysis accuracy. For example, a tremendous number of species have been assembled from short reads, but most of the assemblies are incomplete

and fragmented into several thousands of contigs [2, 3]. To address this issue, the PacBio SMRT sequencing technology, as a representative of third generation sequencing technology, has been attracting more and more attention since its commercial release in 2010 [4]. This technology can currently produce reads of 5-15K bases and some of 100K bases (called long reads) with a cost of approximately $0.4-0.8 per million bases [1, 5]. With this technology, it becomes easier to assemble more complete sequences and perform more accurate analyses [6–8]. Depending on how the long reads are used, sequencing projects can be grouped into two classes.

- *Short and long read hybrid sequencing projects* obtain short reads of sufficient coverage as well as long reads of low or moderate coverage from the same species and assemble them together. When the coverage is low, long reads can fill gaps or form

*Correspondence: baoe@bjtu.edu.cn
Ergude Bao and Lingxiao Lan are joint first authors.
[1]School of Software Engineering, Beijing Jiaotong University, 3 Shangyuan Residence, Haidian District, 100044 Beijing, China
[2]Department of Botany and Plant Sciences, University of California, Riverside, 900 University Ave., 92521 Riverside, CA, USA

scaffolds for the corresponding short read assemblies [9]; when the coverage is moderate, long reads can assemble together with the corresponding short reads [3, 10–12].

- *Long read alone sequencing projects* obtain long reads of high coverage and assemble them alone [7, 13]. These sequencing projects are not as common as the short and long read hybrid sequencing projects because they are more expensive, as the long reads have higher cost than short reads.

Nevertheless, the generated long reads contain 10-15% errors (dominated by insertions and deletions in uniform distribution) [6], so it is important to design efficient algorithms to correct them.

Several error correction algorithms for long reads have been proposed, including PacBioToCA ([6]; the algorithm from the Celera assembler [13]), LSC [8], Proovread [14], CoLoRMap [15], the algorithm from the Cerulean assembler [11], ECTools [16], LoRDEC [17], Jabba [18], DAGCon ([7]; from HGAP assembler), LoRMA [19] and the algorithms from the FALCON and Sprai assemblers (not published). The long read error correction algorithms can be grouped into three classes.

- *Short read based algorithms* PacBioToCA, LSC, Proovread and CoLoRMap align the short reads from the same species to the long reads and use the aligned short reads with low error rate to perform error correction. These algorithms are usually used in short and long read hybrid sequencing projects.
- *Short read assembly based algorithms* the algorithm from Cerulean, ECTools, LoRDEC and Jabba all align the long reads to the de Bruijn graph constructed or contigs assembled from the short reads from the same species to perform error correction. Because of the continuity of the de Bruijn graph or contigs, more error rich regions in the long reads can be aligned and corrected with the de Bruijn graph or contigs. Another benefit of using the de Bruijn graph or contigs is that the alignment of long reads to de Bruijn graph or contigs is much faster than the alignment of short reads to long reads. These algorithms are also usually used in short and long read hybrid sequencing projects.
- *Long read alone algorithms* DAGCon, PacBioToCA in its self-correction mode and the algorithms from FALCON and Sprai find multiple sequence alignments among the long reads, while LoRMA aligns the long reads to the de Bruijn graphs constructed from themselves to perform error correction. These algorithms usually require long read coverage as high as $60-100\times$ and are thus used in the long read alone sequencing projects.

It is worthwhile to note that there are also many short read error correction algorithms for the second generation sequencing technology [20], but they do not work for long reads due to the different error model. The existing long read error correction algorithms could achieve error rates of approximately 1%, but they must discard a large amount of uncorrected bases and thus lead to low throughput. For example, as listed in [7], PacBioToCA and LSC must discard 42.6-87.1% bases in a human brain long read library to achieve the 1% error rate in the corrected and outputted bases. For another example, in [16] and [17], ECTools and LoRDEC also discard 18.2-70.0% bases in *E. Coli* read libraries for error correction. Such a loss of bases is not economical considering the higher cost of long reads compared to short reads, and it may also reduce the completeness of downstream assemblies and the accuracy of analysis. This point was discussed in [14]: "a decrease in throughput could have a strong impact on the further steps of the projects, especially the assembly". Also, as reported in [16], with 4.7-24.2% bases discarded, the lengths of assemblies decrease by 14.3-89.0% for the *S. cerevisiae*, *A. thaliana* and *O. sativa* read libraries.

The low throughput discussed above is because of the following two problems.

- *Error richness problem*: some long read regions are error rich, and it is difficult to align them with sufficient identity to the reference data (short reads for short read based algorithms, short read assembly for short read assembly based algorithms or the other long reads for long read alone algorithms) for correction, or it is difficult to validate and distinguish the true alignments from many false ones aligning them with lower identity.
- *Lack of reference data problem*: some long read regions do not have sufficient reference data for correction, due to low read coverage and/or sequencing gaps.

The short read assembly based algorithms could address the error richness problem to some extent by aligning an error rich long read region with relatively low identity requirements, and then validating the candidate alignments and accepting the one that forms a continuous alignment with its adjacent regions' alignments in the de Bruijn graph or contigs. For example, the algorithm from Cerulean validates long read regions' alignments to contigs of small lengths by first aligning their adjacent regions to contigs of large lengths and then accepting the former alignments adjacent to the latter in the contigs; LoRDEC and Jabba validate long read regions' alignments of low identity to the de Bruijn graph by referencing their adjacent regions' alignments of high identity and then accepting the former alignments adjacent to the latter in the de Bruijn graph. This validation approach is thus called the *adjacent alignment based validation approach*. Some of the remaining algorithms can also address the error richness problem to an extent by making alignments of several passes with different parameter settings [14, 19] or by aligning one pair of paired-end short reads by referencing the alignments of the other pair [15]. However, none of the existing algorithms could address the lack of reference data problem.

To further address the error richness problem and also the lack of reference data problem, in this paper, we propose a novel short read assembly based algorithm called HALC: High throughput Algorithm for Long read error Correction. HALC uses the contigs assembled from the corresponding short reads to correct the long reads. It aligns the long reads to the contigs with a relatively low identity requirement, so that a long read region could be aligned not only to its true genome region but also to the genome region's repeats in the contigs for correction. This novel alignment approach can address the lack of reference data problem and is called the *similar repeat based alignment approach*. It then validates each long read region's alignments with the adjacent alignment based validation approach and also by referencing other long read regions' alignments. This novel validation approach can further address the error richness problem and is called the *long read support based validation approach*.

Implementation
Underlining approaches
Below are the details of HALC's two novel approaches, the similar repeat based alignment approach and the long read support based validation approach, as well as the adjacent alignment based validation approach.

- *Similar repeat based alignment approach* (novel): a long read region could be aligned to its similar repeats in the contigs to guarantee that one long read region is aligned to at least one contig region for correction. Here, a long read region's *similar repeats* are the genome regions of <15% difference to the long read region's true genome region [21]. The similar repeats can be located in the contigs by alignment algorithms with dedicated parameter tunings. By this approach, a long read region of approximately 15% error rate compared to its true genome region can be aligned and converted to its similar repeat of <15% difference from the true genome region. The reduced error rate makes it possible to further refine the long read region with the initial short reads and thus reduce the error rate to <1%. It is worth noting that although the existing error correction algorithms for both second and third generation sequencing technologies try to avoid alignments to repeat regions

[6, 22], our observation and experimental results, in contrast, demonstrate the possibility to make use of some of the alignments (see the "Discussion" section for details).

- *Long read support based validation approach* (novel): the alignments of a long read region and its adjacent regions in the same long read are validated together, and the ones supported by a sufficient number of adjacent regions from the other long reads are accepted. Here, the alignments of two adjacent long read regions are *supported* by another two adjacent long read regions if the latter are aligned to the same contig regions as the former. With this approach, among several aligned contig regions of a long read region, the one corresponding to its true genome region (if exists) is accepted after validation. The prerequisite of this approach is that different long reads should be aligned to a unified set of contig regions, and one alignment of a long read region is to one contig region in the set; otherwise, it is difficult to check if two adjacent long read regions are aligned to the same contig regions as another two.

- *Adjacent alignment based validation approach* (existing): the alignments of a long read region and its adjacent regions in the same long read are validated together, and the ones aligned adjacent to each other in the contigs are accepted. With this approach, among several candidate alignments of a long read region, the one forming the alignment of the highest continuity is accepted after validation.

Figure 1 illustrates these approaches. Combining these approaches, if a long read region has its true genome region in the contigs, several alignments to the true genome region and its similar repeats are obtained through the similar repeat based alignment approach, and the alignment to the true genome region is accepted using the long read support based validation approach as well as the adjacent alignment based validation approach. The long read region can thus be corrected using the true genome region. Otherwise, if a long read region does not have its true genome region in the contigs, the alignment to a similar repeat is obtained and accepted. The long read region can thus be converted to the similar repeat and then refined using the initial short reads.

Algorithm overview
The HALC algorithm consists of the following five major steps, with the long reads, the short reads from the same species and the contigs assembled from the short reads as input.

1. Align the long reads to the contigs with a relatively low identity requirement so that a long read region

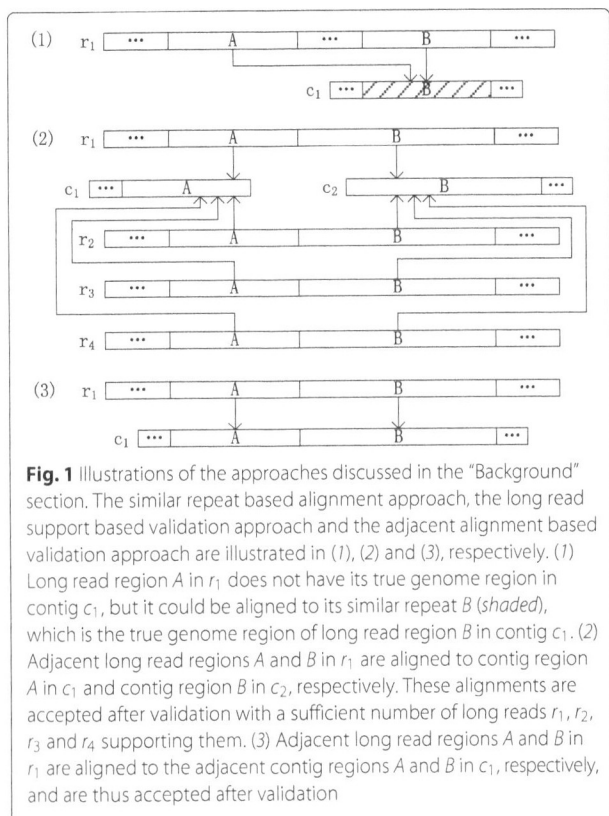

Fig. 1 Illustrations of the approaches discussed in the "Background" section. The similar repeat based alignment approach, the long read support based validation approach and the adjacent alignment based validation approach are illustrated in (1), (2) and (3), respectively. (1) Long read region A in r_1 does not have its true genome region in contig c_1, but it could be aligned to its similar repeat B (shaded), which is the true genome region of long read region B in contig c_1. (2) Adjacent long read regions A and B in r_1 are aligned to contig region A in c_1 and contig region B in c_2, respectively. These alignments are accepted after validation with a sufficient number of long reads r_1, r_2, r_3 and r_4 supporting them. (3) Adjacent long read regions A and B in r_1 are aligned to the adjacent contig regions A and B in c_1, respectively, and are thus accepted after validation

can be aligned to its true genome region or to similar repeats in the contigs.

2. Split the aligned contig regions and the long read regions so that different long reads are aligned to a unified set of contig regions, and one alignment of a long read region is to one contig region in the set.

3. Construct a contig graph from the long read region alignments so that one long read's alternative alignments can be represented by different paths, and the alignment with the highest long read support and continuity has the minimum total edge weight.

   2.1 Construct a graph representing one aligned contig region as a vertex and representing adjacent long read regions' alignments to two contig regions as an edge between their vertices.

   2.2 Assign a small weight to the graph edge between two vertices if the long read regions' alignments are supported by a large number of long read regions, or if the aligned contig regions are adjacent.

4. For each long read, find the paths representing its alternative alignments in the contig graph, and use the one with the minimum total edge weight to correct it.

5. Refine the similar repeat corrected long read regions with the short reads.

Steps 1 and 5 are based on the similar repeat based alignment approach, step 2 guarantees the prerequisite of the long read support based validation approach, and steps 3-4 are based on the long read support based validation approach and the adjacent alignment based validation approach. It is worth noting that the HALC algorithm does not try to maximize the total identity between a long read and the aligned contig regions because considering the high error rate of the long reads, the long read alignment of the maximum total identity may not be the one to the true genome regions. Step 1 is sufficient to guarantee the identity between a long read region and the aligned contig region. Table 1 shows the correspondence between the steps of the algorithm, the approaches the steps are based on, and the problems addressed by the approaches. Figure 2 illustrates the HALC algorithm.

Long read alignment to contigs

In this step, we align the long reads to the contigs with BLASR [23] because (1) it is specifically designed for long read alignment tolerating large numbers of insertions and deletions, and (2) in our experience, the HALC algorithm showed better performance with BLASR than with several other aligners such as BLAST [24], BLAT [25] and MUMMER [26]. The parameter settings of BLASR are *-bestn 20 -minMatch 8 -nCandidates 30 -maxScore 2000 -minAlnLength 15*, with a trade-off between alignment sensitivity and accuracy so that the long read regions are aligned either to their true genome regions or to similar repeats in the contigs. To further improve the alignment sensitivity, we use scaffolds rather than contigs as input because scaffolds contain additional information about contig orientations and orders, and this information could help guide BLASR alignment. For simplicity, we continue using the term contigs rather than scaffolds in the following discussion.

Splitting of contig and long read regions

We split the aligned contig regions and the long read regions following the two rules below. In these two rules, an aligned contig region or long read region is denoted by its starting and ending positions in the underlining

Table 1 Correspondence of algorithm steps, approaches and problems addressed

Steps	Approaches	Problems
1	Similar repeat based alignment	Lack of reference data
3-4	Long read support and adjacent alignment based validation	Error richness
5	Similar repeat based alignment	Lack of reference data

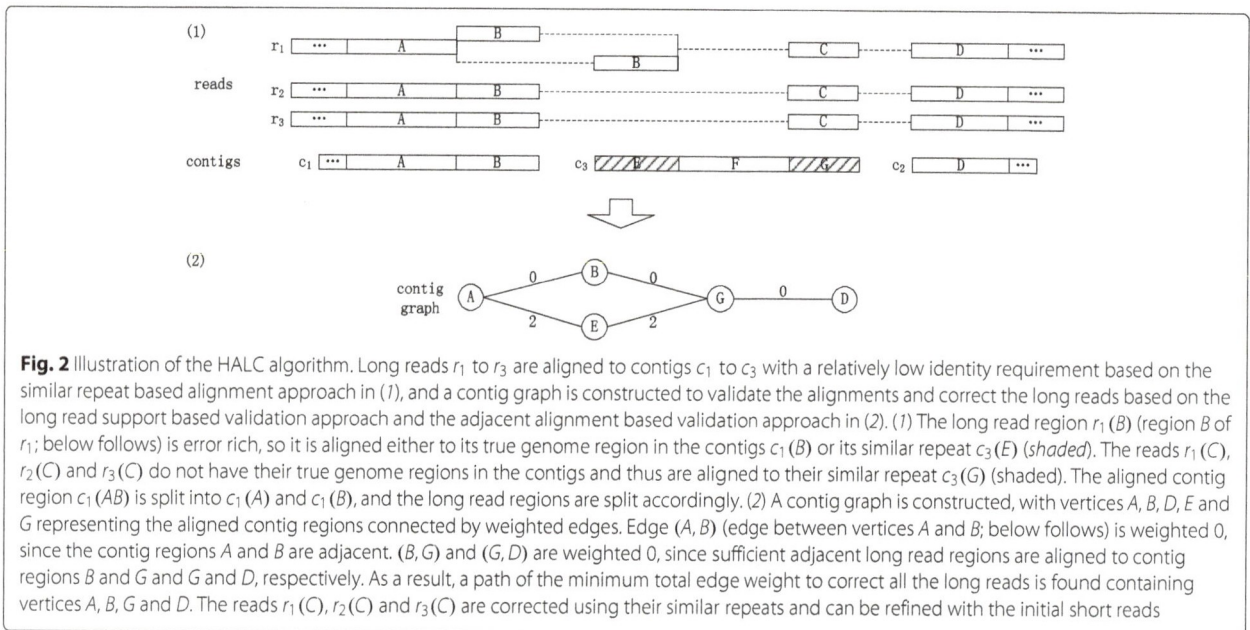

Fig. 2 Illustration of the HALC algorithm. Long reads r_1 to r_3 are aligned to contigs c_1 to c_3 with a relatively low identity requirement based on the similar repeat based alignment approach in (*1*), and a contig graph is constructed to validate the alignments and correct the long reads based on the long read support based validation approach and the adjacent alignment based validation approach in (*2*). (*1*) The long read region r_1 (B) (region B of r_1; below follows) is error rich, so it is aligned either to its true genome region in the contigs c_1 (B) or its similar repeat c_3 (E) (*shaded*). The reads r_1 (C), r_2(C) and r_3(C) do not have their true genome regions in the contigs and thus are aligned to their similar repeat c_3 (G) (shaded). The aligned contig region c_1(AB) is split into c_1 (A) and c_1 (B), and the long read regions are split accordingly. (*2*) A contig graph is constructed, with vertices A, B, D, E and G representing the aligned contig regions connected by weighted edges. Edge (A, B) (edge between vertices A and B; below follows) is weighted 0, since the contig regions A and B are adjacent. (B, G) and (G, D) are weighted 0, since sufficient adjacent long read regions are aligned to contig regions B and G and G and D, respectively. As a result, a path of the minimum total edge weight to correct all the long reads is found containing vertices A, B, G and D. The reads r_1 (C), r_2(C) and r_3(C) are corrected using their similar repeats and can be refined with the initial short reads

genome. For example, an aligned contig region c starting at genome position x and ending at y is denoted as $c(x, y)$.

- Two aligned contig regions $c(x, y)$ and $c(x', y')$ of the same contig are split into three contig regions $c(x, x')$, $c(x', y)$ and $c(y, y')$, if $x < x' < y < y'$.
- Two long read regions $r(x, y)$ and $r(x', y')$ are split into three long read regions $r(x, x')$, $r(x', y)$ and $r(y, y')$, if $r(x, y)$ is aligned to contig regions $c(x, x')$ and $c(x', y)$, and $r(x', y')$ is aligned to contig regions $c(x', y)$ and $c(y, y')$.

These rules are for the general case that two aligned contig regions intersect, while small adjustment can be made to accommodate the case in which one contig region is contained in the other. In practice, long read regions from the same genome region usually contain many differences, so the boundaries of their alignments may be close but different. Therefore, we consider two aligned contig regions $c(x, y)$ and $c(x', y')$ (or two long read regions $r(x, y)$ and $r(x', y')$) as the same contig region (or long read region) without further splitting them if $|x - x'| < \delta$ and $|y - y'| < \delta$, where δ is a small deviation value. The HALC software provides an option -*boundary* to set this value (4 bp by default).

Graph construction

We construct a contig graph with each vertex per aligned contig region and each edge between two vertices if there is at least one pair of adjacent long read regions aligned to the two contig regions. In most of the cases, different pairs of adjacent long read regions can be aligned to the same two contig regions in the same orientation. More

accurately, however, different pairs of adjacent long read regions can be aligned to the same two contig regions in four orientations: forward-forward, forward-reverse, reverse-forward, and reverse-reverse. The contig graph should thus have two vertices for one aligned contig region to represent both the forward and reverse alignments and four edges between the vertices for two aligned contig regions. Therefore, the HALC software provides an option -*accurate* to enable considering the different orientations (yes by default).

Graph weighting

We weight each edge between two vertices in the graph following the two rules below. The first rule guarantees 0 weight for the edges corresponding to the long read regions' alignments to adjacent contig regions, and the second rule guarantees small weights for the edges corresponding to the long read regions' alignments supported by a large number of long read regions.

- If the aligned contig regions of the vertices are adjacent to each other in the initial contig, assign a weight of 0 to the edge.
- If the aligned contig regions are far from each other, assign a weight of $max\{C_0 - C, 0\}$ to the edge, where C_0 is the expected long read coverage on the contigs, and C is the number of adjacent long read regions aligned to the two contig regions.

The expected long read coverage C_0 on the contigs can be calculated automatically by checking the average number of long reads covering a contig base, but the

The BLAT aligner was used to make these alignments because it is a typical aligner for transcriptomic sequences with high sensitivity [25]. We obtained the measurements (1)-(3) above as well as the following measurement: (15) *transcriptome fraction* is the number of transcriptome bases covered by the long reads over the total number of transcriptome bases.

Results on error correction performance

The performance test results on the *E. coli* data set are listed in Table 2(a). A total of 50.4% bases of the initial long reads can be aligned to the corresponding genome with 95.2% identity, indicating a high error rate in the uncorrected long reads. The existing error correction algorithms PacBioToCA, LSC, Proovread, ECTools and LoRDEC can correct and output 23.5-60.8% of the bases. HALC can obtain 7.2–41.1% higher throughput than PacBioToCA, LSC, Proovread and ECTools and is

comparable (<5% difference) to LoRDEC. The alignment ratio, alignment identity and genome fraction of all the algorithms are almost 100% and thus comparable. Except for PacBioToCA and LSC, the average read length of all the algorithms is inversely proportional to the throughput because more but shorter reads can be obtained with higher throughput. The sensitivity and gain of all the algorithms are proportional to the throughput, while the specificity remains comparable.

The performance test results for the *A. thaliana* data set are listed in Table 2(b). HALC can obtain 6.7-24.0% higher throughput than all the existing algorithms. The performance test results on the *Maylandia zebra* data set are listed in Table 2(c). HALC can obtain 7.6% higher throughput than LoRDEC. In both tests, the alignment ratio, alignment identity, genome fraction, sensitivity, gain and specificity of HALC are comparable to or higher than the existing algorithms, and the average read length

Table 2 Evaluation of error correction performance

Method	Throughput	Alignment ratio	Alignment identity	Genome fraction	N reads	Average read length	Sensitivity	Gain	Specificity
				(a) Long reads of E. coli					
Initial	100.0%	50.4%	95.2%	100.0%	75152	2381	-	-	-
PacBioToCA[a]	24.2%	100.0%	100.0%	99.5%	53447	810	-	-	-
LSC	53.5%	98.7%	99.9%	99.7%	115960	825	52.6%	51.7%	99.9%
Proovread	57.4%	100.0%	99.9%	99.7%	44986	2284	57.4%	56.8%	99.9%
CoLoRMap	42.8%	99.7%	100.0%	99.9%	70582	1084	42.7%	42.2%	99.9%
ECTools	23.5%	99.9%	99.2%	99.4%	8095	5211	23.4%	21.8%	99.8%
LoRDEC	60.8%	97.8%	100.0%	99.8%	70164	1549	60.7%	60.5%	100.0%
Jabba	52.8%	99.6%	100.0%	98.6%	26459	3568	52.8%	52.7%	100.0%
HALC	64.6%	98.6%	99.9%	99.8%	78731	1467	64.4%	64.0%	99.9%
				(b) Long reads of A. thaliana					
Initial	100.0%	32.4%	92.4%	82.4%	490418	2645	-	-	-
PacBioToCA[a]	10.7%	99.2%	99.7%	63.9%	260834	535	-	-	-
LSC	25.9%	100.0%	99.5%	71.4%	659123	509	24.2%	22.3%	99.7%
Proovread	27.8%	99.8%	99.7%	79.8%	125786	2864	26.5%	24.9%	99.7%
CoLoRMap	21.4%	99.4%	99.7%	69.3%	230933	1203	20.5%	19.2%	99.8%
ECTools	11.3%	99.8%	99.5%	63.1%	21354	6886	10.8%	9.8%	99.8%
LoRDEC	28.0%	86.4%	99.5%	74.4%	847963	428	25.9%	22.8%	99.6%
Jabba	10.8%	99.6%	99.7%	56.1%	51353	2726	10.5%	9.9%	99.9%
HALC	34.7%	96.5%	99.5%	85.8%	548872	819	33.2%	29.7%	99.3%
				(c) Long reads of Maylandia zebra					
Initial	100.0%	46.9%	91.3%	91.9%	1307812	10082	-	-	-
LoRDEC	33.6%	97.9%	99.7%	89.5%	7372455	601	32.4%	29.8%	99.6%
HALC	41.2%	98.7%	99.6%	90.7%	4833536	1123	40.2%	37.5%	99.4%

The long reads of tests (a)-(c) are from *E.coli*, *A. thaliana* and *Maylandia zebra*, respectively. The initial and error corrected long reads by PacBioToCA, LSC, Proovread, CoLoRMap, ECTools, LoRDEC, Jabba and HALC are compared in the tests. The performance measurements are listed in the "Performance measurements" section.
[a]Some measurements are not available without the correspondence information between a split long read and its initial long read

of HALC is moderate. The results of PaBioToCA, LSC, Proovread, CoLoRMap, ECTools and Jabba are not shown in Table 2(c) because of their very long running time.

The test results in this section indicate that HALC is efficient in correcting and outputting more bases in the initial long reads than the existing algorithms while maintaining sufficient accuracy.

Results on long read assemblies

The assembly results for the error corrected *A. thaliana* long reads are listed in Table 3(a). The number of assembled contigs with HALC corrected long reads is 5.8-29.6% smaller than with most of the existing algorithms, and the N50 value, the largest contig length and the number of covered bases with HALC corrected long reads are 11.4-60.7, 26.6-238.5 and 6.1-141.7% larger than with most of the existing algorithms, respectively. The EPKB value with HALC corrected long reads is 6.8-17.4% smaller than with most of the existing algorithms. Generally, the assembly quality is proportional to the throughput of the algorithms, except for ECTools, with much larger read lengths, and LSC and LoRDEC, with relatively smaller read lengths. The assembly results for the error corrected *Maylandia zebra* long reads are listed in Table 3(b). Even though the number of assembled contigs with HALC corrected long reads is 10.6% larger than with LoRDEC, the N50 value, the largest contig length and the number of covered bases with HALC are 35.9, 33.7 and 58.6% larger than with LoRDEC, respectively. The EPKB value with HALC corrected long reads is 23.6% smaller than with LoRDEC. The results with the initial uncorrected long

reads are not shown because of the limited assembly quality. The assembly results with the *E. coli* long reads are not shown because almost perfect contigs were obtained with the variable long reads, and there is not much difference. These results indicate that HALC corrected long reads can result in more complete assemblies than the existing algorithms with sufficient accuracy.

Results on transcriptome data

For the transcriptome data, the performance test results on the *S. cerevisiae* data set are listed in Table 4. A total of 7.0% bases of the initial long reads can be aligned to the corresponding transcriptome with 99.5% identity, indicating a high error rate in the uncorrected long reads. The existing error correction algorithms LSC, CoLoRMap, LoRDEC and Jabba can obtain 26.0-33.7% throughput, 14.9-29.8% alignment ratio, 99.8-100.0% alignment identity and 15.5-20.6% transcriptome fraction. HALC can obtain 16.1-23.8% higher throughput, 0.7-15.6% higher alignment ratio, and 0.8-5.9% higher transcriptome fraction than all the existing algorithms with comparable alignment identity. The results of the PacBioToCA, Proovread and ECTools are not shown in Table 4 because of their limited performance. It is worth noting that even though some algorithms, such as Proovread, can achieve much better performance by using tailor-made parameters [14], we did not use this procedure to allow a fair comparison. The test results in this section indicate that HALC is also efficient in correcting transcriptome data.

Results with various short read assemblers

With various short read assemblers, HALC exhibits stable results on the *E. coli* and the *S. cerevisiae* data sets, and the difference for all the measurements is below 5% (data not shown). This result indicates that HALC is not very dependent on the upstream short read assemblers and can be used together with various assemblers and for different data types.

Table 3 Evaluation of long read assemblies

Method	N Contigs	N50	Largest contig length	N Covered bases	EPKB
(a) Contigs of A. thaliana					
PacBioToCA	1629	27806	110971	51672726	119.4
LSC	1284	29305	105354	40390383	128.6
Proovread	1193	37828	233854	50379300	123.9
CoLoRMap	1324	35477	150127	52913748	113.1
ECTools	1218	40122	182238	56377176	143.5
LoRDEC	1331	28133	104370	40034274	145.1
Jabba	618	29548	87500	22681404	102.1
HALC	1147	44684	296154	54821730	119.9
(b) Contigs of Maylandia zebra					
LoRDEC	37460	8878	115008	204752772	121.6
HALC	41434	12062	153787	324763656	92.9

The contigs of tests (a)-(b) are for *A. thaliana* and *Maylandia zebra*, respectively. The contigs assembled from the error corrected long reads by PacBioToCA, LSC, Proovread, CoLoRMap, ECTools, LoRDEC, Jabba and HALC are compared in the tests. The performance measurements are listed in the "Performance measurements" section

Table 4 Evaluation of error correction performance on the transcriptomic data set of *S. cerevisiae*

Method	Throughput	Alignment ratio	Alignment identity	Transcriptome fraction
Initial	100.0%	7.0%	99.5%	17.0%
LSC	31.2%	20.2%	99.8%	16.5%
CoLoRMap	26.0%	29.8%	99.8%	20.6%
LoRDEC	33.2%	18.8%	99.9%	16.6%
Jabba	33.7%	14.9%	100.0%	15.5%
HALC	49.8%	30.5%	99.4%	21.4%

The initial and the error corrected long reads by LSC, CoLoRMap, LoRDEC, Jabba and HALC are compared. The performance measurements are listed in the "Performance measurements" section

Running time and memory usage

The running time of HALC on the *E. coli*, A. thaliana, *Maylandia zebra* and *S. cerevisiae* data sets is 0.7h, 7.0h, 53.1h and 0.5h, respectively. The memory usage is 12.3GB, 41.2GB, 33.7GB and 25.9GB, respectively, including the running time and memory usage for short read assemblies by SOAPdenovo2. Compared to the existing algorithms, HALC's running time is much shorter than for PacBioToCA, LSC, Proovread and ECTools and is comparable to or greater than the running time for LoRDEC. Jabba's running time is dependent on the genome sizes. Comparatively, HALC's running time is much smaller on the *Maylandia zebra* data set of large genome size and is larger on the other data sets of small and medium genome sizes. Details of the running time and memory usage are listed in Additional file 1: Table S2. These results indicate that although the main purpose of HALC is to guarantee sufficiently high throughput, it is efficient in running time with acceptable memory usage and can thus scale well for variable project sizes.

Discussion

The most important concern regarding the HALC algorithm is whether the similar repeat based alignment approach introduces false corrections. In theory, false corrections are possible because after a long region is corrected with its similar repeat, it might be refined with the short reads from the similar repeat instead of the ones from the true genome region. However, this problem is not frequent because the refinement algorithm LoRDEC aligns short reads to a long read region by considering not only the long read region's identity but also its adjacent regions in the same long read.

Experimentally, if a similar repeat corrected long read region is a false correction not further refined with the short reads from the true genome region, it will be aligned to its similar repeat in the corresponding genome instead of its true genome region. In other words, it will be aligned to a genome region included in the short read contigs instead of the genome region not included in the contigs. Therefore, we refer to the short read contigs to check for false corrections. We aligned both the HALC corrected long reads and the short read contigs to the genomes on the *E. coli*, *A. thaliana* and *Maylandia zebra* data sets used above and calculated the percentage of genome bases not covered by the contigs and the percentage of long read bases aligned to these genome bases. A much larger value of the former than the latter would indicate that many long read regions are aligned to their similar repeats instead of the true genome regions and are thus false corrections. For the *E. coli* data set, the two values are 4.4 and 3.9%; for the *A. thaliana* data set, the two values are 21.5 and 23.6%; for the *Maylandia zebra* data set, the two values are 42.8 and 39.3%, all comparable. This result indicates

limited false corrections by HALC. Furthermore, since many similar repeat corrected long read regions are not false corrections and are aligned to the genome regions not included in the contigs, we calculated the identity between the genome bases not covered by the contigs and the long read bases aligned to these bases to see the accuracy of the similar repeat corrected long read regions. The identity values are 99.9, 99.3 and 99.6% for the three data sets, respectively. This result indicates high accuracy of the similar repeat corrected long read regions.

In addition, we also refer to the error corrected long reads produced by the existing algorithms to check for false corrections. We aligned the error corrected long reads by various algorithms to the genomes on all three data sets and calculated the percentage of genome bases above the various long read coverages. A much smaller percentage of genome bases above the small long read coverages for HALC than the existing algorithms, or a much larger percentage of genome bases above the large long read coverages for HALC, would indicate that many long read regions are aligned to their similar repeats instead of the true genome regions and are thus false corrections. The plot for the *E. coli* data set is shown in Fig. 3, and the curve of HALC exhibits a similar switch to the existing algorithms from high coverage to low at 25×. The plots of the *A. thaliana* and the *Maylandia zebra* data sets are also available in Additional file 1: Figures S1 and S2, respectively, and similar results can be observed. This result also indicates limited false corrections by HALC. Indeed, the amount of errors contained

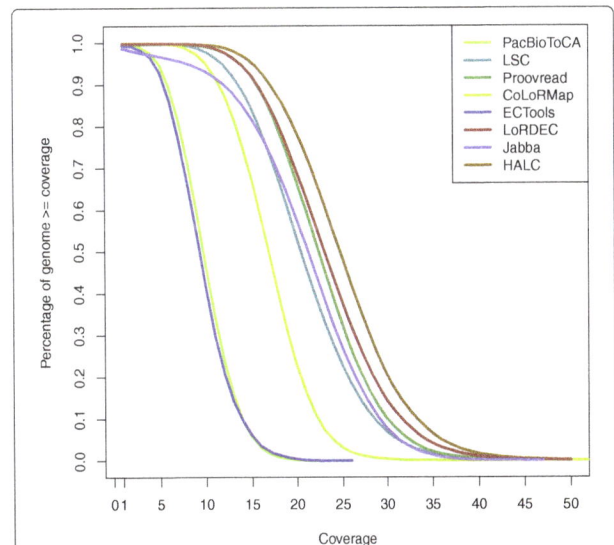

Fig. 3 Percentage of genome covered above various long read coverages on the *E. coli* data. The percentage of genome bases (y-axis) is plotted with long read coverage from 1× to 50× (x-axis), corresponding to the error correction results of different algorithms in Table 2(a)

in the long read assemblies with HALC is another reflection of the false corrections. Table 3 shows that the EPKB values with HALC are smaller or comparable to the ones with most of the existing algorithms, also indicating limited false corrections by HALC (see the "Results on long read assemblies" section).

Conclusions

This study introduces HALC, a high throughput algorithm for PacBio long read error correction. With the similar repeat based alignment approach, the long read regions without true genome regions in the contigs can be aligned; with the long read support based approach, the long read regions' alignments with the highest long read support and continuity can get accepted. Hence, more long read bases can be corrected with accuracy. The experimental results indicate that HALC can correct more bases in the long reads than the existing error correction algorithms while achieving comparable or higher accuracy. As a result, HALC can help to obtain more complete assemblies by providing the error corrected long reads.

Abbreviations
CI: Number of correct bases in the initial long reads; CD: Number of correct bases in the discarded bases; EI: Number of errors in the initial long reads; EPKB: Number of errors per 100K bp in the initial contigs; FN: False negative; FP: False positive; TH: Throughput; TN: True negative; TP: True positive

Acknowledgements
We thank Thomas Girke and Tao Jiang from the University of California, Riverside for their very helpful suggestions during the improvement of this work. We also thank Leena Salmela and Eric Rivals, the authors of LoRDEC, for providing us with the Error Correction Evaluation Toolkit for full long reads. We acknowledge the support of the core facilities at the Institute for Integrative Genome Biology (IIGB) at the University of California, Riverside.

Funding
This work was supported by grants from the National Science Foundation of China [61502027 to EB] and the Fundamental Research Funds for the Central Universities [2015RC045 to EB]. The funders had no role in study design, data collection and analysis, decision to publish, or preparation of the manuscript.

Authors' contributions
EB designed the algorithm and wrote the manuscript. LL implemented and tested the algorithm. Both authors read and approved the final manuscript.

Competing interests
The authors declare that they have no competing interests.

References

1. Rhoads A, Au KF. Pacbio sequencing and its applications. Genomics, proteomics & bioinformatics. 2015;13(5):278–89.
2. Luo R, Liu B, Xie Y, Li Z, Huang W, Yuan J, He G, Chen Y, Pan Q, Liu Y, et al. Soapdenovo2: an empirically improved memory-efficient short-read de novo assembler. GigaScience. 2012;1(1):18.
3. Gnerre S, MacCallum I, Przybylski D, Ribeiro FJ, Burton JN, Walker BJ, Sharpe T, Hall G, Shea TP, Sykes S, et al. High-quality draft assemblies of mammalian genomes from massively parallel sequence data. Proc Natl Acade Sci. 2011;108(4):1513–1518.
4. Eid J, Fehr A, Gray J, Luong K, Lyle J, Otto G, Peluso P, Rank D, Baybayan P, Bettman B, et al. Real-time dna sequencing from single polymerase molecules. Science. 2009;323(5910):133–8.
5. Lee H, Gurtowski J, Yoo S, Nattestad M, Marcus S, Goodwin S, McCombie WR, Schatz M. Third-generation sequencing and the future of genomics. bioRxiv. 2016. https://doi.org/10.1101/048603.
6. Koren S, Schatz MC, Walenz BP, Martin J, Howard JT, Ganapathy G, Wang Z, Rasko DA, McCombie WR, Jarvis ED, et al. Hybrid error correction and de novo assembly of single-molecule sequencing reads. Nature Biotechnol. 2012;30(7):693–700.
7. Chin CS, Alexander DH, Marks P, Klammer AA, Drake J, Heiner C, Clum A, Copeland A, Huddleston J, Eichler EE, et al. Nonhybrid, finished microbial genome assemblies from long-read smrt sequencing data. Nature methods. 2013;10(6):563–9.
8. Au KF, Sebastiano V, Afshar PT, Durruthy JD, Lee L, Williams BA, van Bakel H, Schadt EE, Reijo-Pera RA, Underwood JG, et al. Characterization of the human esc transcriptome by hybrid sequencing. Proc Natl Acad Sci. 2013;110(50):4821–830.
9. English AC, Richards S, Han Y, Wang M, Vee V, Qu J, Qin X, Muzny DM, Reid JG, Worley KC, et al. Mind the gap: upgrading genomes with pacific biosciences rs long-read sequencing technology. PloS ONE. 2012;7(11): 47768.
10. Bankevich A, Nurk S, Antipov D, Gurevich AA, Dvorkin M, Kulikov AS, Lesin VM, Nikolenko SI, Pham S, Prjibelski AD, et al. Spades: a new genome assembly algorithm and its applications to single-cell sequencing. J Comput Biol. 2012;19(5):455–77.
11. Deshpande V, Fung ED, Pham S, Bafna V. Cerulean: a hybrid assembly using high throughput short and long reads. In: Algorithms in Bioinformatics. Berlin Heidelberg: Springer-Verlag; 2013. p. 349–63.
12. Ye C, Hill C, Ruan J, et al. Dbg2olc: Efficient assembly of large genomes using the compressed overlap graph. 2014. https://arxiv.org/abs/1410.2801.
13. Myers EW, Sutton GG, Delcher AL, Dew IM, Fasulo DP, Flanigan MJ, Kravitz SA, Mobarry CM, Reinert KH, Remington KA, et al. A whole-genome assembly of drosophila. Science. 2000;287(5461): 2196–204.
14. Hackl T, Hedrich R, Schultz J, Förster F. proovread: large-scale high-accuracy pacbio correction through iterative short read consensus. Bioinformatics. 2014;30(21):3004–11.
15. Haghshenas E, Hach F, Sahinalp SC, Chauve C. Colormap: Correcting long reads by mapping short reads. Bioinformatics. 2016;32(17):545–51.
16. Lee H, Gurtowski J, Yoo S, Marcus S, McCombie WR, Schatz M. Error correction and assembly complexity of single molecule sequencing reads. BioRxiv. 2014. https://doi.org/10.1101/006395.
17. Salmela L, Rivals E. Lordec: accurate and efficient long read error correction. Bioinformatics. 2014;30(24):3506–14.
18. Miclotte G, Heydari M, Demeester P, Rombauts S, Van de Peer Y, Audenaert P, Fostier J. Jabba: hybrid error correction for long sequencing reads. Algoritm Mol Biol. 2016;11(1):1.
19. Salmela L, Walve R, Rivals E, Ukkonen E. Accurate selfcorrection of errors in long reads using de bruijn graphs. Bioinformatics. 2016;33(6):799–806.
20. Yang X, Chockalingam SP, Aluru S. A survey of error-correction methods for next-generation sequencing. Brief Bioinform. 2013;14(1):56–66.
21. Smit AF. The origin of interspersed repeats in the human genome. Curr Opin Genet Dev. 1996;6(6):743–8.
22. Yang X, Dorman KS, Aluru S. Reptile: representative tiling for short read error correction. Bioinformatics. 2010;26(20):2526–533.
23. Chaisson MJ, Tesler G. Mapping single molecule sequencing reads using basic local alignment with successive refinement (blasr): application and theory. BMC bioinforma. 2012;13(1):238.
24. Altschul SF, Gish W, Miller W, Myers EW, Lipman DJ. Basic local alignment search tool. J Mol Biol. 1990;215(3):403–10.
25. Kent WJ. Blat–the blast-like alignment tool. Genome research. 2002;12(4): 656–64.
26. Delcher AL, Kasif S, Fleischmann RD, Peterson J, White O, Salzberg SL. Alignment of whole genomes. Nucleic Acids Res. 1999;27(11):2369–376.
27. Salzberg SL, Phillippy AM, Zimin A, Puiu D, Magoc T, Koren S, Treangen TJ, Schatz MC, Delcher AL, Roberts M, et al. Gage: A critical evaluation of genome assemblies and assembly algorithms. Genome Res. 2012;22(3): 557–67.
28. Grabherr MG, Haas BJ, Yassour M, Levin JZ, Thompson DA, Amit I, Adiconis X, Fan L, Raychowdhury R, Zeng Q, et al. Full-length transcriptome assembly from rna-seq data without a reference genome. Nature Biotechnol. 2011;29(7):644–52.
29. Zerbino DR, Birney E. Velvet: algorithms for de novo short read assembly using de bruijn graphs. Genome Res. 2008;18(5):821–9.

30. Simpson JT, Wong K, Jackman SD, Schein JE, Jones SJM, Birol I. Abyss: a parallel assembler for short read sequence data. Genome Res. 2009;19(6): 1117–1123.
31. Schulz MH, Zerbino DR, Vingron M, Birney E. Oases: Robust de novo rna-seq assembly across the dynamic range of expression levels. Bioinformatics. 2012;28(8):1086-92.
32. Robertson G, Schein J, Chiu R, Corbett R, Field M, Jackman SD, Mungall K, Lee S, Okada HM, Qian JQ, et al. De novo assembly and analysis of rna-seq data. Nature methods. 2010;7(11):909–12.
33. Conte MA, Kocher TD. An improved genome reference for the african cichlid, metriaclima zebra. BMC genomics. 2015;16(1):1.
34. Li H, Durbin R. Fast and accurate long-read alignment with burrows–wheeler transform. Bioinformatics. 2010;26(5):589–95.
35. Gurevich A, Saveliev V, Vyahhi N, Tesler G. Quast: quality assessment tool for genome assemblies. Bioinformatics. 2013;29(8):1072–75.

"gnparser": a powerful parser for scientific names based on Parsing Expression Grammar

Dmitry Y. Mozzherin[1][*][†] ⓘ, Alexander A. Myltsev[2][†] and David J. Patterson[3]

Abstract

Background: Scientific names in biology act as universal links. They allow us to cross-reference information about organisms globally. However variations in spelling of scientific names greatly diminish their ability to interconnect data. Such variations may include abbreviations, annotations, misspellings, etc. Authorship is a part of a scientific name and may also differ significantly. To match all possible variations of a name we need to divide them into their elements and classify each element according to its role. We refer to this as 'parsing' the name. Parsing categorizes name's elements into those that are stable and those that are prone to change. Names are matched first by combining them according to their stable elements. Matches are then refined by examining their varying elements. This two stage process dramatically improves the number and quality of matches. It is especially useful for the automatic data exchange within the context of "Big Data" in biology.

Results: We introduce Global Names Parser (*gnparser*). It is a Java tool written in Scala language (a language for Java Virtual Machine) to parse scientific names. It is based on a Parsing Expression Grammar. The parser can be applied to scientific names of any complexity. It assigns a semantic meaning (such as genus name, species epithet, rank, year of publication, authorship, annotations, etc.) to all elements of a name. It is able to work with nested structures as in the names of hybrids. *gnparser* performs with ≈ 99% accuracy and processes 30 million name-strings/hour per CPU thread. The *gnparser* library is compatible with Scala, Java, R, Jython, and JRuby. The parser can be used as a command line application, as a socket server, a web-app or as a RESTful HTTP-service. It is released under an Open source MIT license.

Conclusions: Global Names Parser (*gnparser*) is a fast, high precision tool for biodiversity informaticians and biologists working with large numbers of scientific names. It can replace expensive and error-prone manual parsing and standardization of scientific names in many situations, and can quickly enhance the interoperability of distributed biological information.

Keywords: Biodiversity, Biodiversity informatics, Scientific name, Parser, Semantic parser, Names-based cyberinfrastructure, Scala, Parsing Expression Grammar

Background

Conventions

Throughout the paper we use the terms "name", "scientific name", and "name-string" in particular ways. "Name" refers to one or several words that act as a label for a taxon. A "scientific name" is a name formed in compliance with a nomenclatural code (Code) or, if beyond the scope of the Codes, is consistent with the expectations of a Code.

The term "name-string" is the sequence of characters (letters, numbers, punctuation, spaces, symbols) that forms the name. A name can be expressed in the form of many name-strings (for example, see Fig. 1). There are about two and a half million currently accepted names for extinct and extant species. There are approximately ten million of legitimately formed scientific names and hundreds of millions of possible name-strings for them. We use the term "elements" for the components of a name-string. Traditionally, in biological literature, scientific names for genera and taxa below genus are presented in *italics*. In this paper, where we wish to emphasize examples of name-strings, we use **bold font**.

---

*Correspondence: mozzheri@illinois.edu

[†]Equal contributors

[1]University of Illinois, Illinois Natural History Survey, Species File Group, 1816 South Oak St., Champaign, IL, 61820, USA

Full list of author information is available at the end of the article

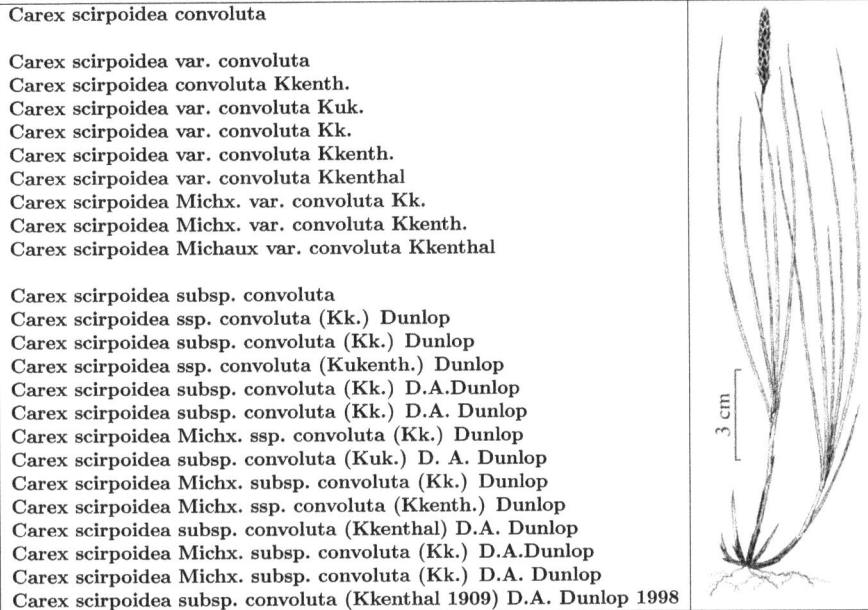

Carex scirpoidea convoluta
Carex scirpoidea var. convoluta
Carex scirpoidea convoluta Kkenth.
Carex scirpoidea var. convoluta Kuk.
Carex scirpoidea var. convoluta Kk.
Carex scirpoidea var. convoluta Kkenth.
Carex scirpoidea var. convoluta Kkenthal
Carex scirpoidea Michx. var. convoluta Kk.
Carex scirpoidea Michx. var. convoluta Kkenth.
Carex scirpoidea Michaux var. convoluta Kkenthal
Carex scirpoidea subsp. convoluta
Carex scirpoidea ssp. convoluta (Kk.) Dunlop
Carex scirpoidea subsp. convoluta (Kk.) Dunlop
Carex scirpoidea ssp. convoluta (Kukenth.) Dunlop
Carex scirpoidea subsp. convoluta (Kk.) D.A.Dunlop
Carex scirpoidea subsp. convoluta (Kk.) D.A. Dunlop
Carex scirpoidea Michx. ssp. convoluta (Kk.) Dunlop
Carex scirpoidea subsp. convoluta (Kuk.) D. A. Dunlop
Carex scirpoidea Michx. subsp. convoluta (Kk.) Dunlop
Carex scirpoidea Michx. ssp. convoluta (Kkenth.) Dunlop
Carex scirpoidea subsp. convoluta (Kkenthal) D.A. Dunlop
Carex scirpoidea Michx. subsp. convoluta (Kk.) D.A.Dunlop
Carex scirpoidea Michx. subsp. convoluta (Kk.) D.A. Dunlop
Carex scirpoidea subsp. convoluta (Kkenthal 1909) D.A. Dunlop 1998

Fig. 1 Some legitimate versions of the scientific name for the 'Northern Bulrush' or 'Singlespike Sedge'. The genus (*Carex*), species (*scirpoidea*), and subspecies (*convoluta*) may be annotated (var., subsp., and ssp.) or include or omit the name of the original authority for the infraspecies (Kükenthal), or for the species (Michaux), or for the current infraspecific combination (Dunlop). The name of the authority is sometimes abbreviated, sometimes differently spelled, and may be with or without initials and dates. This list is not complete. Image courtesy of [42]

Introduction

Biology is entering a "Big Data" age, where global and fast access to all knowledge is envisaged. Progress towards this vision is still limited in scope. One impediment, especially for the long tail of smaller sources (of which some are not yet digital), is the absence of devices to inter-connect distributed data. The names of organisms are invaluable in "Big Data" biology because they can be treated as metadata and as such can be used to discover, index, organize, and interconnect distributed information about species and other taxa [1]. The use of names for informatics purposes is not straightforward because, for example, there may be many legitimate spellings for a name (Fig. 1). A cyberinfrastructure that uses names to manage information about organisms must determine which name-strings are variant forms of the same scientific name.

Figure 1 presents some of the different legitimate variants of a scientific name in order to make the point that there is not a single correct way to spell scientific names. Because of these variations, fewer than 15% of the names in comparisons of large biological databases could be matched based on exact spellings of name-strings [2]. In order to improve this simple metric for interoperability, we need to identify variants of the same name. We refer to the process of addressing variant spellings (there being other causes of different names for the same

taxon) as "lexical reconciliation". Lexical reconciliation involves linking the alternative spelling variants for the same taxon into a "lexical group". Most biologists do this intuitively — they recognize that the name-strings in Fig. 1 refer to the same taxon. They do so by "parsing" the name-strings into elements (genus name, species name, authors, ranks etc.) and mentally discarding less significant elements such as annotations and authorship. It then becomes clear all of name-strings are formed around the Latin elements **Carex scirpoidea convoluta**. We refer to the form of the scientific name without authority or annotations as the "canonical form". Further analysis of the name-strings reveals two different lexical groups (separated in Fig. 1 by a line break) for, probably, one taxonomic concept:

- **Carex scirpoidea var. convoluta** description by **Kükenthal**
- **Carex scirpoidea subsp. convoluta** rank determination by **Dunlop**.

In the past, the need to parse scientific names to form normalized names has mostly been achieved manually. A person familiar with rules of botanical nomenclature would be able to analyse the 24 name-strings in this example with relative ease, but not thousands or millions of name-strings - especially if they include scientific names to which more than one nomenclatural code may be applied. The manual splitting of names into even only two parts — the latinized elements of taxon names that make up the canonical form and the authorship — is slow and therefore expensive. To scale this exercise up requires an algorithmic solution, a scientific name parser!

The strategy of the algorithmic approach is to identify which combinations of the most atomic parts of a name-string (i.e. the UTF-8 encoded characters) represent words (such as genus name, species name, authors, annotations) or dates. An early algorithmic approach to parsing scientific names was with "regular language" implemented as regular expression [3]. A regular expression is a sequence of characters that describes a search pattern [4]. For example, a regular expression "[A-Z][a-z]{2}" recognizes a word that starts from a capital letter followed by two small letters (e.g. "Zoo"). Scientific names almost universally follow patterns that are influenced by the Codes of Nomenclature: such as the use of spaces to separate words, capitalization of generic names and authors, or the inclusion of four digit dates between the middle of the 18th century and the present. This makes most names amenable to parsing by regular expressions. Current examples of scientific name parsers based on regular expressions are GBIF's *name-parser* [5], and *YASMEEN* [6].

While regular expression is a powerful approach to string parsing, it has limitations. It cannot elegantly deal with name-strings where an authorship element is present in the middle of the name (for example **Carex scirpoidea Michx. subsp. convoluta (Kük.) D.A.Dunlop**). Indeed, regular expressions are not well suited to any targets with recursive (nested) elements [7], such as hybrid formulae (e.g. **Brassica oleracea L. subsp. capitata (L.) DC. con- var. fruticosa (Metzg.) Alef. × B. oleracea L. subsp. cap- itata (L.) var. costata DC.**). Name parsing built on regular expressions is impractical for complex name-strings.

Another limitation with most regular expression software tools is that they are "black boxes" that allow developers very limited interaction with the parsing process. They do not reveal much information about the parsing context and developers cannot call a procedure during a parsing event. As a result, complex regular expression-based parsers are difficult to implement and maintain, and functions such as error recovery, detailed warnings, descriptions of errors are missing.

We wanted to deal with scientific names across a very broad range of complexity and to give more flexibility than can be achieved with a regular expression approach. We believe that a scientific name parser should satisfy the following requirements.

1. **High Quality.** A parser should be able to break names into their semantic elements to the same standards that can be achieved by a trained nomenclaturalist

or better. This will give users confidence in the automated process and allow them to set aside tedious and expensive manual parsing.

2. **Global Scope.** A parser should be able to parse all types of scientific names, inclusive of the most complex name-strings such as hybrid formulae, multi-infraspecific names, names with multilevel authorships and so on. No name-strings should be left unparsed, otherwise biological information attached to them may remain undiscoverable.

3. **Parsing Completeness.** All information included in a name-string is important, not just the canonical form of the scientific name. Authorship, year, rank information allow us to distinguish homonyms, similar names, synonyms, spelling mistakes, or chresonyms. Access to such information improves the performance of subsequent reconciliation (the mapping of all alternative name-strings for the same taxon against each other).

4. **Speed.** Users, especially large-scale aggregators of biodiversity data, are more satisfied with speedy processing of data as it allows them to move forward to more purposeful value-adding tasks. Speed reduces the purchasing/operating costs of the hardware used for production parsing.

5. **Accessibility.** To be available to the widest possible audience, a parser should be released as a stand-alone program, have good documentation, be able to work as a library, to function as a command line tool, as a tool within a graphical interface, to run as a socket or as RESTful services.

These requirements became our design goals. Based on our experience with prototype systems, we chose to use Parsing Expression Grammar and Scala language.

Adoption of Parsing Expression Grammar

Parsing Expression Grammar (PEG) [8] have been introduced for parsing strings. PEG allows developers to define the rules ("grammar") that describe the general structure of target strings. Such rules can be used to deconstruct scientific names. The rules are built from the ground up, starting from the simplest — such as a combination of "characters" separated by "spaces". That 'rule' identifies most "words". Digits and other characters make dates identifiable. Further rules can be applied, such as a "genus" rule can describe a part of a polynomial name-string in which the first word begins with combination of a "capital_character" followed by several "lower_case_characters" that fall within a relatively small spectrum of allowed characters; "authorship" would consist of one or more capitalized words and followed perhaps by a "year". Within some instances of authorship, authors may be grouped to form "author-teams". PEG rules are designed to be recursive. They can be expanded to deal with increasingly complex name-strings, or address errors such as absent or extra spaces, or OCR errors. Each rule can have programmatic logic attached, making the PEG approach very flexible. We believe that PEG suits our goals better than regular expressions for the following reasons:

- PEG is better suited than regular expressions for strings with a recursive structure;
- the syntax of scientific names is formal enough to be closer to an algebraic structure rather than to a natural language. Inconsistencies and ambiguities in scientific name-strings are relatively rare because they usually comply with the requirements and conventions of nomenclatural codes;
- scientific name-strings are short enough to avoid problems with computational complexity and memory consumption;
- programming a parser with PEG can describe parsing rules in a domain-specific language;
- domain-specific languages offer great flexibility for logic within the rules, for example to report errors in name-strings.

The Global Names project created a specialized parsing library *biodiversity* in 2008 [9]. It was written in Ruby and based on PEG. It uses the *TreeTop* Ruby library [10] as the underlying PEG implementation.

The PEG approach allowed us to deal with complex scientific names gracefully. It gave us flexibility to incorporate edge cases and to detect common mistakes during the parsing process. The *biodiversity* library has enjoyed considerable popularity. At the time of writing, it had been downloaded more than 150,000 times [11], it is used by many taxon name resolution projects (e.g. Encyclopedia of Life [12], Canadian Register of Marine Species (CARMS) [13], the iPlant TNRS [14], and World Registry of Marine Species (WoRMS) [15]. According to statistics compiled by BioRuby, *biodiversity*, at the time of writing, has been the most popular bio-library in the Ruby language [16].

We were pleased with PEG approach for parsing scientific names, but regard the *biodiversity* parser library as a working prototype. It has allowed us to make further improvements and deliver a better, faster production-grade parser.

Other approaches
There is a growing number of algorithms and tools in machine learning and natural language processing that aim to recognize parts of texts. They include statistical parsing [17], context-free grammars [18], fuzzy context-free grammars [19], and named entity recognition [20]. Unsupervised deep learning [21, 22] increases the quality of entity recognition without extensive curation and programming efforts by people. We chose not to use these approaches for the following reasons.

- The limited scope of a parser. A parser of scientific names very rarely needs to work with name-strings of more than 15 words.
- There is no need for recognition. A scientific name-string parser is usually applied to preexisting lists of scientific names. There is no requirement to recognize scientific names in larger bodies of text. Other scientific name recognition and discovery tools are available.
- Formal grammar. Scientific names are formed in compliance with well-defined and formal codes of nomenclature. They have predictable structures making the requirements for a scientific name-string parser to be more similar to parsers of programming languages than to tools designed to work with natural languages.
- Scale and throughput. We created the parser to serve the needs of biodiversity aggregators. A core design requirement was to develop a lightweight library for inputs of millions of scientific name-strings per second, and to be processed locally.
- Stand-alone approach. We did not wish the parser to rely on local or remote previously known information of genera, species, author names, or other scientific names. *gnparser* relies instead on morphological features of scientific name-strings.
- Determinism. Biologists know that there is only a single correct parsed version of a scientific name. A scientific names parser must produce a single "correct" result for each input string. A parser should provide meta information on every part of the string.

Adoption of Scala
The pre-existing *biodiversity* package is not speedy and cannot scale because it uses Ruby as its programming language. Ruby is one of the best languages for rapid prototyping, but it is an interpreted dynamic language with, originally, a single-threaded runtime during execution. This makes it slow and inappropriate for "Big Data" tasks. We concluded that we needed a replacement language environment with the following properties:

- a mature technology;
- multithreaded, with high performance and scalability;
- an active support community with an Open source friendly culture;
- a wide range of libraries: utilities, web frameworks, etc.;
- a powerful development environment with IDEs, testing frameworks, debuggers, profilers and the like;
- mature libraries for search and cluster computations;
- interoperable with languages popular in scientific community (R, Python, Matlab);

- natural support of domain specific languages embedded in the hosted language.

While many of the properties are true for Ruby, other properties, such as high performance, scalability and interoperability, are not. To meet all requirements, and exploiting what we had learned from *biodiversity*, we rewrote the code using Scala (a Java virtual machine programming language [23]), and the Open source *parboiled2* library [24] which we improved [25]. The *parboiled2* library implements PEG in Scala. An alternative to *parboiled2* is the Scala combinators library [26]. We did not use it because it is slow and has memory consumption problems.

The functional programming features of Scala allowed us to build a domain specific language that describes the grammar's rules to parse scientific names. This produces a Parsing Expression Grammar with considerably more flexibility than external lexers such as Bison or Yacc. As this domain specific language is within *parboiled2*, it can take advantage of the Macro capacity of Scala [27] to optimize the compilation of the code and the subsequent running of the program. As a result, the software performs with high efficiency. The resulting *gnparser* library is faster, more scalable and more flexible than its predecessor.

We limited this version to work with scientific names that comply with the botanical, zoological, and prokaryotic codes of nomenclature, but not with names of viruses because they are formed in different ways [2, 28] and need a different PEG. We intend to add this later.

Implementation

The *gnparser* project is entirely written in Scala. It supports two major Scala versions: 2.10.6+ and 2.11.x. The code is organized into four modules:

1. *"parser"* is the core module used by all other modules. It parses scientific names from the most atomic components of a name-string to semantically-defined terms. It includes the parsing grammar, an abstract syntax tree (AST) composed of the elements of scientific names, warning and error facilities. When the parsing is complete and semantic elements of name-strings have been assigned to AST nodes, the elements can be recombined and formatted to meet further needs. For example:
   - *normalizer* converts input name-strings into a consistent style;
   - *canonizer* creates canonical forms of the latinized elements of names;
   - *JSON renderer*, the parsing result is converted to JSON [29] to allow developers to work with the output using other languages. The output (Fig. 2, also see Results and discussion) has the following information: **'details'** contains the

JSON-representation of a parsed scientific name; **'quality_warnings'** describes potential problems if names are not well-formed; **'quality'** depicts a quality level of the parsed name; and **'positions'** maps the positions of every element in a parsed name to the semantic meaning of the element. Full and formal explanation of all parser fields is given as a JSON schema and can be found online [30] [also see Additional file 1].

2. The *"spark-python"* module contains facilities to use *"gnparser"* with Apache Spark scripts written in Python. Apache Spark is a highly distributive and scalable development environment for processing massive sets of data. Spark is written in Scala, but can also be used with Python, R and Java languages. Spark programs written in Java and Scala are able to run *"parser"* in a distributed fashion natively.

3. The *"examples"* module contains examples to assist developers in adding *"parser"* functionality into other popular programming languages such as Java, Scala, Jython, JRuby, and R.

4. The *"runner"* module contains the code that allows users to run *"parser"* from a command line as a standalone tool or to run it as a TCP/IP socket or HTTP web server. It depends on the *"parser"* module. The core part is the launch script *"gnparse"* (for Linux/Mac and Windows) that creates a JVM instance and runs *"parser"* on multiple threads against the input provided via a socket or file. This module also contains a web application and a RESTful interface to offer simpler ways to access *"parser".* *"web"* achieves interactions with *"parser"* via HTTP protocol. It works both with simple web (HTML) and REST API interfaces. Figure 2 illustrates a parsing example using the web-interface. Socket and REST services use Akka framework which makes them highly concurrent and scalable.

"parser" and *"examples"* can run in JVM 1.6+. *"runner"* requires JVM 1.8+. Documentation is available in a README file [see Additional file 2].

Parsing rules

gnparser v0.3.1 contains 76 PEG rules. In turn, these rules make use of more elementary rules provided by the *parboiled2* library. The rules are domain-specific based on hours of conversations with leading taxonomists, study of nomenclatural codes, and feedback of the users.

As an example, the *yearNumber* rule is given below. It detects the year in which a name was published. *Rule[Year]* is a type of the returning value of the rule. Using domain-specific language and elementary rules of *parboiled2* we capture the start and the end positions of a year substring (lines #1 and #2). This matches a substring that represents a year in scientific name-strings. A

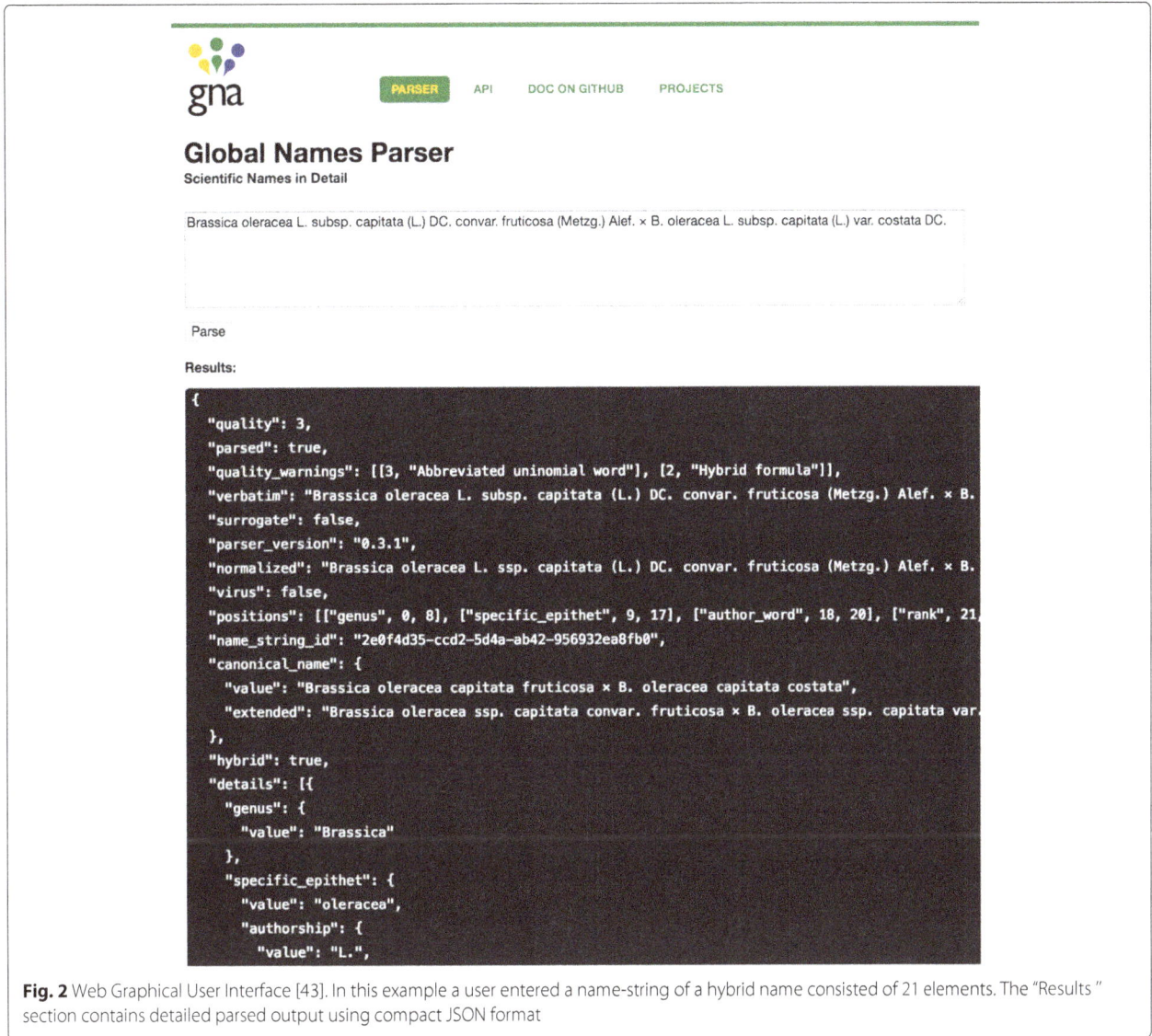

Fig. 2 Web Graphical User Interface [43]. In this example a user entered a name-string of a hybrid name consisted of 21 elements. The "Results" section contains detailed parsed output using compact JSON format

publication year is usually a number between 1753 [31] and the present. A year substring might have one or two digits substituted with question marks if the exact year of a publication is unknown. The capture is then passed as a parameter to a parser action (line #3). Parser action, a Scala function, might produce warnings or a class instance of defined type (*Rule[Year]*).

```
def yearNumber: Rule[Year] = rule { capturePos( // #1
    CharPredicate("12") ~ CharPredicate("0789") ~
    Digit ~ (Digit|'?') ~ '?'.? // #2
) ~> { (yPos: CapturePosition) => // #3
    FactoryAST.year(yPos) // #4
  }
}
```

We then assemble more complex inter-dependent rules (lines #5 to #10), and finally combine all of them into the

rule *year* on line #11 that consists of prioritized alternatives of all previously defined rules.

```
def yearWithChar = rule { yearNumber ~ capturePos(Alpha)} // #5
def yearWithParens = rule { '(' ~ (yearWithChar |
    yearNumber) ~ ')'} // #6
def yearWithPage = rule { (yearWithChar | yearNumber) ~
    ':' ~ oneOrMore(Digit)} // #7
def yearApprox = rule { '[' ~ yearNumber ~ ']'} // #8
def yearWithDot = rule { yearNumber ~ '.'} // #9
def yearRange = rule { yearNumber ~ '-' ~
    capturePos(Digit.+) ~ (Alpha ++ "?").*} // #10
def year = rule { yearRange | yearApprox |
    yearWithParens | yearWithPage | yearWithDot |
    yearWithChar | yearNumber // #11
}
```

This enables the incorporation of the *year* rule into all cases where it might be needed. For example on line #12 we indicate that *year* must be present in the matcher for the *authorsYear* rule.

```
def authorsYear: RuleNodeMeta[AuthorsGroup] = rule {
authorsGroup ~ softSpace ~ (',' ~ softSpace).? ~ year ~> { // #12
(aM: NodeMeta[AuthorsGroup], yM: NodeMeta[Year]) =>
val a1 = for { a <- aM; y <- yM} yield a.copy(year = y.some)
a1.changeWarningsRef((aM.node, a1.node))
}
}
```

Installation

"*gnparser*" is available for launch in three bundles.

- A *parser* artifact is provided via the Maven central repository of Java code [32]. Physically it is a relatively small jar file without embedded external dependencies. The artifact can be accessed in custom projects by a build system such as Maven, Gradle, or SBT. The build system identifies and provides access to all dependent jars.

- A Zip-archived "fat jar" is located at the project's GitHub repository. The jar contains the compiled files of *gnparser* along with all necessary dependencies to launch it within JVM. The archive is also bundled with a launch script (for Windows, OS X and Linux) that can run a command line interface to *gnparser*.

- The project's Docker container image is located at Docker Hub [33]. Docker provides an additional layer of abstraction and automation of operating-system-level virtualization on Linux. It can be thought of as a lightweight virtualization technology within a Linux OS host. When it is setup properly, everything — starting from JVM and ending with Scala and SBT — can be run with simple commands that will, for example, pull the *gnparser*'s Docker image from the DockerHub, and run the socket or web server on an appropriate port.

Testing methods

Data for our tests were sets of 1000 and 100,000 name-strings randomly chosen from 24 million unique name-strings of the Global Names Index (GNI) [34]. The name-strings in GNI are collected from a large variety of biodiversity data sources and are pre-identified as scientific names. While GNI contains some incorrectly classified strings, it is the largest compilation of name-strings representing scientific names. It is not biased towards any particular taxon or particular variant of name, and so the extracted datasets are believed to represent naturally occurring data quite well. The datasets are randomly chosen and are therefore mixtures of well-formed names, lexical variants of names, names with formatting and spelling mistakes, and name-strings that were misrepresented as names. Name-strings in the sets are independent of each other. An evaluation dataset with 1000 names is included as Additional file 3.

We compared the performance of *gnparser* with two other projects: *biodiversity* parser [9, 35] (also developed by Global Names team), and the GBIF *name-parser* [5]. The following versions were used: *gnparser* v. 0.2.0, GBIF

name-parser v. 0.1.0, *biodiversity* v. 3.4.1. To make comparisons, we calculated *Precision*, *Recall* and *Accuracy* (as described below) using a dataset consisting of 1000 name-strings. We also tested the YASMEEN parser from iMarine [6]. With our dataset, YASMEEN generated many more mistakes than other parsers (*Precision* 0.534, *Recall* 1.0, *F1* 0.6962), and was unable to finish a full dataset without crashing. We excluded it from further tests.

To estimate the quality of the parsers, we relied on their performance in representing canonical forms and terminal authorships. A canonical form represents the latinized elements of taxon names, while the terminal authorship refers to the author of the lowest subtaxon found in the scientific name. For example, with **Oriastrum lycopodioides Wedd. var. glabriusculum Reiche**, the canonical form is **Oriastrum lycopodioides glabriusculum** and the terminal authorship is **Reiche**, not **Wedd**.

When both the canonical form and the terminal authorship were determined correctly we marked the result as true positive (N_{tp}). If one or both of them were determined incorrectly, the result was marked as a false positive (N_{fp}). Name-strings correctly discarded from parsing were marked as true negatives (N_{tn}). False negatives (N_{fn}) were name-strings which should have been parsed, but were not. The results of the tests are summarized in Table 1:

Accuracy — the proportion of all results that were correct. It is calculated as:

$$Accuracy = \frac{N_{tp} + N_{tn}}{N_{tp} + N_{tn} + N_{fp} + N_{fn}}$$

Precision — the proportion of name-strings parsed correctly compared to all detected name-strings. It is calculated as:

$$Precision = \frac{N_{tp}}{N_{tp} + N_{fp}}$$

Recall — the proportion of correctly detected name-strings relative to all parseable name-strings and is calculated as:

Table 1 Precision/Recall for parsers applied to 1000 name-strings

	gnparser	gbif-parser	Biodiversity
True positive	978	955	971
True negative	13	12	13
False positive	9	32	16
False positive	0	1	0
Precision	0.989	0.968	0.984
Recall	1.0	0.999	1.0
F1	0.994	0.983	0.992
Accuracy	0.989	0.967	0.984

$$Recall = \frac{N_{tp}}{N_{tp} + N_{fn}}$$

The $F1 - measure$ is a balanced harmonic mean (where *Precision* and *Recall* have the same weight). When *Precision* and *Recall* differ, $F1 - measure$ allows results to be compared. It is calculated as

$$F1 = \frac{2 \times Precision \times Recall}{Precision + Recall}$$

Some names in the dataset were not well-formed. If a human could extract the canonical form and the terminal authorship from them, we included them in our assessment. Examples of such name-strings are **"Hieracium nobile subsp. perclusum (Arv. -Touv.) O. Bolòs & Vigo"** (the problem for the parser here is an introduced space within an author's name), **"Campylium gollanii C. M?ller ex Vohra 1970 [1972]"** (with a miscoded UTF-8 symbol and an additional year in square brackets), **"Myosorex muricauda (Miller, 1900)."** (with a period after the authorship).

Parsers analyze the structure of name-strings, but they cannot determine if a string is a "real" name. For example, in the case of a name-string that has the same form as a subspecies such as **"Example name Word var. something Capitalized Words, 1900"**. In such a case, the identification of a canonical form as **"Example name something"** and terminal authorship as **"Capitalized Words, 1900"** would be considered a true positive. Clearly, it will be important for name-management services to distinguish between name-strings of scientific names, names of viruses, surrogate names, and non-names. To find out how well parsers distinguished strings which are not scientific names, we calculated *Accuracy* for discarded/non-parsed strings. If the parser worked well, non-parsed strings would include only names of viruses and terms that do not comply with the codes of zoological, prokaryotic, and botanical nomenclature.

We processed 100,000 name-strings with each parser. Each parser discarded close to 1,000 name-strings as non-parseable. *Accuracy*, in this case, provided the percentage of correctly discarded names out of all discarded by the parser names. We do not know *Recall*, as it was not reasonable to manually determine this for 100,000 names. To get a sense of names which should be discarded but were parsed instead, we analysed intersections and differences of the results between the three parsers as shown in Table 2.

To establish the throughput of parsing we used a computer with an Intel i7-4930K CPU (6 cores, 12 threads, at 3.4 GHz), 64GB of memory, and 250GB Samsung 840 EVO SSD, running Ubuntu version 14.04. Throughput was determined by processing 1,000,000 random name-strings from Global Names database.

Table 2 Accuracy of non-parseable names detection out of 100,000 name-strings

	gnparser	gbif-parser	Biodiversity
True discarded	1131	1082	1161
Correctly discarded	1129	940	1152
Incorrectly discarded	2	142	9
Accuracy	0.998	0.869	0.992

To study the effects of parallel execution on throughput we used the *ParallelParser* class from *biodiversity* parser. We used '*gnparse file –simple*' (a command line-based script set to return simplified output) for *gnparser*. For GBIF *name-parser*, we created a thin wrapper with multithreaded capabilities [36]. The following versions had been used for throughput benchmarks: *gnparser* v. 0.3.1, GBIF *name-parser* v. 0.1.0, *biodiversity* v. 3.4.1.

Results and discussion

We discuss and compare *gnparser*, GBIF *name-parser* and *biodiversity* parser in the context of our requirements for quality, global scope, parsing completeness, speed, and accessibility.

High quality parsing

Quality is the most important of the 5 requirements. GBIF *name-parser* uses regular expressions approach, while *gnparser* and *biodiversity* parsers use the PEG approach. Results for quality measurements are shown in Tables 1 and 2. We include the 1,000 tested names as Additional file 3.

If test data contain a large proportion of true negatives (N_{tn}) *Accuracy* will not be a good measure as it favors algorithms that distinguish negative results rather than finding positive ones. We manually checked our test datasets and established that $\approx$ 1% were not scientific names. Given that true negatives are rare, they will have very limited influence on *Accuracy*. *Recall* for all parsers was high, hence false negatives are not important.

Accuracy is probably the best measure for our tests. All 3 parsers performed very well, with *Accuracy* values higher than 95%. Both *gnparser* and *biodiversity* parser approached the 99% mark which we regard as the metric for production quality. Most of the false positives came from name-strings with mistakes. For example, out of 11 false positives (below) that *gnparser* found in the 1000 name-string test data set, only 2 (the first 2) were well-formed names.

Eucalyptus subser. Regulares Brooker
Jacquemontia spiciflora (Choisy) Hall. fil.
Acanthocephala declivis variety guianensis Osborn, 1904
Atysa (?) frontalis

Bumetopia (bumetopia) quadripunctata Breuning, 1950

Cyclotella kã ¹/₄tzingiana Thwaites

Elaphidion (romaleum) tæniatum Leconte, 1873

Hieracium nobile subsp. perclusum (Arv. -Touv.) O. Bolòs & Vigo

Leptomitus vitreus (Roth) Agardh?

Myosorex muricauda (Miller, 1900).

Papillaria amblyacis (M<81>ll.Hal.) A.Jaeger

We do expect a parser to deal with names that are not well-formed. That means overcoming problems such as aberrant characters which might arise from Unicode character miscodings, inappropriate annotations, or other mistakes. To alert users, *gnparser* generates a warning when it identifies a problem in a name-string. The other parsers do not have this feature.

When parsers reach ≈ 80% *Accuracy*, they hit a "long tail" of problems where each particular type of a problem is rare. Every new manual check of additional test sets of 1,000–10,000 name-strings reveals new issues. Examples of these challenges are given elsewhere [2]. For all three parsers, developers have to perform the meticulous task of adding new rules to address each rare case. That is, parsers need to be subject to continuous improvement. The problems found during preparation of this paper are being addressed in the next version of *gnparser*. As the parsing rules improve, we believe that *gnparser* can reach > 99.5% *Accuracy* without diminishing *Recall*.

As we incorporate new rules to increase *Recall*, we have to consider the risks of reducing *Precision* by introducing new false positives. For example, the GBIF *name-parser* allows the genus element of a name-string to start with a lowercase character. As a result the name-strings below were parsed as if they were scientific names, while the other parsers ignored them:

acid mine drainage metagenome

agricultural soil bacterium CRS5639T18-1

agricultural soil bacterium SC-I-8

algal symbiont of Cladonia variegata MN075

alpha proteobacterium AP-24

anaerobic bacterium ANA No.5

anoxygenic photosynthetic bacterium G16

archaeon enrichment culture clone AOM-SR-A23

bacterium endosymbiont of Plateumaris fulvipes

bacterium enrichment culture DGGE band 61_3_FG_L

barley rhizosphere bacterium JJ-220

bovine rumen bacterium niuO17

Strategies like these may increase *Recall* with certain low-quality datasets, but they decrease *Precision*. Many "dirty" datasets contain recurring problems. As an example, DRYAD contains many name-strings in which elements of scientific names are concatenated with an interpolated character such as '_' (e.g. "Homo_sapiens" and "Pinoyscincus_jagori_grandis") [2]. For them, our solution was to include a "preparser" script which "normalizes" known problems that are inherent within particular datasets and then apply a high quality parser to the result.

Our testing also revealed differences between regular expressions and PEG approaches. Both can achieve high quality results with canonical forms of scientific names, but the regular expressions are less suitable for more complex name-strings. The recursive or nested nature of some scientific names can cause problems which become insurmountable for regular expressions.

Global scope

If we want to connect biological data using scientific names, no name-strings should be missed or rejected, no matter how complex they are. During our testing we found that *Accuracy* of GBIF's *name-parser* was depressed because, in part, the parser did not recognize hybrid formulae and infrasubspecific names with more then one infraspecific epithet. This case underscores the limitations of the regular expression approach. As examples, the following were not parsed by the GBIF *name-parser*:

Erigeron peregrinus ssp.callianthemus var. eucallianthemus (a name-string with two infraspecificx epithets)

Polyporus varius var. nummularius f. undulatus (Pilát) Domanski, Orlos & Skirg. (two infraspecific epithets)

Salvelinus fontinalis x Salmo gairdneri (hybrid formula)

Echinocereus fasciculatus var. bonkerae × E. fasciculatus var. fasciculatus (hybrid formula)

The PEG approach supports nested parsing rules to create progressively more complex rules that manage such cases. The capacity to address recursion allows *gnparser* to handle the full spectrum of scientific names that we have presented to it.

Parsing Completeness

The extraction of canonical forms from name-strings representing scientific names is the most beneficial and widely used parsing goal. Sometimes, however, this may not be sufficient because the canonical form does not always distinguish a name completely.

In the example in Fig. 1 **Carex scirpoidea convoluta** is a canonical form for **Carex scirpoidea var. convoluta Kükenthal** and **Carex scirpoidea ssp. convoluta (Kük.) Dunlop.** The first non-parsed name-string refers to the variety **convoluta** of **Carex scirpoidea** that had been described by **Kükenthal**. The second captures Dunlop's reclassification of **convoluta** as a subspecies. We are not able to distinguish between these two different names without knowing the rank and/or the corresponding authorship. Furthermore, it is useful to see in the second example that **(Kük.)** was the original author and **Dunlop** was the author of the new combination. Also, canonical forms do not distinguish between homonyms. The heather, *Pieris japonica* (Thunb.) D. Don ex G. Don and the butterfly, *Pieris japonica* Shirôzu, 1952 have the same canonical form **Pieris japonica**.

After matching by canonical form, rank, authors, and "types" of authorship allow us to distinguish name-strings with similar or identical canonical elements. The name-string **Carex scirpoidea Michx. var. convoluta Kükenth.** adds the information that the species **Carex scirpoidea** was described by **Michx** but is not evident in the examples in the paragraph above.

Another area in which parsers with limited abilities can give misleading results is with negated names [2]. In these cases, the name-string includes some annotation or marks to indicate that the information associated with the name does NOT refer to the taxon with the scientific name that is included. Examples include **Gambierodiscus aff toxicus** or **Russula xerampelina-like sp.**

All components of a name may be important and need to be parsed and categorized. With *gnparser*, we describe the meaning of every element in the parsed name-string and present the results in JSON format. Parsing of **Carex scirpoidea Michx. subsp. convoluta (Kük.) D.A. Dunlop** gives the following JSON output

The output includes the semantic meaning of all parsed elements in a name-string, indicates if the name-string was parsed successfully, if it is a virus name, a hybrid, or a surrogate. Surrogates are name-strings that are alternatives to names (such as acronyms) and they may or may not include part of a scientific or colloquial name (e.g. **Coleoptera sp. BOLD:AAV0432**). The output also includes a statement of the position of each element in the name-string. Last, but not least, the JSON output contains UUID version 5 calculated from the verbatim name-string. This UUID is guaranteed to be the same for the same name-string, promoting its use to globally connect information and annotations.

The output usually covers every semantic element in the name-string. The fields in the output illustrated above have the following meanings.

name_string_id: UUID v5 identifier;

parsed: whether a name-string was successfully parsed (true/false);

quality: how well-formed a name-string is (range from 1 to 3, 1 is the best);

parser_version: version of a parser used;

verbatim: name-string as was submitted to *gnparser*;

normalized: name-string modified by the parser to give a normalized style;

canonical_name: a special form of normalization that includes only the scientific elements of the name, this form is contained within most name-strings relating to scientific names;

```
1  {
2    "name_string_id" : "203213f3-99d1-5f5e-810a-4453c4d220cb",
3    "parsed" : true, "quality" : 1, "parser_version" : "0.3.1",
4    "verbatim" : "Carex scirpoidea Michx. subsp. convoluta (Kük.) D.A. Dunlop",
5    "normalized" : "Carex scirpoidea Michx. ssp. convoluta (Kük.) D. A. Dunlop",
6    "canonical_name" : {
7    "value" : "Carex scirpoidea convoluta", "extended" : "Carex scirpoidea ssp. convoluta"
8    },
9    "hybrid" : false, "surrogate" : false, "virus" : false,
10   "details" : [ {
11   "genus" : { "value" : "Carex" },
12   "specific_epithet" : {
13   "value" : "scirpoidea",
14   "authorship" : {
15   "value" : "Michx.",
16   "basionym_authorship" : { "authors" : [ "Michx." ] }
17   }
18   },
19   "infraspecific_epithets" : [ {
20   "value" : "convoluta", "rank" : "ssp.",
21   "authorship" : {
22   "value" : "(Kük.) D. A. Dunlop",
23   "basionym_authorship" : { "authors" : [ "Kük." ] },
24   "combination_authorship" : { "authors" : [ "D. A. Dunlop" ] }
25   }
26   } ]
27   } ],
28   "positions" : [ [ "genus", 0, 5 ], [ "specific_epithet", 6, 16 ], [ "author_word", 17, 23 ],
29   [ "rank", 24, 30 ], [ "infraspecific_epithet", 31, 40 ], [ "author_word", 42, 46 ],
30   [ "author_word", 48, 50 ], [ "author_word", 50, 52 ], [ "author_word", 53, 59 ] ]
31  }
```

hybrid: whether the name-string refers to a hybrid (true/false);

surrogate: whether a name-string is a surrogate name (true/false);

details: describes the semantic elements within the name-string inclusive of the following;

genus: reports the genus part of the name (in this case Carex);

specific epithet: reports the species epithet (scirpoidea);

authorship: reports the authorship of the combination (Michx.);

basionym authorship: reports the authorship of the basionym (Michx.)

infraspecific epithets: reports the infraspecies name if present (convoluta) with rank (ssp.)

authorship: reports the authors of the infraspecies name ((Kük.) D. A. Dunlop)

basionym authorship: reports the author of the basionym of infraspecies name element (["Kük."]);

combination authorship: reports the author of the infraspecies name combination (D. A. Dunlop); and

positions: identifies each name element and where it starts and ends.

The complete list of fields for the *gnparser*'s output exists as a JSON Schema file [30] [see Additional file 1].

Parsing speed

In the areas of performance discussed above, there is little difference between *biodiversity* parser and *gnparser*. There is, however, a dramatic difference in their parsing speed and ability to scale. Parsing tasks that took 20 hours with earlier *biodiversity* parsers can now be completed in a few minutes on a multithreaded computer. Parsing is a key to other services such as name-reconciliation and subsequent resolution. Improvements to the speed of the parser will increase user satisfaction elsewhere.

Results on the speed performance are given in Fig. 3. The performance depends on the number of CPU threads used. On 1 thread *gnparser* was 7 times faster than *biodiversity*, 10 times faster on 4 threads, and 14 times faster on 12 threads.

gnparser displays functionality not presented in the GBIF *name-parser* as described in previous sections. In spite of this additional functionality *gnparser* outperformed other tested parsers.

Accessibility

By 'accessibility' we refer to the ability of the software code to be used by a wide audience. For Open source projects, accessibility is very important. If more people use a software, the more cost-effective is its development.

Parsing scientific names is essential for organizing biodiversity data. Many biodiversity database environments and projects include a parsing algorithm. Examples are

Threads	gnparser	gbif-parser	biodiversity	Ratio		
				gn	gbif	bio
1	8178	6389	1111	1	0.78	0.14
2	14125	12638	1722	1	0.89	0.12
4	25125	21994	2556	1	0.88	0.10
8	33541	30972	2777	1	0.92	0.08
12	36369	31833	2527	1	0.88	0.07

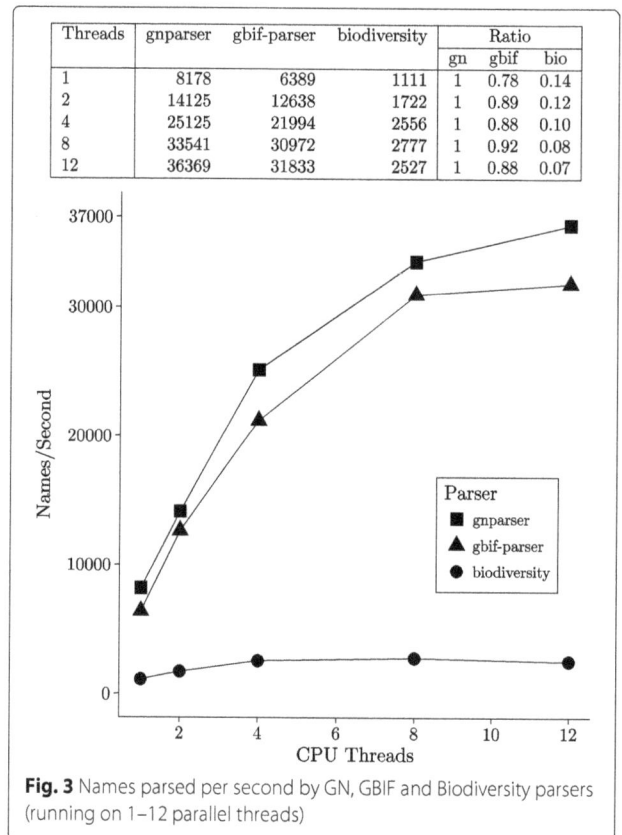

Fig. 3 Names parsed per second by GN, GBIF and Biodiversity parsers (running on 1–12 parallel threads)

uBio [37], the Botanical Society of Britain and Ireland [38], FAT [39], NetiNeti [40], and Taxonome [41]. A modular approach offers an option of re-use and avoids replication of effort. *biodiversity* was the first biodiversity parser to be released as a stand-alone package that could be used as a module — as it was with the iPlant project [35]. The same approach has now been adopted with the GBIF *name-parser* [5], YASMEEN [6], and *gnparser*.

We designed *gnparser* with accessibility in mind from the outset. Scala language allows the use of *gnparser* as a library in Scala, Java, Jython, JRuby and a variety of other languages based on Java Virtual Machine it can also be used natively in R and Python via JVM-binding libraries. Apache Spark, a "Big Data" framework, is also supported. The following example illustrates how a client written in Jython can access the *gnparser* functionality.

```
from org.globalnames.parser import
ScientificNameParser
snp = ScientificNameParser.instance()
result = snp.fromString("Homo sapiens
L.").renderCompactJson()
print result
```

If programmers want to use *gnparser* in some JVM-incompatible language they can connect to the parser via a socket server interface. There is also a command line tool, a web interface, and a RESTful API. In 2016, Encyclopedia of Life started to parse name-strings using *gnparser* socket server.

We pay close attention to documentation, trying to keep it detailed, clear, and up to date. We have an extensive test suite [see Additional file 4] that describes the parser's behavior and contains examples of *gnparser* functionality and output format.

This commitment to accessibility creates a larger potential audience for the parser, and will help many researchers and programmers deal with the problems that arise from variant forms of scientific names.

Conclusions

The performance of the scientific names parsers is summarised in Table 3. The two PEG-based parsers — *biodiversity* and *gnparser* are similar. They are based on the same algorithmic approach and follow similar design goals. While we had the option of modifying the rules for *biodiversity* to improve *Accuracy*, we preferred to create a new tool from scratch to overcome limitations in speed, scalability and accessibility. We needed to address speed at Global Names because existing software took too long to parse or reparse 24 million name-strings. *gnparser* can be used natively by larger variety of programming languages than *biodiversity*, because JVM-based languages and tools are so widely used. Our first goal for *gnparser* was complete coverage of the *biodiversity*'s test suite. We continue to improve *gnparser* while *biodiversity* entered maintenance mode. That explains a slight difference in *Accuracy* by these two parsers.

gbif-parser is a high quality product. However, its regular expressions-based algorithm limits its usability. The recursive nature of some scientific names creates significant obstacles for intrinsically non-recursive algorithms such as regular expressions. Coverage of multi-infraspecific names and hybrids, both with recursive patterns, is prohibitively expensive for such an approach.

Table 3 Summary comparison of Scientific Name Parsers

	gnparser	gbif-parser	Biodiversity
Accuracy	98.9%	96.7%	98.4%
Hybrid formulas support	Yes	No	Yes
Infrasubspecies support	Yes	No	Yes
Throughput (names/s/thread)	8178	6389	1111
Parsing details	Complete	Partial	Complete
Library for the same languages	Yes	Yes	Yes
Library for other languages	Yes	Yes	No
Command line tool	Yes	No	Yes
Socket server	Yes	No	Yes
Web interface	Yes	Yes	Yes
RESTful service	Yes	Yes	Yes

In conclusion, this paper describes *gnparser*, a powerful tool for working with biodiversity information. It transforms names of taxa into their semantic elements. This allows standardization of names by, for example, representing them as canonical forms. This step dramatically improves name matching within and among data sources, and this increases the amount of data on a single taxon that can be integrated. Parsing can be used to improve the discovery of names in sources, and creating a common taxonomic index to multiple sources. Parsing allows users to extract, compare and analyse metadata within the name-strings, and allowing comparisons of the efforts of individuals or to map trends over time. The *gnparser* tool is released under MIT Open source license, contains command line executable, socket, web, and REST services, and is optimized for use as a library in languages like Scala, Java, R, Jython, JRuby.

Additional files

Additional file 1: Includes a full and formal explanation of all parser fields as a JSON schema.

Additional file 2: README.rst file that is converted to HTML format. It is also available at project home page [44].

Additional file 3: 1,000 name-strings randomly selected from GNI and used to determine *Accuracy*, *Precision* and *Recall* data (Table 1).

Additional file 4: Extensive test suite that describes the parser's behavior. It is also a source of examples of parser functionality and output format. Test suite consists of a pipe delimited input (scientific name) and parsed output in JSON format.

Abbreviations
AAM: Alexander A. Myltsev; API: Application program interface; AST: Abstract Syntax Tree; BHL: Biodiversity heritage library; DJP: David J. Patterson; DYM: Dmitry Y. Mozzherin; GBIF: Global Biodiversity information facility; GNA: Global names architecture; GNI: Global names index; JSON: JavaScript object notation; JVM: Java virtual machine; PEG: Parsing expression grammar; REST: Representational state transfer

Acknowledgements
The authors thank David Mark Welch (Josephine Bay Paul Center, Marine Biological Laboratory) for the leadership at the beginning of the *gnparser* project. The authors also thank administrators of the Species File Group for the much needed support during the transfer of the GNA grant from the Marine Biological Laboratory to University of Illinois.

Funding
This work is supported by the National Science Foundation (NSF DBI-1356347). The Species File Group of the University of Illinois provided an additional funding. The funding bodies had no role in the study design, data collection and analysis, decision to publish, or preparation of the manuscript.

Authors' contributions
DYM and AAM designed *gnparser*. DYM created requirements, test suite and the original version of *gnparser*. AAM optimized *gnparser* for speed, refactored it into three internal subprojects. DYM set Docker containers and Kubernetes scripts. DYM and AAM wrote online documentation and JSON schema to formalize output. DJP corrected parser's results, calibrated quality output and errors output. DYM and AAM drafted manuscript and DJP edited its final version. All authors read and approved the final manuscript.

Competing interests
The authors declare that they have no competing interests.

Declarations
All authors have gone through the manuscript and contents of this article have not been published elsewhere.

Author details
[1] University of Illinois, Illinois Natural History Survey, Species File Group, 1816 South Oak St., Champaign, IL, 61820, USA. [2] IP Myltsev, Kaslinskaya St., Chelyabinsk, 454084, Russia. [3] University of Sydney, Sydney, Australia.

References
1. Patterson DJ, Cooper J, Kirk PM, Pyle RL, Remsen DP. Names are key to the big new biology. Trends Ecol Evol. 2010;25(12):686–91. doi:10.1016/j.tree.2010.09.004.
2. Patterson D, Mozzherin D, Shorthouse D, Thessen A. Challenges with using names to link digital biodiversity information. Biodiversity Data J. 2016;4:8080. doi:10.3897/BDJ.4.e8080.
3. Leary PR, Remsen DP, Norton CN, Patterson DJ, Sarkar IN. uBioRSS: Tracking taxonomic literature using RSS. Bioinformatics. 2007;23(11): 1434–6. doi:10.1093/bioinformatics/btm109.
4. Aho AV, Ullman JD. Foundations of Computer Science vol. 2. USA: Computer Science Press New York; 1992.
5. GBIF name-parser. https://github.com/gbif/name-parser/releases/tag/name-parser-2.10. Accessed 18 Apr 2017.
6. Vanden Berghe E, Coro G, Bailly N, Fiorellato F, Aldemita C, Ellenbroek A, Pagano P. Retrieving taxa names from large biodiversity data collections using a flexible matching workflow. Ecol Inform. 2015;28:29–41. doi:10.1016/j.ecoinf.2015.05.004.
7. Yu S. Handbook of formal languages, regular languages. New York: Springer Verlag; 1997.
8. Ford B. Parsing Expression Grammars: A Recognition-Based Syntactic Foundation. In: Proceedings of the 31st ACM SIGPLAN-SIGACT, ACM, New York, 2004. Symposium on Principles of Programming Languages; 2004. p. 111–22.
9. GlobalNamesArchitecture/biodiversity: Scientific Name Parser. https://github.com/GlobalNamesArchitecture/biodiversity. Accessed 18 Apr 2017.
10. Treetop. https://github.com/cjheath/treetop. Accessed 18 Apr 2017.
11. RubyGems — biodiversity search. https://rubygems.org/search?query=biodiversity. Accessed 18 Apr 2017.
12. Encyclopedia of Life. http://eol.org/. Accessed 18 Apr 2017.
13. Canadian Register of Marine Species. http://www.marinespecies.org/carms/. Accessed 18 Apr 2017.
14. iPlant Taxonomic Name Resolution Service. http://tnrs.iplantcollaborative.org/. Accessed 18 Apr 2017.
15. WoRMS - World Register of Marine Species. http://www.marinespecies.org/. Accessed 18 Apr 2017.
16. Ruby Libraries for Biology. http://biogems.info/. Accessed 18 Apr 2017.
17. Charniak E. Statistical Language Learning. USA: MIT Press; 1996.
18. Aho AV, Ullman JD. The Theory of Parsing, Translation, and Compiling. Upper Saddle River: Prentice-Hall, Inc.; 1972.
19. Asveld PRJ. A fuzzy approach to erroneous inputs in context-free language recognition. In: Proceedings of the Fourth International Workshop on Parsing Technologies IWPT'95. Prague, Czech Republic: Institute of Formal and Applied Linguistics, Charles University; 1995. p. 14–25.
20. Nadeau D, Sekine S. A survey of named entity recognition and classification. Lingvisticae Investigationes. 2007;30(1):3–26. doi:10.1075/li.30.1.03nad.
21. Mikolov T, Sutskever I, Chen K, Corrado GS, Dean J. Distributed representations of words and phrases and their compositionality. In: Advances in Neural Information Processing Systems 26. USA: Curran Associates, Inc.; 2013. p. 3111–9.
22. Schmidhuber J. Deep learning in neural networks: An overview. Neural Netw. 2015;61:85–117.
23. Odersky M, Altherr P, Cremet V, Emir B, Maneth S, Micheloud S, Mihaylov N, Schinz M, Stenman E, Zenger M. An overview of the Scala programming language. Technical report. 2004.
24. Myltsev A, Doenitz M. parboiled2: a macro-based approach for effective generators of parsing expressions grammars in Scala 2017. in preparation. http://myltsev.com/papers/parboiled2.pdf. Accessed 10 May 2017.
25. GlobalNamesArchitecture/parboiled2: A macro-based PEG parser generator for Scala 2.10+ doi:10.5281/zenodo.50340. https://github.com/GlobalNamesArchitecture/parboiled2. Accessed 18 Apr 2017.
26. Moors A, Piessens F, Odersky M. Parser combinators in Scala. Department of Computer Science, KU Leuven, Leuven, Belgium. 2008.
27. Burmako E. Scala macros: Let our powers combine! On how rich syntax and static types work with metaprogramming. In: Proceedings of the 4th Workshop on Scala SCALA '13. New York: ACM; 2013. p. 3–1310. doi:10.1145/2489837.2489840.
28. King AMQ, Adams MJ, Carstens EB, Lefkowitz EJE. Virus Taxonomy: Classification and Nomenclature of Viruses: Ninth Report of the International Committee on Taxonomy of Viruses. Amsterdam: Elsevier Academic Press; 2012, pp. 1–1338.
29. Bray T. The JavaScript object notation (JSON) data interchange format. Google Inc. Online. 2014.
30. JSON schema for gnparser output. http://globalnames.org/schemas/gnparser.json. Accessed 18 Apr 2017.
31. Linne CV. Plantarum: Exhibentes Plantas Rite Cognitas Ad Genera Relatas Cum Differentiis Specificis, Nominibus Trivialibus, Synonymis Selectis, Locis Natalibus Secundum; Holmiae, Impensis Laurentii Salvii, Stockholm, 1753, p. 583.
32. Maven Central: Global Names Artifacts. https://search.maven.org/#search|ga|1|globalnames. Accessed 18 Apr 2017.
33. Global Names Parser Docker Image. https://hub.docker.com/r/gnames/gnparser/. Accessed 18 Apr 2017.
34. Global Names Index. http://gni.globalnames.org. Accessed 18 Apr 2017.
35. Boyle B, Hopkins N, Lu Z, Raygoza Garay JA, Mozzherin D, Rees T, Matasci N, Narro ML, Piel WH, McKay SJ, Lowry S, Freeland C, Peet RK, Enquist BJ. The taxonomic name resolution service: an online tool for automated standardization of plant names,. BMC Bioinformatics. 2013;14(1):16. doi:10.1186/1471-2105-14-16.
36. gbifparser: v0.1.0 2015. doi:10.5281/zenodo.34848. http://dx.doi.org/10.5281/zenodo.34848. Accessed 18 Apr 2017.
37. uBio Name Parser. http://www.ubio.org/tools/explode.php. Accessed 18 Apr 2017.
38. Botanical Society of Britain and Ireland Taxon Name Parser. http://bsbidb.org.uk/taxonnameparser.php. Accessed 18 Apr 2017.
39. Sautter G, Böhm K, Agosti D. A combining approach to Find All Taxon names (FAT) in legacy biosystematics literature. Biodivers Inform. 2006;3: 46–58. doi:10.2307/1216144.
40. Akella LM, Norton CN, Miller H. NetiNeti: discovery of scientific names from text using machine learning methods. BMC Bioinformatics. 2012;13(1):211. doi:10.1186/1471-2105-13-211.
41. Kluyver TA, Osborne CP. Taxonome: a software package for linking biological species data. Ecol Evol. 2013;3(5):1262–5. doi:10.1002/ece3.529.
42. Flora of North America Editorial Committee E. Flora of North America. Vol. 23, Magnoliophyta: Commelinidae (in Part): Cyperaceae. New York and Oxford: Oxford University Press; 2002, p. 551.
43. Global Names Parser Web App. http://parser.globalnames.org. Accessed 18 Apr 2017.
44. GlobalNamesArchitecture/gnparser: Split scientific names to meaningful elements with meta information. https://github.com/GlobalNamesArchitecture/gnparser. Accessed 18 Apr 2017.

16

Comparison of different cell type correction methods for genome-scale epigenetics studies

Akhilesh Kaushal[1], Hongmei Zhang[1*] (iD), Wilfried J. J. Karmaus[1], Meredith Ray[1], Mylin A. Torres[2,3], Alicia K. Smith[2,4] and Shu-Li Wang[5*]

Abstract

Background: Whole blood is frequently utilized in genome-wide association studies of DNA methylation patterns in relation to environmental exposures or clinical outcomes. These associations can be confounded by cellular heterogeneity. Algorithms have been developed to measure or adjust for this heterogeneity, and some have been compared in the literature. However, with new methods available, it is unknown whether the findings will be consistent, if not which method(s) perform better.

Results: *Methods*: We compared eight cell-type correction methods including the method in the minfi R package, the method by Houseman et al., the Removing unwanted variation (RUV) approach, the methods in FaST-LMM-EWASher, ReFACTor, RefFreeEWAS, and RefFreeCellMix R programs, along with one approach utilizing surrogate variables (SVAs). We first evaluated the association of DNA methylation at each CpG across the whole genome with prenatal arsenic exposure levels and with cancer status, adjusted for estimated cell-type information obtained from different methods. We then compared CpGs showing statistical significance from different approaches. For the methods implemented in minfi and proposed by Houseman et al., we utilized homogeneous data with composition of some blood cells available and compared them with the estimated cell compositions. Finally, for methods not explicitly estimating cell compositions, we evaluated their performance using simulated DNA methylation data with a set of latent variables representing "cell types".

Results: Results from the SVA-based method overall showed the highest agreement with all other methods except for FaST-LMM-EWASher. Using homogeneous data, minfi provided better estimations on cell types compared to the originally proposed method by Houseman et al. Further simulation studies on methods free of reference data revealed that SVA provided good sensitivities and specificities, RefFreeCellMix in general produced high sensitivities but specificities tended to be low when confounding is present, and FaST-LMM-EWASher gave the lowest sensitivity but highest specificity.

Conclusions: Results from real data and simulations indicated that SVA is recommended when the focus is on the identification of informative CpGs. When appropriate reference data are available, the method implemented in the minfi package is recommended. However, if no such reference data are available or if the focus is not on estimating cell proportions, the SVA method is suggested.

Keywords: Cell-type composition, CpG sites, Genome-scale DNA methylation, Surrogate variables

* Correspondence: hzhang6@memphis.edu; slwang@nhri.org.tw
[1]Division of Epidemiology, Biostatistics, and Environmental Health, University of Memphis, Memphis 38152, TN, USA
[5]National Institute of Environmental Health Sciences, National Health Research Institutes, Miaoli, Taiwan
Full list of author information is available at the end of the article

Background

Whole blood is frequently utilized in genome-wide association studies of DNA methylation patterns in relation to environmental exposures or clinical outcomes. However, for DNA methylation assessed from whole blood, the association between DNA methylation and an exposure of interest could be confounded by cellular heterogeneity [1, 2]. In larger epidemiological studies, it is not feasible to isolate and profile every individual cell subset. Thus, several algorithms have been developed to measure and adjust for cellular heterogeneity in whole blood.

Houseman et al. proposed a method to infer the cell mixture proportions based on a regression calibration technique, which uses an external validation dataset to calibrate the model and correct for the bias [3]. Jaffe and Irizarry [4] modified the Houseman et al.'s algorithm and tailored it to predict cell mixture composition of DNA-methylation profiles obtained from a different Illumina platform. This cell type correction method is implemented in Bioconductor [5] package minfi [6]. The above two approaches require external validation datasets and are designed to identify cell mixtures in tissues such as whole blood.

Apart from these two reference-based techniques, non-reference-based methods have also been developed. An advantage of these non-reference-based methods is that they can be applied to other tissues in addition to blood. Zou et al. developed a non-reference-based method, FaST-LMM-EWASher. This approach is built upon linear mixed models with top principal components as the covariates. Another set of methods infer latent variables for cell type compositions, which are then included in association assessments. These methods include RefFreeEWAS and its recently improved version (RefFreeCellMix), surrogate variable analysis (SVA), and ReFACTor [7–10]. RefFreeEWAS [7] and RefFreeCellMix [8] both utilize singular value decompositions (SVDs) and extract latent subject and cell-specific effects, but RefFreeCellMix incorporated additional constraints and utilities aiming to reduce the occurrence of false positives. Surrogate variable analysis (SVA) [9], based on SVDs of residuals in linear regressions, uses permutations to identify statistically significant eigen-vectors and consequently infer potential confounding factors (surrogate variables). A Bioconductor package is available to estimate surrogate variables using this approach [9]. Finally, ReFACTor [10] is based on principal component analyses on a set of potentially informative CpG sites.

Removing unwanted variation (RUV) is an approach different from the aforementioned methods and it was designed to estimate cell type heterogeneity and built upon factor analyses. This approach utilizes reference CpGs inferred from a reference database, based on which factor analyses are conducted. The factors are then included in subsequent analyses for the purpose of adjusting for cell type effects. Although a reference database is needed, this method does not estimate cell type proportions as done in the minfi package and in the Houseman et al. method.

In our earlier work Kaushal et al. [11], we compared five methods, Houseman et al., minfi, FaST-LMM-EWASher, RefFreeEWAS, and a method by use of SVA. McGregor et al. [12] compared the methods noted above except for ReFACTor and RefFreeCellMix and focused on assessment of associations between DNA methylation and a variable of interest by taking cell type compositions into account. With the additional methods included (ReFACTor and RefFreeCellMix), it was unclear whether the findings would be consistent with those in McGregor et al., Kaushal et al., and if not, which method(s) might perform better. To this end, we first applied each cell type correction method (Houseman et al., minfi, RUV, FaST-LMM-EWASher, ReFACTor, RefFreeEWAS, and RefFreeCellMix) as well as the surrogate variable analyses (SVA) to two real data sets. For these real data sets, we evaluated the association between genome-scale DNA methylation and a variable of interest adjusting for cell type compositions. We assessed the agreement within each data set in terms of identified CpGs between different methods and the consistency of findings of each method between different data sets. We also qualitatively compared different approaches based on existing knowledge about the sparsity of informative genes, pathways and genetic functions. For the method implemented in the minfi package and that proposed by Houseman et al., we utilized a homogeneous real data set with some blood cells composition available and compared the true cell counts with the estimated cell compositions. For methods free of reference groups (FaST-LMM-EWASher, RefFreeEWAS, RefFreeCellMix, ReFACTor, and SVA), we further utilized simulated data generated under different scenarios to compare different methods, which, combined with findings from the real data, enabled us to comprehensively assess each method.

Results

Findings from prenatal arsenic exposure and DNA methylation data

We used genome-scale DNA methylation data from a birth cohort study consisting of 64 cord blood samples examining multiple prenatal factors in relation to child health outcomes, pilot of the nationwide Taiwan Maternal and Infant Cohort Study [13, 14].

Via linear regressions, we assessed the association of DNA methylation at each CpG site across the whole genome with prenatal urinary arsenic exposure levels (a continuous measure), adjusting for cell-type effects with cell type information inferred from one of the eight methods. For each method, we recorded the number of CpGs showing statistically significant associations with prenatal urinary arsenic exposure after adjusting for multiple testing by controlling false discovery rate (FDR) at 0.05. ReFACTor identified the largest number of CpGs (~60,000) and no CpGs were detected by FaST-LMM-EWASher (Table 1). RefFreeCellMix also identified a large number of CpGs (~3000). SVA and RefFreeEWAS detected more CpGs compared to the remaining methods. (Table 1). Next, we assessed the number of identified CpGs that overlapped between different methods. The diagram in Fig. 1 shows the overlap of CpG sites from four approaches (Houseman et al., minfi, RefFreeEWAS, and SVA) as well as the analyses without adjusting for cell types (we did not include all eight methods in this Venn diagram for clarity). Results from SVA showed the best agreement with findings from the other four analyses (Fig. 1). Two identified CpG sites cg06434480 and cg10662395 were common to all these five analytical methods labeled in Fig. 1. Further comparisons indicated that CpG site cg10662395 was also identified by RefFree-CellMix and RUV, and was the only CpG site to overlap among all seven analyses (Houseman et al., minfi, RefFreeEWAS, SVA, RefFreeCellMix and RUV, as well as the analyses without adjusting for cell types). Although ReFACTor identified the largest number of CpGs, they did not overlap with the joint findings from the aforementioned seven analyses. Overall, CpGs identified via SVA overlapped with those from the Houseman et al. method, minfi and RefFreeEWAS (p-value < 0.0001, Table 1, Fig. 1. The definition of percentage overlap is given in the Methods section. One of the two CpGs (cg06434480 and cg10662395), cg06434480, is located within 200 base pairs

of the transcription start site of gene *HMGCR* (3-hydroxy-3-methylglutaryl-CoA reductase) which is known to be associated with inorganic arsenic exposure [15]. In a study conducted in humans, Mono-methylated arsenic (MMA) was found to downregulate the gene expression of *HMGCR*, a gene involved in cholesterol biosynthesis [16]. The other CpG, cg10662395, is located in the body region of gene *HCN2* (hyperpolarization activated cyclic nucleotide gated potassium channel 2). This gene was not found to be directly associated with arsenic exposure in the literature, but *HCN2* has been known to regulate pacemaker activity in the heart and the brain of mice and humans [17, 18]. Arsenic has been found to induce QT interval (i.e., time between initial deflection of QRS complex to the end of T wave) prolongation probably by altering potassium ion channel [19].

The motivation of adjusting for cell types was due to the potential confounding effects of cell type compositions with respect to the association of arsenic exposure with DNA methylation, caused by the association of arsenic exposure with cell type compositions [20–23]. Our assessment on the correlations between total arsenic exposure and estimated cell type proportions also supported the potential confounding-role of cell types (Additional file 1: Material S1). To support the existence of such confounding effects, we assessed the associations with and without adjusting for cell type proportions at all CpG sites. We found that at more than 99% of all the CpGs the effects (regression coefficients) of prenatal arsenic exposure changed by more than 10% when not adjusting for cell type (the median of the coefficients was 2.32 with 5th percentile of 0.40 and 95th percentile of 3.46) to adjusting for cell type (the corresponding statistics were 0.080, 0.0073 and 0.25), indicating a need of adjusting for cell types.

Overall, the analysis based on SVA identified CpG sites that had better overlap with the CpGs identified by other methods. To acquire the biological relevance of CpGs uniquely identified by use of SVA, we implemented DAVID to perform Gene Ontology (GO) analysis and to identify KEGG pathways. The 455 (out of 498) significant CpGs identified uniquely by SVA were mapped to genes using Illumina annotation file for 450 K DNA methylation array. Of great interest, GO categories related to transcription and regulation of RNA metabolic process were enriched after controlling FDR at 0.05, as well as three KEGG pathways, endocytosis, cancer pathway and MAPK signaling pathway (a complete list is included in Additional file 2: Material S2). A discussion on the connection of arsenic exposures and the identified GO categories and KEGG pathways is presented in the Discussion section.

Table 1 Number of significant CpG sites with and without cell type correction and overlap with the SVA method (data on prenatal arsenic exposure and DNA methylation)

Method	Identified CpGs (N)[#]	Overlap with SVA (%)	p-value[##]
Houseman et al.	10	1.20	<0.0001
minfi	57	4.62	<0.0001
SVA	498	–	–
RefFreeEWAS	133	6.01	<0.0001
RefFreeCellMix	2932	0.60	1.0
ReFACTor	58,871	13.03	1.0
EWASher [a]	0	0.0	–
RUV	356	0.20	1.0
Unadjusted [b]	3	0.60	<0.0001

[#]The selection of CpG sites is based on FDR-adjusted p-values (FDR is controlled at 0.05)
[##]P-value is based on Fishers exact test for overlap with results from SVA. The null hypothesis is that there is no overlap with the CpGs identified based on SVA
[a]The FasT-LMM-EWASher method
[b]Unadjusted: cell type compositions were not included in the analyses

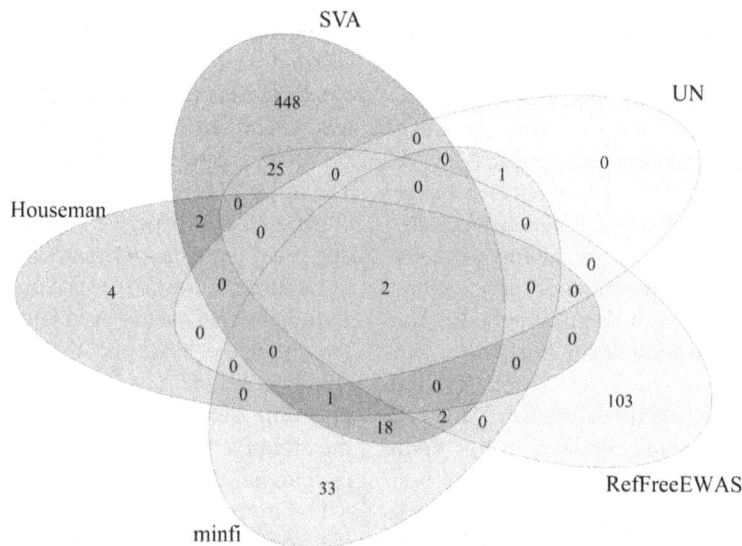

Fig. 1 Venn diagram illustrating the overlap of identified CpG sites that were associated with prenatal arsenic exposure at FDR level of 0.05 after incorporating estimated cell type compositions by different methods for the association study of prenatal arsenic exposure with DNA-methylation. Results from Houseman et al., *minfi*, *RefFreeEWAS*, and *SVA* as well as the analyses without adjusting for cell types are displayed (Results from other methods are in the text). *"UN"*: results from an analysis without adjusting for cell type compositions

Findings from example data on cancer

We repeated the same analysis on an example data set provided by the FasT-LMM-EWASher package. A tutorial website for applying all the cell type composition inference methods to this example data is available at https://akhilesh362.wordpress.com/. This data set includes DNA methylation from the Illumina 27 K array and measures of a binary variable (cancer status) for 204 subjects. In total, 7648 CpGs were included in our study based on initial screening done by the FasT-LMM-EWASher package. The purpose of the initial screening

is to exclude probes that are essentially not methylated or completely methylated. In this example data, cell type proportions were likely to be different on average between subjects with cancer and those without cancer, based on two-sample t-tests applied to logit-transformed sample proportions (Additional files 3 and 4: Materials S3 and S4), explaining the potential need to adjust for their confounding effects. Since Illumina 27 K focuses more on cancer genes, DNA methylation at a large number of CpG sites showed statistically significant associations with cancer status (Table 2). Some similar findings

Table 2 Number of significant CpG sites with and without cell-correction methods and overlap of CpG sites with those from the SVA method (example data from FasT-LMM-EWASher package)

Method	Identified CpGs (N)[#]	Overlap with SVA (%)	p-value[##]	J-index [c]
Houseman et al.	1835	54.71	<0.0001	0.40
minfi	3589	84.59	<0.0001	0.40
SVA	1888	–	–	–
RefFreeEWAS	788	30.51	<0.0001	0.30
RefFreeCellMix	1006	18.38	<0.0001	0.10
ReFACTor	4224	87.45	<0.0001	0.40
EWASher [a]	3	0.16	<0.0001	0
RUV	6008	99.95	<0.0001	0.30
Unadjusted [b]	3768	82.89	<0.0001	0.40

[#]The selection of CpG sites is based on FDR-adjusted p-values (FDR is controlled at 0.05)
[##] P-value is based on Fishers exact test for overlap. The null hypothesis is that there is no overlap with the CpGs identified based on SVA
[a]The FasT-LMM-EWASher method
[b]Unadjusted: cell type compositions were not incorporated into the analyses
[c]J-index is Jaccard index

as in Table 1 were observed. ReFACTor identified a large number of CpGs, Fast-LMM-EWASher identified the least number of CpG sites, and SVA agreed nicely with minfi (Jaccard similarity index = 0.4). A unique observation from this analysis is that RUV identified the largest number of CpGs (6008 CpGs, close to the number of CpGs in the candidate pool, 7648 CpGs). Since the original Houseman et al. method was designed specifically for Illumina 27 K platform, it is understandable that SVA showed a better overlap with results from this approach (Jaccard similarity index = 0.4) compared to the results in the prenatal arsenic exposure and DNA methylation data. In total, 3 identified CpGs (cg22029275 located in the 1st Exon of *FAM123A* gene, cg07080358 located in 1st Exon of *CNRIP1*, and cg15202954 located within 200 base pair of transcription start site of *NALCN* gene) were common to all the eight cell correction methods as well as to the analyses without cell type composition adjusted. There is evidence that these three genes (*FAM123A, CNRIP1 and NALCN*) are associated with the risk of colorectal cancer [24–26].

DAVID analysis of genes associated with the significant CpGs identified uniquely by SVA led to the identification of three GO categories related to plasma membrane at FDR of 0.05 (integral to plasma membrane, intrinsic to plasma membrane, and plasma membrane part), as well as KEGG pathways such as pathways in cancer and signaling pathways (Additional file 2: Material S2), which indicates that genes corresponding to these CpG sites may play a role in the regulation of cancer.

Findings from breast cancer status and DNA-methylation data

This analysis uses a data set discussed in Smith et al. [27]. Breast cancer status, DNA-methylation, and cell counts for granulocytes, monocyte, and lymphocytes for 61 subjects at baseline and a subset of 39 subjects at 6 months follow up are implemented in the analyses. Among all the methods discussed, the method implemented in the minfi package and the original Houseman et al. method were able to estimate cell proportions. We used minfi and the Houseman et al. approach to estimate the proportions of granulocyte, monocyte and lymphocyte cells. Lymphocyte proportions were derived by adding the proportions of B cell, T cell and Natural Killer (NK) cells. For the three cell types (granulocyte, monocyte and lymphocyte), Pearson correlations between estimated (minfi) and true cell proportions were 0.85, 0.79, 0.88 at baseline and 0.84, 0.78, 0.87 at the 6 month follow up, respectively. For the correlations based on the Houseman et al. method, they were 0.84, 0.78 and 0.88 at baseline and 0.78, 0.73 and 0.83 at the 6 month follow up, respectively. All the correlations showed statistically significant difference from zero (*p*-value < 0.05).

Findings from simulated data

We simulated data applying two scenarios with the first scenario focusing on latent variable effects (comparable to effects of cell composition) and the second focusing on latent variable effects with confounding (comparable to effects of cell composition as well as confounding effects). In total, 100 data sets were simulated under each scenario. Details of the simulation scenarios are given in the Methods section. The simulated data were used to evaluate the five methods that do not estimate cell proportions nor need reference databases, specifically, FaST-LMM-EWASher, RefFreeEWAS, RefFreeCellMix, ReFACTor, and SVA.

For data under all scenarios, we applied each of the five methods to each simulated data to draw information on cell compositions. We then incorporated the information to assess the associations of "DNA methylation" with the variable of interest at each pseudo CpG site, and compared each method by assessing the sensitivity and specificity of the selected CpG sites across all 100 data sets. Regardless of the number of important CpGs, FaST-LMM-EWASher resulted in the lowest sensitivity but the highest specificity for both scenarios, consistent with findings from real data (Table 3). Findings from RefFreeEWAS, RefFreeCellMix, ReFACTor, and SVA are, in general, comparable for data simulated under scenario 1, but SVA gives consistently higher sensitivity and specificity in all settings (Table 3). For data simulated under scenario 2 with high correlations ($\rho = 0.7$), SVA outperformed FaST-LMM-EWASher, RefFreeEWAS, RefFreeCellMix and ReFACTor and had higher sensitivity and specificity. Compared with RefFreeEWAS, overall RefFreeCellMix outperformed when confounding effects were present, showing much higher sensitivities with relatively lower specificities. Results from ReFACTor indicated extremely low specificity under scenario 2, which is consistent with the rather large numbers of CpGs identified in real data. The performance of FaST-LMM-EWASher was similar between the two scenarios and was inferior to all other methods. On the other hand, the SVA method performed well under both scenarios, followed by RefFreeEWAS and RefFreeCellMix with RefFreeEWAS being weaker in capturing confounding effects. We also considered a situation with $\rho = 0.3$, mimicking a situation of moderate confounding (Additional file 5: Material S5), and similar patterns observed as those from the relatively two extreme cases ($\rho = 0$ and $\rho = 0.7$).

In the above simulations, we fixed the regression coefficients of the important CpGs. To demonstrate the pattern of sensitivity and specificity, we implemented receiver operating characteristic (ROC) plots. In total, 100 data sets were simulated under scenario 1 with regression coefficients for the variable of interest ranged

Table 3 Summary of sensitivity, specificity of Unadjusted, FaST-LMM-EWASher, RefFreeEWAS, SVA, ReFACTor and RefFreeCellMix for 100 simulated data across three settings

	Sensitivity (Median, 95% interval)		Specificity (Median, 95% interval)	
	Scenario 1 ($\rho = 0$)	Scenario 2 ($\rho = 0.7$)	Scenario 1 ($\rho = 0$)	Scenario 2 ($\rho = 0.7$)
	Number of Important CpGs =50			
Unadjusted	0.960 (0.470, 1.000)	1.000 (1.000, 1.000)	1.000 (0.987, 1.000)	0.000 (0.000, 0.000)
Ewasher [a]	0.000 (0.000, 0.000)	0.000 (0.000, 0.000)	1.000 (0.999, 1.000)	1.000 (0.999, 1.000)
RefEWAS [b]	1.000 (0.960, 1.000)	0.000 (0.000,0.494)	0.997 (0.994, 0.999)	0.579 (0.055,1.000)
CellMix [c]	1.000 (0.980, 1.000)	1.000 (1.000, 1.000)	0.997 (0.993, 0.999)	0.546 (0.199, 0.923)
ReFACTor	1.000 (0.960, 1.000)	1.000 (1.000, 1.000)	0.996 (0.825, 1.000)	0.000 (0.000, 0.000)
SVA [d]	1.000 (0.980, 1.000)	1.000 (0.960, 1.000)	0.998 (0.996, 1.000)	0.998 (0.996, 1.000)
	Number of Important CpGs =100			
Unadjusted	0.980 (0.664, 1.000)	1.000 (1.000, 1.000)	0.999 (0.976, 1.000)	0.000 (0.000, 0.000)
Ewasher [a]	0.000 (0.000, 0.000)	0.000 (0.000, 0.000)	1.000 (0.999, 1.000)	1.000 (0.999, 1.000)
RefEWAS [b]	1.000 (0.965,1.000)	0.000 (0.000,0.403)	0.995 (0.991, 0.998)	0.520 (0.014,1.000)
CellMix [c]	1.000 (0.975, 1.000)	1.000 (1.000, 1.000)	0.988 (0.968, 0.996)	0.211 (0.047, 0.525)
ReFACTor	1.000 (0.965, 1.000)	1.000 (1.000, 1.000)	0.994 (0.808, 0.998)	0.000 (0.000, 0.000)
SVA [d]	1.000 (0.990,1.000)	0.990 (0.965, 1.000)	0.996 (0.993, 0.999)	0.996 (0.993, 0.999)
	Number of Important CpGs =150			
Unadjusted	0.993 (0.723, 1.000)	1.000 (1.000, 1.000)	0.999 (0.965, 1.000)	0.000 (0.000, 0.000)
Ewasher [a]	0.000 (0.000,0.000)	0.000 (0.000,0.000)	1.000 (0.999,1.000)	1.000 (0.999,1.000)
RefEWAS [b]	0.993 (0.973,1.000)	0.000 (0.000,0.294)	0.992 (0.986, 0.997)	0.496 (0.013,1.000)
CellMix [c]	1.000 (0.983, 1.000)	1.000 (1.000, 1.000)	0.975 (0.929, 0.993)	0.098 (0.022, 0.293)
ReFACTor	1.000 (0.980, 1.000)	1.000 (1.000, 1.000)	0.989 (0.794, 0.997)	0.000 (0.000, 0.000)
SVA [d]	1.000 (0.993,1.000)	0.993 (0.970, 1.000)	0.992 (0.988, 0.996)	0.993 (0.988, 0.996)

ρ = correlation between primary covariate and latent variables
$\rho = 0$ corresponds to data simulated from Scenario 1, while $\rho = 0.7$ corresponds to data simulated from Scenario 2
[a]FaST-LMM-EWASher
[b]RefFreeEWAS
[c]RefFreeCellMix
[d]Surrogate variable analysis

from 0.01 to 0.3. For each data set, we calculated sensitivity and specificity of selected CpGs, based on which we estimated the ROC curves. Sensitivities from FaST-LMM-EWASher were substantially low and were not considered in this demonstration. The performance of RefFreeEWAS, RefFreeCellMix, and ReFACTor was comparable under scenario 1 (Table 3). We therefore only presented ROC curves for ReFreeEWAS and SVA for the purpose of comparison (Fig. 2). The findings are consistent with what we observed from Table 2 for scenario 1, that is, SVA performed better than RefFreeEWAS. In addition, the results indicated that both SVA and RefFreeEWAS have high specificity regardless of the underlying regression coefficients, indicating the conservativism when selecting informative CpGs.

Discussion

We compared eight cell-type correction methods using real and simulated data. Based on DNA methylation in a cohort study, the methods in ReFACTor identified the largest number of CpGs (~60 K CpGs), none of which overlapped with the common CpGs detected by other methods including the analysis without adjusting for cell type compositions (but excluding the method in FaST-LMM-EWASher). The method in FaST-LMM-EWASher did not identify any CpG sites. Except for ReFACTor and FaST-LMM-EWASher, at least one detected CpG was shared between all the other methods. More than 50% of CpGs identified by the Houseman et al. method and by the approach implemented in minfi were also detected by the SVA method; The overlap in CpGs was much less between these two methods and the remaining methods. The genes associated with CpGs uniquely identified by using SVA with prenatal urinary arsenic as primary exposure led to the enrichment of GO categories and KEGG pathways that were consistent with our understanding with respect to the effect of arsenic on DNA methylation. Arsenic exposure leads to generation of

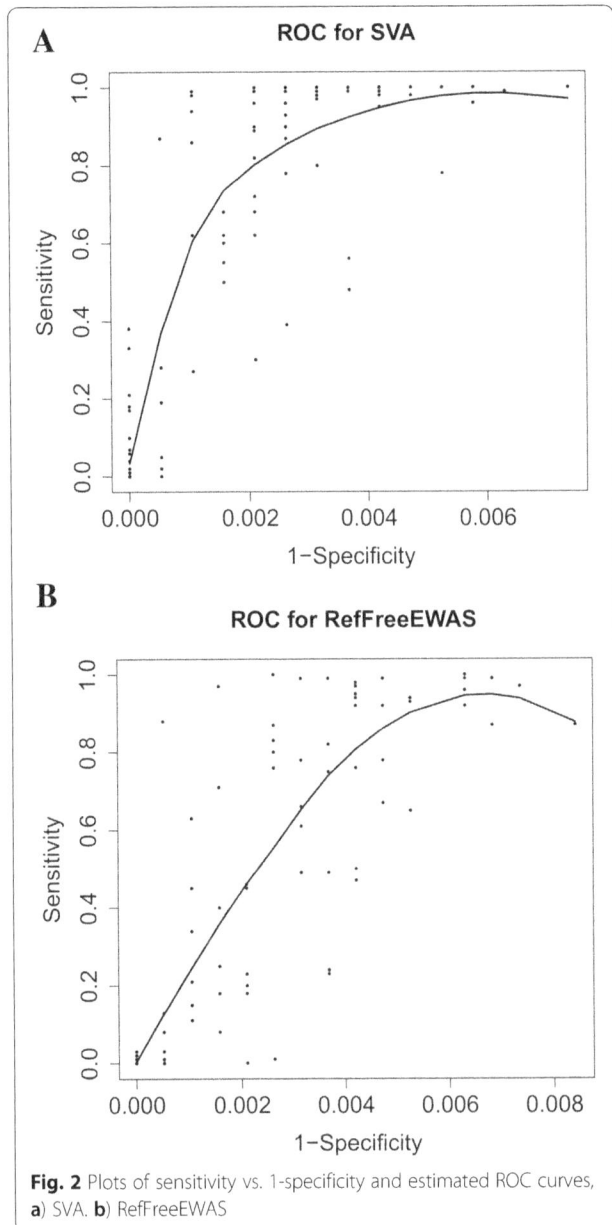

Fig. 2 Plots of sensitivity vs. 1-specificity and estimated ROC curves, **a**) SVA. **b**) RefFreeEWAS

reactive oxygen species (ROS) which induces DNA damage [28]. This reactive oxygen species play a crucial role in signal transduction pathways, transcription factor regulation [29], and mitogen activated protein kinases (MAPKs) signal transduction pathway is one such pathway that is affected by ROS [30]. DAVID analysis of genes associated with the CpGs uniquely identified by SVA for FaST-LMM-EWASher example dataset led to enrichment of KEGG pathways in cancer. All these imply that the CpGs uniquely identified by using SVA are potentially informative. Using the example dataset provided by FasT-LMM-EWASher method, we found that all methods except for FasT-LMM-EWASher identified a large number of CpG

sites. This was likely due to the platform used to measure DNA methylation levels (Illumina 27 K), which is centered more on cancer genes. However, CpGs identified based on ReFACTor and RUV were close to the number of CpGs in the pool of candidate CpGs, indicating possible inflations. On the other hand, results from minfi showed the greatest overlap with the SVA method (Table 2). Based on these two real data sets, results from the method in the minfi package and those from SVA were most agreeable. However, for real data, the underlying truth was unknown, which was the motivation of incorporating a data set with cell counts known and the use of a series of simulation studies. Findings from these data were further discussed in this section.

Using the available cell counts in the cancer status and DNA methylation dataset we observed agreements between cell types estimated by Houseman et al. and minfi, but minfi showed a better agreement. The Houseman et al. approach was designed for the Illumina 27 K bead-chip array, which may not fit the 450 K array as noted in the literature [4]. The modification of the Houseman et al. approach implemented in the minfi package, on the other hand, is suitable for both 27 K and 450 K array. The reference data were from six adult white European males. It has been shown that DNA methylation patterns vary by sex, age and ancestry [31–35]. Generalizing the cell mixtures estimated by minfi to studies with both genders and non-Europeans of different age groups may potentially introduce bias.

Further simulations investigating reference-free methods supported the findings from real data. Regardless of the number of important CpGs, FaST-LMM-EWASher showed the lowest sensitivity, indicating low power to identify truly important CpGs if using that method to adjust for cell type compositions. ReFACTor produced the lowest specificity when confounding effects were present, supporting the rather low overlapping with findings from other methods. On the other hand, findings from ReFACTor, RefFreeEWAS, RefFreeCellMix and SVA were in general comparable for data simulated under scenario 1 (no-confounding effects) but SVA gave consistently higher sensitivity and specificity when cofounding effects present.

The SVA approach does not provide estimates on cell type compositions; however, our ultimate goal was not to estimate cell counts. The goal was to identify an approach that best assesses DNA methylation differentiation due to exposure or diseases, corrected for a potential cell type bias. From this viewpoint and the findings from real data and the high sensitivities and specificities from simulations (under both scenarios, confounding and no confounding), using SVA to adjust for cell type compositions seems to be an appropriate method and may perform better than other existing methods. After including two additional methods (ReFACTor and RefFreeCellMix) and implementing more stringent conditions of confounding in simulations (by assuming higher correlations and dynamic correlations in nearby CpGs), we reached the same conclusion as in McGregor et al. [12] when the focus was on assessing associations. We would like to point out that since the method in minfi focuses on estimating cell type compositions, it does not have the ability to address variations in any unknown factors. Thus, it is expected that this approach will not outperform the SVA approach in association studies, as found in McGregor et al., although the method in minfi provides better estimates on cell type proportions. It is also worth noting that information included in the surrogate variables produced by the SVA method may also include other information in addition to cell type compositions. There is a potential of over-adjustment by use of this approach. Furthermore, we note that all these reference-free methods can be directly applied to genome-wide bisulfite sequencing data and we expect similar findings in terms of their ability in inferring cell type compositions.

Conclusions

When appropriate reference data are available and if inferences on cell type compositions are needed, the method implemented in the minfi package is recommended. However, if no such reference data are available or if the focus is not on estimating cell proportions, the SVA method is suggested to correct for bias resulting from varying cell mixtures, the same conclusion given by McGregor et al. [12].

Methods

In this section we briefly describe the existing techniques for estimating cell proportions or inferring latent variables due to cell compositions, data sets (real and simulated) to assess these methods, and statistical methods used in the analyses. All the analyses were programmed in R and a tutorial website including all the programs demonstrating the methods is available at https://akhilesh362.wordpress.com/, and can also be accessed via http://www.memphis.edu/sph/people/faculty_profiles/zhang.php.

Existing methods for cell compositions

Reference-based methods: Houseman et al. [3] developed a method for cell type correction that capitalizes on the idea that differentially methylated regions (DMRs) can serve as a signature for the distribution of different types of white blood cells. It uses these DMRs as a surrogate in a regression calibration based technique to identify the cell mixture distribution. Regression calibrations can lead to bias estimates, thus an external validation data is used to calibrate the model and to correct for the bias [36]. Their method was specifically for the Illumina 27 k beadchip array (Illumina, Inc., San Diego, CA, USA).

The method by Jaffe and Irizarry [4] was adapted from the Houseman et al. [3] method and is tailored for Illumina450k along with 27 k array. The algorithm in Houseman et al. identified 500 CpG sites used to estimate cell mixture proportions from the Illumina 27 k array. The modification of Jaffe and Irizarry was motivated because of the existence of probe SNPs in the 500 CpG sites and the inconsistency of CpG sites between the 27 k and 450 k arrays. In addition, the flow-sorted data of the six adult male subjects were used as references [37] when DNA methylation was measured in peripheral blood. For DNA methylation in cord blood, cord blood reference data were used [38].

The method of removing unwanted variation (RUV) uses information from a reference database, but it does not estimate cell type proportions. Instead, this approach is based on the information on negative control probes and performs factor analysis on these probes to identify factors due to unmeasured confounders. These factors are then included in subsequent analyses to adjust for cell type effects. The negative control probes were chosen as the top 500 CpG sites from the reference database of DNA methylation known to be correlated with the cell types [39].

Reference-free methods: In total, four commonly used or recently developed reference-free methods are implemented in our study, FaST-LMM-EWASher, RefFreeEWAS, RefFreeCellMix, and ReFACTor. These methods do not need any external validation datasets and have the potential to adjust for cell mixture arising from any tissue, including blood. FaST-LMM-EWASher [40] applies the maximum likelihood (ML) approach in linear mixed models and optimizes spectral decomposition to estimate cell types [41]. RefFreeEWAS utilizes singular value decomposition (SVD) to decompose the residuals of unadjusted linear models along with unadjusted linear coefficient estimates, and estimates latent subject and cell-specific effects. Bootstrap estimates for coefficient standard errors are used to account for the correlation in the error structure. Surrogate variable analysis (SVA) estimates potential confounding factors from a singular value decomposition (SVD) of residuals and was initially applied to gene expression data [42]. SVA utilizes the concept of expression heterogeneity while estimating surrogate variables. Expression heterogeneity (EH) refers to certain plausible biological profiles of the subject, which may not be captured by the covariates in study. Compared to the method in RefFreeEWAS, SVA decomposes the residual matrix and utilizes permutations to identify statistically significant eigenvectors which serve as a representative of EH (the so-called eigengenes), and then infers surrogate variables based on theses "eigengenes". Surrogate variables from SVA have the potential to cover information on cell types in DNA methylation from blood cells.

The method built in the R package RefFreeCellMix is improved from that in RefFreeEWAS. It uses a variant of non-negative matrix factorization to decompose the total methylation sites into CpG-specific methylation states for a pre-specified number of cell types and subject-specific cell-type distributions [8]. Another approach in the R package, ReFACTor, implements a variant form of principal component analysis (PCA) to adjust for the cell type effects. This method assumes that a small number of methylation sites are affected by underlying cell mixtures. It filters out CpGs if the variation is not large enough (the default cutoff is standard deviation = 0.02). To avoid filtering out too many CpGs, in our analyses, we excluded CpGs such that their standard deviations were in the lower 5th percentile. By default this method searches for top 500 most informative methylation sites and performs PCA with a fixed number of components on these CpG sites to obtain the components. These ReFACTor components can be used as a covariate in epigenome wide association study or can be added one at time to remove the inflation due to cell type composition [10].

Three real data sets used to compare the approaches

These three data sets include data on prenatal arsenic exposure and DNA methylation, an example data from FaST-LMM-EWASher, and data on breast cancer status and DNA methylation. The first two data sets were utilized to demonstrate each of the five methods for cell type compositions and their agreement in terms of identified CpGs potentially associated with a variable of interest. The third data set was used to assess the agreement between the estimated cell type proportions (using the Houseman et al. method and the method in minfi) and the physical counts of the cells. This data set served as a benchmark and was critical for the comparison between the Houseman et al. method and the method in minfi. The benchmark data used to compare reference-free

methods were simulated data, as discussed in the next section.

Prenatal arsenic exposure and DNA methylation data: The data were from a birth cohort study examining multiple prenatal and postnatal factors in relation to child health outcomes, part of the nationwide Taiwan prenatal and infant cohort study [13, 14] established in Taiwan in 2000–2001. In total, 64 subjects with genome-scale DNA methylation measured in cord blood and levels of prenatal arsenic exposure were included in our study. DNA methylation data were pre-processed including quantile normalization, probe-type correction, and probe SNPs exclusion. After pre-processing, in total, 385,183 CpG sites were included in the analyses. All the five methods were applied to this data set. This and the following example data set were used to compare the performance of the five methods.

An example cancer data from FaST-LMM-EWASher: This is an example data provided by the FaST-LMM-EWASher package [43]. It was originally used to illustrate the method incorporated in FaST-LMM-EWASher. In total, 204 subjects with cancer status and DNA methylation from Illumina 27 K array on 25,978 CpG sites are available.

Breast cancer status and DNA methylation data: This data set has been previously described [27] and has genome-scale DNA methylation and breast cancer status available on 61 subjects at baseline and on 39 subjects at 6 month follow-up along with complete blood counts. After pre-processing, 484,489 CpG sites were included in the study. In this article, we focus on granulocytes, monocyte and lymphocytes cells since proportions of these cells can be estimated by use of the minfi package and the original Houseman et al. approach. In our study, proportions of these cells from the physical counts were compared to the cell proportions estimated by minfi and the Houseman et al. method.

Both studies (the arsenic and DNA methylation related study and the breast cancer and DNA methylation related study) were approved by internal review board. Nurses and doctors were involved in the data collections. None of the authors were involved in data collection and handling. The data used in this analysis were de-identified.

Simulated data sets to compare the approaches

To further evaluate the three reference-free methods (FaST-LMM-EWASher, RefFreeEWAS, RefFreeCell-Mix, ReFACTor, and SVA), we simulated DNA methylation data under different settings with "latent" variables representing "cell types". These data sets served as benchmark data for comparing reference-methods because the underlying truth was known. Two simulation scenarios were employed to evaluate the methods.

Scenario 1: We simulated DNA methylation data at 2000 CpG sites across 600 samples, of which the first n CpG sites were associated with covariates of interest (e.g., level of arsenic exposure) and a set of latent variables, and the remaining CpG sites were only associated with the latent variables. The set of latent variables represent "cell types". One covariate of interest was considered and generated from a Normal distribution with mean 0 and variance 1 ($N(0, 1)$), The coefficient of this covariate was set at 0.3 and the intercept in the regressions was set to 0.5. Five "latent" variables were used and generated from five different Normal distributions: $N(0,5)$, $N(3,1)$, $N(0,1)$, $N(2,4)$, $N(0,3)$, respectively. The association of DNA methylation and the latent variables was assumed linear and the coefficients were generated from $N (0.5, 0.01)$. The distribution of random errors in the linear regressions was assumed to be Normal with mean 0 and variance 1.2 for the n CpGs, mean 0 and variance of 1.2 for the next 100 CpGs, and mean 0 and variance 2 for the remaining CpGs. The last setting with larger variance in random errors was for situations where the influence of cell types on DNA methylation was weaker.

We took three values of n, $n = 50$, 100, and 150, representing different sparsity levels (from high to low) of informative CpGs. In total, 100 data sets for each n were simulated. Note that under this scenario, the covariates and latent variables were generated separately and had no correlations.

Scenario 2: Latent variables generated under this scenario have potential confounding effects. The overall setting is the same as in scenario 1, except that the covariate of interest and the five latent variables (6 variables in total) were correlated such that correlation is equal to $0.7^{|i-j|}$, $i, j = 1, 2, 3, 4, 5, 6$. For instance, the correlation of the continuous covariate with the first latent variable was 0.7, and with the second latent variable was $0.7^2 = 0.49$.

The flow of the analyses plan

The overall flow of the analyses plan is as follows: Step 1. We applied all the eight methods to two real data sets (the prenatal arsenic exposure and DNA methylation data and the example cancer data from FaST-LMM-EWASher) to assess the agreement within each data set in terms of identified CpGs between different methods, assess the consistency of each method between different data sets, and use existing knowledge and tools to qualitatively compare each approach. Step 2. We used the breast cancer status and DNA methylation data which had cell counts available to quantitatively compare the two reference-based approaches (the method in minfi and the method by Houseman et al.) in terms of their agreement with the true cell counts. Step 3. We implemented simulated data to compare the reference-free methods based on sensitivities and specificities. The inability of quantitative comparison in Step 1 (where the underlying truth was unknown) motivated the subsequent comparisons in Steps 2 and 3 (underlying truth was known).

Statistical analyses

Linear regression-based analyses were used to assess the associations of DNA methylation with variables of interest with cell type heterogeneity adjusted using eight different methods. In the analyses of the two real data sets (the arsenic and DNA methylation data, and the FaST-LMM-EWASher example data), we recorded CpG sites showing statistically significant association with variables of interest (i.e., arsenic exposure and cancer) after implementing different cell type heterogeneity inference methods. We also inferred the number of statistically significant CpG sites without adjusting for cell type heterogeneity. To compare the eight cell type heterogeneity inference methods (Houseman et al., minfi, FaST-LMM-EWASher, RefFreeEWAS, RefFreeCellMix, ReFACTor, RUV, and SVA), we assessed the percentage of overlap between different methods in the number of identified CpG sites that showed statistical significance, and calculated a similarity index, Jaccard index (J-index) [44]. The percentage of overlap is calculated as the number of identified CpGs overlapped with that from SVA divided by the number of CpGs identified by SVA. We used Fisher exact test to assess the significance of overlap. Jaccard index measures the similarity between two finite sample sets. We used a Bioconductor package GeneOverlap to calculate this index. To assess whether the CpGs uniquely identified by the SVA approach were informative, we used the Database for Annotation, Visualization and Integrated Discovery (DAVID) [45, 46] to analyze the enrichment in Gene ontology (GO) [47] categories and Kyoto Encyclopedia of Genes and Genomes (KEGG) [48, 49] pathways.

As for each simulated data set, we calculated sensitivity and specificity of the selected CpG sites for each cell type heterogeneity inference method. They were calculated by comparing the detected CpGs with the truly important CpGs. For each of the five methods, median of sensitivity and specificity along with 95% empirical intervals across 100 data sets were recorded for each setting under each simulation scenario.

Additional files

Additional file 1: Supplemental Material S1. Figure A. Pearson correlations between inorganic arsenic levels (in log10 scale) and cell type proportions. Figure B. Pearson correlations between total arsenic levels (in log10 scale) and cell type proportions.

Additional file 2: Supplemental Material S2. Functional annotation of genes related to CpGs identified by SVA method for the Taiwanese and example data in the FasT-LMM-EWASher package.

Additional file 3: Supplemental Material S3. T-test results for differences in cell proportions for six cell types across cases and control (cancer status).

Additional file 4: Supplemental Material S4. Boxplots depicting the difference in proportions for six cell types across cases and control (cancer status).

Additional file 5: Supplemental Material S5. Summary of sensitivity, specificity of Unadjusted, FaST-LMM-EWASher, RefFreeEWAS, SVA, ReFACTor and RefFreeCellMix for 100 simulated data across three settings for $\rho = 0, 0.3$ and 0.7.

Abbreviations
CpG: Cytosine phosphate guanine; DAVID: Database for Annotation, Visualization and Integrated Discovery; DMRs: Differentially methylated regions; FDR: False discovery rate; GO: Gene ontology; KEGG: Kyoto Encyclopedia of Genes and Genomes; ROS: Reactive oxygen species; SNPs: Single nucleotide polymorphism; SVA: Surrogate variable analysis

Acknowledgements
The authors gratefully acknowledge the cooperation of the gynecologists and pediatricians who participated in this study, and appreciate the hard work of the National Health Research Institutes, Miaoli, Taiwan, in recruiting the subjects and specimen processing.

Funding
We thank the National Science Council, Taiwan, (grants no.: MOST 103-2314-B-400-006, 104-2314-B-400-001) for the subject recruitment, specimen processing and generation of the methylation data. The work of Hongmei Zhang and Wilfried Karmaus was supported by R01AI091905 (PI: Wilfried Karmaus) and R01AI121226 (MPI: Hongmei Zhang). None of the funding agencies played a role in the design or conclusions of the present study.

Authors' contributions
HZ conceived the study. AK and HZ wrote the manuscript, MR provided detailed editing on the manuscript, HZ provided guidance on data simulation, analytical and statistical aspects. WK motivated the analyses and contributed to the manuscript. MR provided simulation codes for scenario 1. AK performed the statistical analyses. SW, MAT, and AKS provided data and contributed to the manuscript. All authors were involved in editing and revising the manuscript.

Competing interests
The authors declare that they have no competing interests.

Author details
[1]Division of Epidemiology, Biostatistics, and Environmental Health, University of Memphis, Memphis 38152, TN, USA. [2]Winship Cancer Institute, Emory University, 1365 Clifton Rd. NE, Atlanta 30322, GA, USA. [3]Department of Radiation Oncology, Emory University School of Medicine, 1365 Clifton Rd. NE, Atlanta 30322, GA, USA. [4]Department of Psychiatry and Behavioral Sciences, Emory University School of Medicine, 101 Woodruff Circle, Suite 4000, Atlanta 30322, GA, USA. [5]National Institute of Environmental Health Sciences, National Health Research Institutes, Miaoli, Taiwan.

References
1. Adalsteinsson BT, Gudnason H, Aspelund T, Harris TB, Launer LJ, Eiriksdottir G, Smith AV, Gudnason V. Heterogeneity in white blood cells has potential to confound DNA methylation measurements. Plos One. 2012;7(10):e46705.
2. Talens RP, Boomsma DI, Tobi EW, Kremer D, Jukema JW, Willemsen G, Putter H, Slagboom PE, Heijmans BT. Variation, patterns, and temporal stability of DNA methylation: considerations for epigenetic epidemiology. FASEB J. 2010;24(9):3135–44.
3. Houseman EA, Accomando WP, Koestler DC, Christensen BC, Marsit CJ, Nelson HH, Wiencke JK, Kelsey KT. DNA methylation arrays as surrogate measures of cell mixture distribution. BMC Bioinformatics. 2012;13:86.
4. Jaffe AE, Irizarry RA. Accounting for cellular heterogeneity is critical in epigenome-wide association studies. Genome Biol. 2014;15(2):R31.
5. Gentleman RC, Carey VJ, Bates DM, Bolstad B, Dettling M, Dudoit S, Ellis B, Gautier L, Ge Y, Gentry J, et al. Bioconductor: open software development for computational biology and bioinformatics. Genome Biol. 2004;5(10):R80.
6. Aryee MJ, Jaffe AE, Corrada-Bravo H, Ladd-Acosta C, Feinberg AP, Hansen KD, Irizarry RA. Minfi: a flexible and comprehensive Bioconductor package for the analysis of Infinium DNA methylation microarrays. Bioinformatics. 2014;30(10):1363–9.
7. Houseman EA, Molitor J, Marsit CJ. Reference-free cell mixture adjustments in analysis of DNA methylation data. Bioinformatics. 2014;30(10):1431–9.
8. Houseman EA, Kile ML, Christiani DC, Ince TA, Kelsey KT, Marsit CJ. Reference-free deconvolution of DNA methylation data and mediation by cell composition effects. BMC Bioinformatics. 2016;17:259.
9. Leek JT, Storey JD. Capturing heterogeneity in gene expression studies by surrogate variable analysis. Plos Genet. 2007;3(9):1724–35.
10. Rahmani E, Zaitlen N, Baran Y, Eng C, Hu D, Galanter J, Oh S, Burchard EG, Eskin E, Zou J, et al. Sparse PCA corrects for cell type heterogeneity in epigenome-wide association studies. Nat Methods. 2016;13(5):443–5.
11. Kaushal A, Zhang H, Karmaus WJJ, Wang JSL. Which methods to choose to correct cell types in genome-scale blood-derived DNA methylation data? BMC Bioinformatics. 2015;16 Suppl 15:7.
12. McGregor K, Bernatsky S, Colmegna I, Hudson M, Pastinen T, Labbe A, Greenwood CM. An evaluation of methods correcting for cell-type heterogeneity in DNA methylation studies. Genome Biol. 2016;17(1):84.
13. Lin L-C, Wang S-L, Chang Y-C, Huang P-C, Cheng J-T, Su P-H, Liao P-C. Associations between maternal phthalate exposure and cord sex hormones in human infants. Chemosphere. 2011;83(8):1192–9.
14. Wang S-L, Su P-H, Jong S-B, Guo YL, Chou W-L, Päpke O. In utero exposure to dioxins and polychlorinated biphenyls and its relations to thyroid function and growth hormone in newborns. Environ Health Perspect. 2005;113:1645–50.
15. Liu S, Guo X, Wu B, Yu H, Zhang X, Li M. Arsenic induces diabetic effects through beta-cell dysfunction and increased gluconeogenesis in mice. Sci Rep. 2014;4:6894.
16. Guo L, Xiao Y, Wang Y. Monomethylarsonous acid inhibited endogenous cholesterol biosynthesis in human skin fibroblasts. Toxicol Appl Pharmacol. 2014;277(1):21–9.
17. Small EM, Frost RJ, Olson EN. MicroRNAs add a new dimension to cardiovascular disease. Circulation. 2010;121(8):1022–32.
18. Elinder F, Mannikko R, Pandey S, Larsson HP. Mode shifts in the voltage gating of the mouse and human HCN2 and HCN4 channels. J Physiol. 2006;575(Pt 2):417–31.

19. Mumford JL, Wu K, Xia Y, Kwok R, Yang Z, Foster J, Sanders WE. Chronic arsenic exposure and cardiac repolarization abnormalities with QT interval prolongation in a population-based study. Environ Health Perspect. 2007;115(5):690–4.

20. Hernandez-Castro B, Doniz-Padilla LM, Salgado-Bustamante M, Rocha D, Ortiz-Perez MD, Jimenez-Capdeville ME, Portales-Perez DP, Quintanar-Stephano A, Gonzalez-Amaro R. Effect of arsenic on regulatory T cells. J Clin Immunol. 2009;29(4):461–9.

21. Biswas D, Banerjee M, Sen G, Das JK, Banerjee A, Sau TJ, Pandit S, Giri AK, Biswas T. Mechanism of erythrocyte death in human population exposed to arsenic through drinking water. Toxicol Appl Pharmacol. 2008;230(1):57–66.

22. Andrew AS, Jewell DA, Mason RA, Whitfield ML, Moore JH, Karagas MR. Drinking-water arsenic exposure modulates gene expression in human lymphocytes from a U.S. population. Environ Health Perspect. 2008; 116(4):524–31.

23. Soto-Pena GA, Luna AL, Acosta-Saavedra L, Conde P, Lopez-Carrillo L, Cebrian ME, Bastida M, Calderon-Aranda ES, Vega L. Assessment of lymphocyte subpopulations and cytokine secretion in children exposed to arsenic. FASEB J. 2006;20(6):779–81.

24. Bethge N, Lothe RA, Honne H, Andresen K, Troen G, Eknaes M, Liestol K, Holte H, Delabie J, Smeland EB, et al. Colorectal cancer DNA methylation marker panel validated with high performance in Non-Hodgkin lymphoma. Epigenetics. 2014;9(3):428–36.

25. Sjoblom T, Jones S, Wood LD, Parsons DW, Lin J, Barber TD, Mandelker D, Leary RJ, Ptak J, Silliman N, et al. The consensus coding sequences of human breast and colorectal cancers. Science. 2006;314(5797):268–74.

26. Blot-Chabaud M, Wanstok F, Bonvalet JP, Farman N. Cell sodium-induced recruitment of Na(+)-K(+)-ATPase pumps in rabbit cortical collecting tubules is aldosterone-dependent. J Biol Chem. 1990;265(20):11676–81.

27. Smith AK, Conneely KN, Pace TW, Mister D, Felger JC, Kilaru V, Akel MJ, Vertino PM, Miller AH, Torres MA. Epigenetic changes associated with inflammation in breast cancer patients treated with chemotherapy. Brain Behav Immun. 2014;38:227–36.

28. Li D, Morimoto K, Takeshita T, Lu Y. Arsenic induces DNA damage via reactive oxygen species in human cells. Environ Health Prev Med. 2001;6(1):27–32.

29. Martindale JL, Holbrook NJ. Cellular response to oxidative stress: signaling for suicide and survival. J Cell Physiol. 2002;192(1):1–15.

30. Son Y, Kim S, Chung HT, Pae HO. Reactive oxygen species in the activation of MAP kinases. Methods Enzymol. 2013;528:27–48.

31. El-Maarri O, Becker T, Junen J, Manzoor SS, Diaz-Lacava A, Schwaab R, Wienker T, Oldenburg J. Gender specific differences in levels of DNA methylation at selected loci from human total blood: a tendency toward higher methylation levels in males. Hum Genet. 2007;122(5):505–14.

32. Boks MP, Derks EM, Weisenberger DJ, Strengman E, Janson E, Sommer IE, Kahn RS, Ophoff RA. The relationship of DNA methylation with age, gender and genotype in twins and healthy controls. Plos One. 2009;4(8):e6767.

33. Teschendorff AE, West J, Beck S. Age-associated epigenetic drift: implications, and a case of epigenetic thrift? Hum Mol Genet. 2013;22(R1):R7–R15.

34. Barfield RT, Almli LM, Kilaru V, Smith AK, Mercer KB, Duncan R, Klengel T, Mehta D, Binder EB, Epstein MP, et al. Accounting for population stratification in DNA methylation studies. Genet Epidemiol. 2014;38(3):231–41.

35. Fraser HB, Lam LL, Neumann SM, Kobor MS. Population-specificity of human DNA methylation. Genome Biol. 2012;13(2):R8.

36. Carroll RJ, Ruppert D, Stefanski LA, Crainiceanu CM. Measurement error in nonlinear models: a modern perspective. CRC press. 2006.

37. Reinius LE, Acevedo N, Joerink M, Pershagen G, Dahlen SE, Greco D, Soderhall C, Scheynius A, Kere J. Differential DNA methylation in purified human blood cells: implications for cell lineage and studies on disease susceptibility. Plos One. 2012;7(7):e41361.

38. Bakulski KM, Feinberg JI, Andrews SV, Yang J, Brown S, L Mckenney S, Witter F, Walston J, Feinberg AP, Fallin MD. DNA methylation of cord blood cell types: applications for mixed cell birth studies. Epigenetics. 2016;11(5):354–62.

39. Gagnon-Bartsch JA, Speed TP. Using control genes to correct for unwanted variation in microarray data. Biostatistics. 2012;13(3):539–52.

40. Zou J, Lippert C, Heckerman D, Aryee M, Listgarten J. Epigenome-wide association studies without the need for cell-type composition. Nat Methods. 2014;11(3):309–11.

41. Lippert C, Listgarten J, Liu Y, Kadie CM, Davidson RI, Heckerman D. FaST linear mixed models for genome-wide association studies. Nat Methods. 2011;8(10):833–5.

42. Leek JT, Storey JD. Capturing heterogeneity in gene expression studies by surrogate variable analysis. Plos Genet. 2007;3(9):e161.

43. Zou JY. Correcting for Sample Heterogeneity in Methylome-Wide Association Studies. Methods Mol Biol. 2015;1589:107–14.

44. Jaccard P. The distribution of the flora in the alpine zone. New phytologist. 1912;11:37–50.

45. da Huang W, Sherman BT, Lempicki RA. Systematic and integrative analysis of large gene lists using DAVID bioinformatics resources. Nat Protoc. 2009; 4(1):44–57.

46. da Huang W, Sherman BT, Lempicki RA. Bioinformatics enrichment tools: paths toward the comprehensive functional analysis of large gene lists. Nucleic Acids Res. 2009;37(1):1–13.

47. Ashburner M, Ball CA, Blake JA, Botstein D, Butler H, Cherry JM, Davis AP, Dolinski K, Dwight SS, Eppig JT, et al. Gene ontology: tool for the unification of biology. The gene ontology consortium. Nat Genet. 2000;25(1):25–9.

48. Kanehisa M, Goto S, Sato Y, Kawashima M, Furumichi M, Tanabe M. Data, information, knowledge and principle: back to metabolism in KEGG. Nucleic Acids Res. 2014;42(Database issue):D199–205.

49. Kanehisa M, Goto S. KEGG: kyoto encyclopedia of genes and genomes. Nucleic Acids Res. 2000;28(1):27–30.

quantGenius: implementation of a decision support system for qPCR-based gene quantification

Špela Baebler[*], Miha Svalina, Marko Petek, Katja Stare, Ana Rotter, Maruša Pompe-Novak and Kristina Gruden

Abstract

Background: Quantitative molecular biology remains a challenge for researchers due to inconsistent approaches for control of errors in the final results. Due to several factors that can influence the final result, quantitative analysis and interpretation of qPCR data are still not trivial. Together with the development of high-throughput qPCR platforms, there is a need for a tool allowing for robust, reliable and fast nucleic acid quantification.

Results: We have developed "quantGenius" (http://quantgenius.nib.si), an open-access web application for a reliable qPCR-based quantification of nucleic acids. The quantGenius workflow interactively guides the user through data import, quality control (QC) and calculation steps. The input is machine- and chemistry–independent. Quantification is performed using the standard curve approach, with normalization to one or several reference genes. The special feature of the application is the implementation of user-guided QC-based decision support system, based on qPCR standards, that takes into account pipetting errors, assay amplification efficiencies, limits of detection and quantification of the assays as well as the control of PCR inhibition in individual samples. The intermediate calculations and final results are exportable in a data matrix suitable for further statistical analysis or visualization. We additionally compare the most important features of quantGenius with similar advanced software tools and illustrate the importance of proper QC system in the analysis of qPCR data in two use cases.

Conclusions: To our knowledge, quantGenius is the only qPCR data analysis tool that integrates QC-based decision support and will help scientists to obtain reliable results which are the basis for biologically meaningful data interpretation.

Keywords: Quantitative molecular biology, Quantitative PCR, Nucleic acid quantification, Web application, Decision support system

Background

The immense potential of quantitative molecular biology in life sciences is challenged by inconsistent approaches for control of errors in the final results. Due to its performance characteristics and general applicability, quantitative PCR (qPCR) has become the golden standard method for the quantification of nucleic acids. Although with the help of laboratory automation, qPCR data generation has become easy and fast, quantitative data analysis and interpretation is still not trivial due to several factors that can influence the final result. To ensure high quality of results and allow for potential reproduction of experiment, the Minimum Information for Publication of Quantitative Real-Time PCR Experiments (MIQE) guidelines have been proposed [1].

qPCR is used to measure the quantity of target DNAs in a given sample through repeated cycles of DNA amplification. The cycle at which the observed amplification-derived fluorescence first exceeds a certain threshold is called the quantification cycle (Cq). The analysis starts with the examination of the amplification curves and initial assessment of their quality, followed by the determination of the C_q values, which are further used for the quantification of the nucleic acids. It can be performed by either a standard curve or a comparative approach (formerly referred to as "absolute" and "relative" quantification, respectively). Both approaches are relative, but each is based on its own assumptions [1]. In

* Correspondence: spela.baebler@nib.si
Department of Biotechnology and Systems Biology, National Institute of Biology, Ljubljana 1000, SI, Slovenia

the standard curve approach, the number of target DNA molecules in the sample is calculated using a calibration curve of serially diluted DNA standards of known concentrations. The calibration curve presents a linear relationship between the C_q and the logarithm of the initial amount of template DNA. Test sample copy numbers are calculated from the linear regression of the standard curve, assuming equal amplification efficiencies for the standard and test samples [2]. When reference materials with known contents are available, the outcomes are absolute copy numbers [3] whereas when the copy numbers of the targets in the standards are not known, relative standard curves can be used to determine copy numbers ratios between different samples [4]. The second approach, comparative quantification, is based on determining the fold-differences in the expression of the target in relation to the reference gene. The most popular, comparative threshold cycle method ($\Delta\Delta C_q$) relies on a direct comparison of the C_q values and assumes equal and 100% efficiencies of the target and the reference gene. However, the amplification efficiencies between different genes analysed can differ which makes the $\Delta\Delta C_q$ method unsuitable in many cases [5]. Consequently, modifications that allow for amplification efficiency correction have been developed [1]. Although they do not perform as accurate as the standard curve approach [3], they can be applicable in research applications where high accuracy is not needed.

The efficiency of PCR amplification is considered as one of the most important parameters in qPCR analysis, as it strongly influences the final result [5, 6]. The efficiency is defined as the fraction of target molecules that are copied in one PCR cycle. Deviations from an optimal 100% efficiency are observed as inhibition, caused by the presence of inhibitory components, or over-amplification, caused by compound or structural conformation changes during the PCR [6]. Most common and broadly accepted way of efficiency determination is from the slope of a standard curve using linear least squares regression [1] where the preciseness of the efficiency estimate is affected by qPCR platform, the number of replicate reactions and serial dilution volume [7]. Recently, robust regression methods were shown to present a reliable alternative because they are less affected by outliers [8]. Alternatively, the efficiency can be calculated from the fluorescent increment in single amplification curves which were shown to be less accurate (reviewed in [9]) and they also require an additional step in the analysis that is sometimes cumbersome and impractical. The efficiency of the amplification is highly dependent on primer sequence and therefore the assumption of most quantification algorithms is that PCR efficiency is assay-dependent and sample-independent [10]. Yet, it is not uncommon that individual samples

originating from different or even same matrix have different amplification efficiencies [6] which can result in quantification inaccuracies [11]. A simple control of efficiency in individual samples can be performed by analysing two dilutions of the same sample [6].

Normalization controls for variations in extraction yield, reverse-transcription yield and efficiencies of amplification, thus enabling comparisons of nucleic acid concentrations across different samples. Various normalization strategies and reference genes selection algorithms have been proposed with the common guideline that several validated reference genes should be used for normalization (reviewed in [12]).

Although numerous commercial and open-access software tools for the analysis of qPCR data exist (see [13] for a recent review), they lack quality control (QC) of the final result that would aid the researcher in interpreting it. We have developed the web application quantGenius (http://quantgenius.nib.si), the only qPCR data analysis solution that integrates a QC-based decision support system (DSS). Among other features, it includes a control of inhibition in individual samples which is extremely useful when working with difficult samples, such as environmental or plant samples. In this way, it helps the scientist to obtain reliable results in a fast and high-throughput manner and thus provides the basis for biologically meaningful data interpretation.

Implementation

Front-end of the web application is built in HTML, CSS and JavaScript. Back end is written in PHP with extensive use of Laravel framework. The data is stored and managed using MySQL relational database management system (Additional file 1). The application is fully functional in most popular web browsers (Chrome, Internet Explorer 9+ and Firefox) with enabled JavaScript.

The most recent quantGenius release is available at http://quantGenius.nib.si. The source code for quantGenius is freely available under the GNU General Public License version 3.0. All the application functionalities are freely available without login or registration. Nevertheless, registration and login option have been implemented for users that wish to keep their datasets for later analysis.

In the application, data are organised as experiments, containing data for all the assays that were analysed in a sample set (see screenshot in the Additional file 1).

The quantGenius workflow features three main steps: 1) data import, 2) interactive calculation of target and reference genes copy numbers and normalization to reference genes with implemented QC-DSS and 3) export of final results in a gene-sample matrix format (Fig. 1). quantGenius enables a transparent overview of all calculations, including intermediate values and mathematical

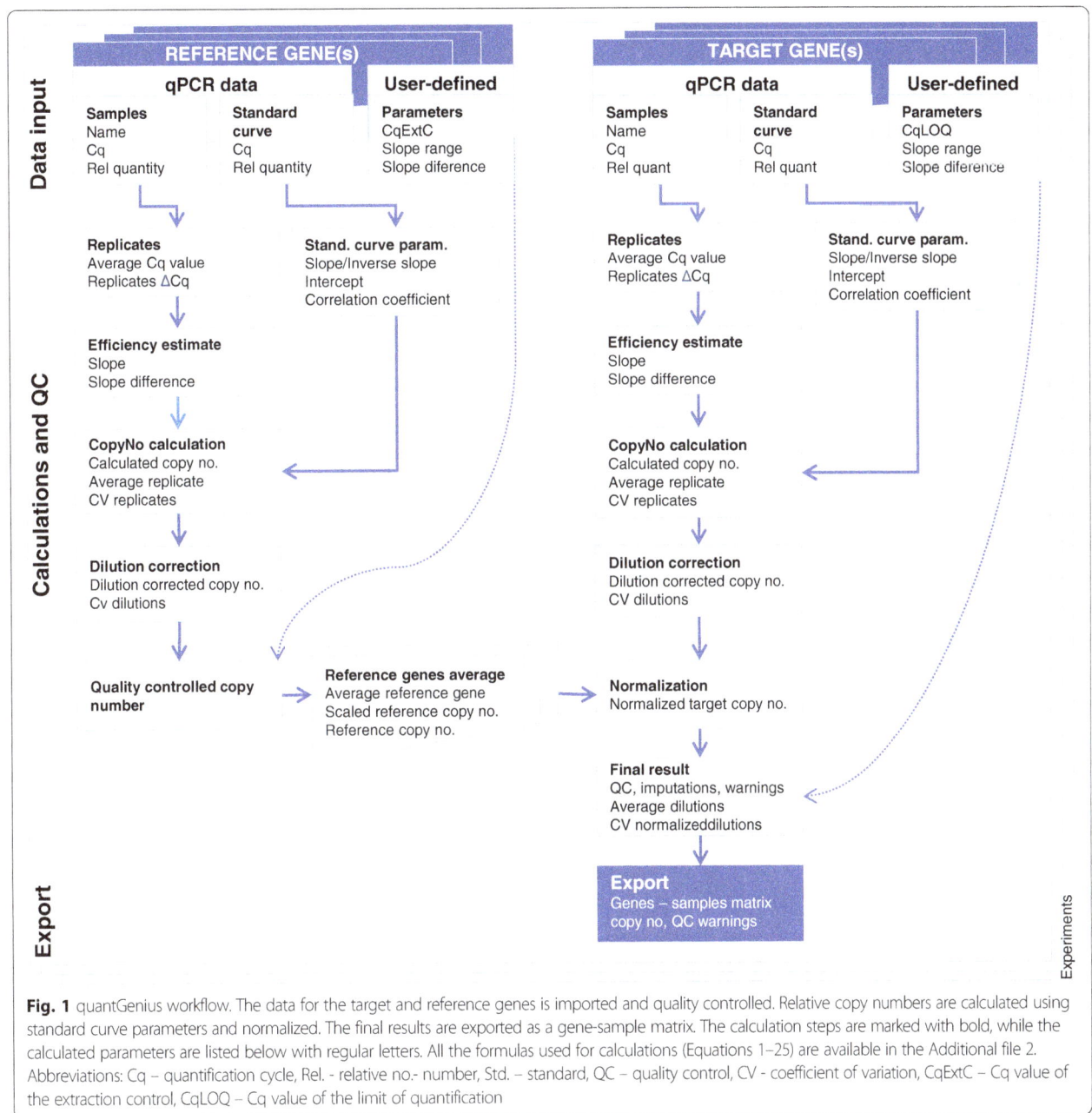

Fig. 1 quantGenius workflow. The data for the target and reference genes is imported and quality controlled. Relative copy numbers are calculated using standard curve parameters and normalized. The final results are exported as a gene-sample matrix. The calculation steps are marked with bold, while the calculated parameters are listed below with regular letters. All the formulas used for calculations (Equations 1–25) are available in the Additional file 2. Abbreviations: Cq – quantification cycle, Rel. - relative no.- number, Std. – standard, QC – quality control, CV - coefficient of variation, CqExtC – Cq value of the extraction control, CqLOQ – Cq value of the limit of quantification

formulas used as well as QC-based decisions. All the formulas used for calculations are available in the Additional file 2 (Equations 1–25) and a detailed user manual is available on the application website.

Results

Platform-independent and consistency-checked data input

The application's input is qPCR machine- and chemistry- independent. For each tested assay, sample names, C_q values and relative copy numbers (based on sample dilutions) are imported by pasting pre-formatted tab-delimited data into the input form. In this way, sample data analysed with one or two sample dilutions and any

number of technical replicates can be processed. Data for the standard curve, which can be either actual or relative copy numbers, are imported separately (Additional file 3).

Standard output files of the microfluidic qPCR platform BioMark (Fluidigm) can be converted to a format suitable for the import using the "Fluidigm data prep tool", available on the quantGenius website.

All imported data are automatically checked for consistency (i.e., that the sample names, replicates and a number of dilutions are consistent between the target and reference genes) to prevent wrong calculations due to incorrect imports (for example copy-pasting errors).

Copy number calculation and normalization to reference genes

In quantGenius, a standard curve quantification approach is implemented, which allows for the calculation of comparable copy numbers on multi-plate experiments, when the same standard curve is used on all plates. For optimal transparency of the process, the calculations are performed in several steps (Fig. 1, Additional file 2), differing slightly whether simple (one-dilution) or two-dilution analysis is selected. Based on the standard curve parameters (Additional file 2, Equations 3-8), sample target and reference gene copy numbers are calculated (Additional file 2, Equation 11). In the next steps, replicate copy numbers are averaged and sample dilution is taken into account (Additional file 2, Equations 13, 14).

Target gene copy numbers are normalized to reference gene copy numbers, or in the case of several reference genes, to their average (Fig. 2; Additional file 2, Equations 17–20). To avoid unequal contribution of the individual reference genes and to allow for quantification in the cases where data for one of the reference genes is missing due to QC issues, all the reference gene copy numbers are scaled to the average of the reference gene that was imported first (Additional file 2, Equation 18).

User-guided quality control-based decision support system

The unique and novel feature of the presented application, quantGenius, is the implementation of an easy-to-use QC-based DSS that enables robust analyses of quantitative biology data. It includes all critical parameters of qPCR QC, such as technical pipetting errors, nucleic acid extraction and reverse transcription yields, estimations of the detection and quantification ranges of the assays as well identification presence of inhibitors in the individual samples [3]. Several QC parameters are calculated at different steps of the workflow (Fig. 1, Additional file 2, Equations 2, 4, 5, 8–10, 13, 15, 22). The QC stringency is user-controlled, based on the level of accuracy required for particular application (Fig. 3). By changing the QC parameters all the data are instantly recalculated. Moreover, the "clone experiment" option allows for analysis of the same experiment with different QC parameters and thus direct comparison of the effects that the parameter settings changes have on the final results.

quantGenius enables a transparent overview of all QC-related issues and decisions. Highlighting of the values that fall out of the pre-defined QC thresholds enables the identification of the pipetting errors in the standard curve or target sample reactions as well as standard curve dilutions that are out of the quantification range which should all be manually removed by the user (Fig. 3).

Based on the implemented DSS, the final result will be, in the case that the data is out of the quantification range, imputed or, if the calculated numbers are considered unreliable, not given. In both cases, warnings are issued, notifying the user about the QC issues. The decision tree slightly differs depending on whether simple (one-dilution) or two-dilution calculation approach is chosen (Fig. 4) and hierarchically takes into account the following factors:

Extraction control. For each reference gene assay, C_{qExtC}, a C_q value indicating a valid nucleic acid (DNA

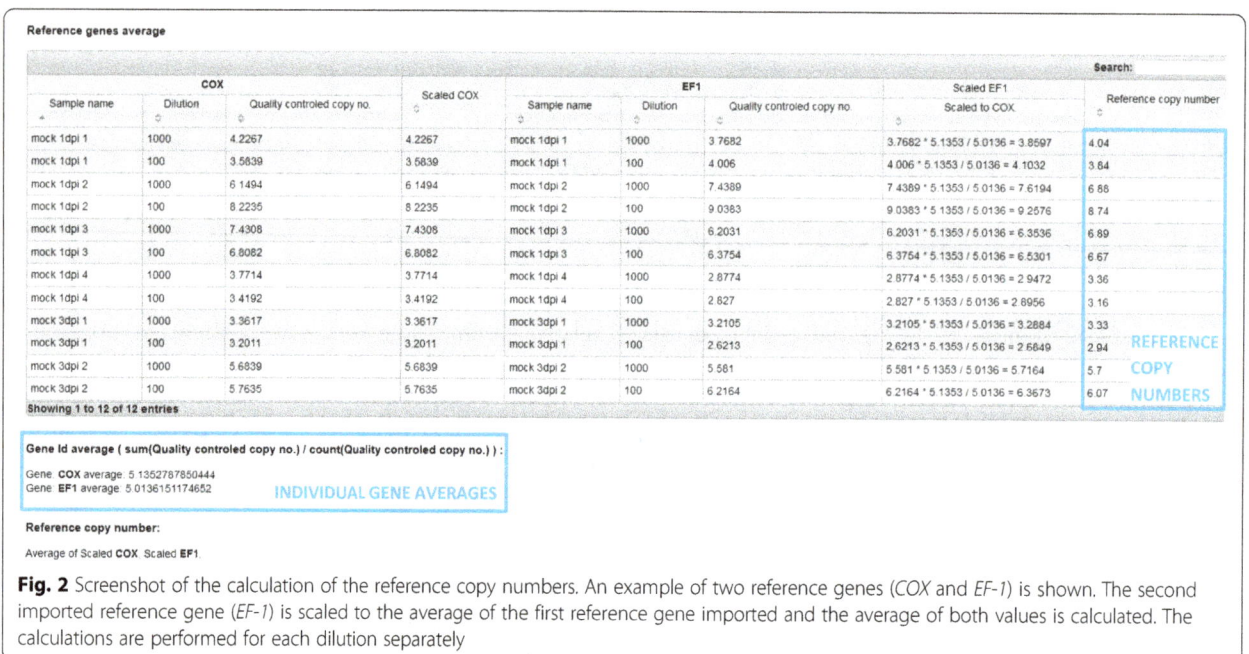

Fig. 2 Screenshot of the calculation of the reference copy numbers. An example of two reference genes (*COX* and *EF-1*) is shown. The second imported reference gene (*EF-1*) is scaled to the average of the first reference gene imported and the average of both values is calculated. The calculations are performed for each dilution separately

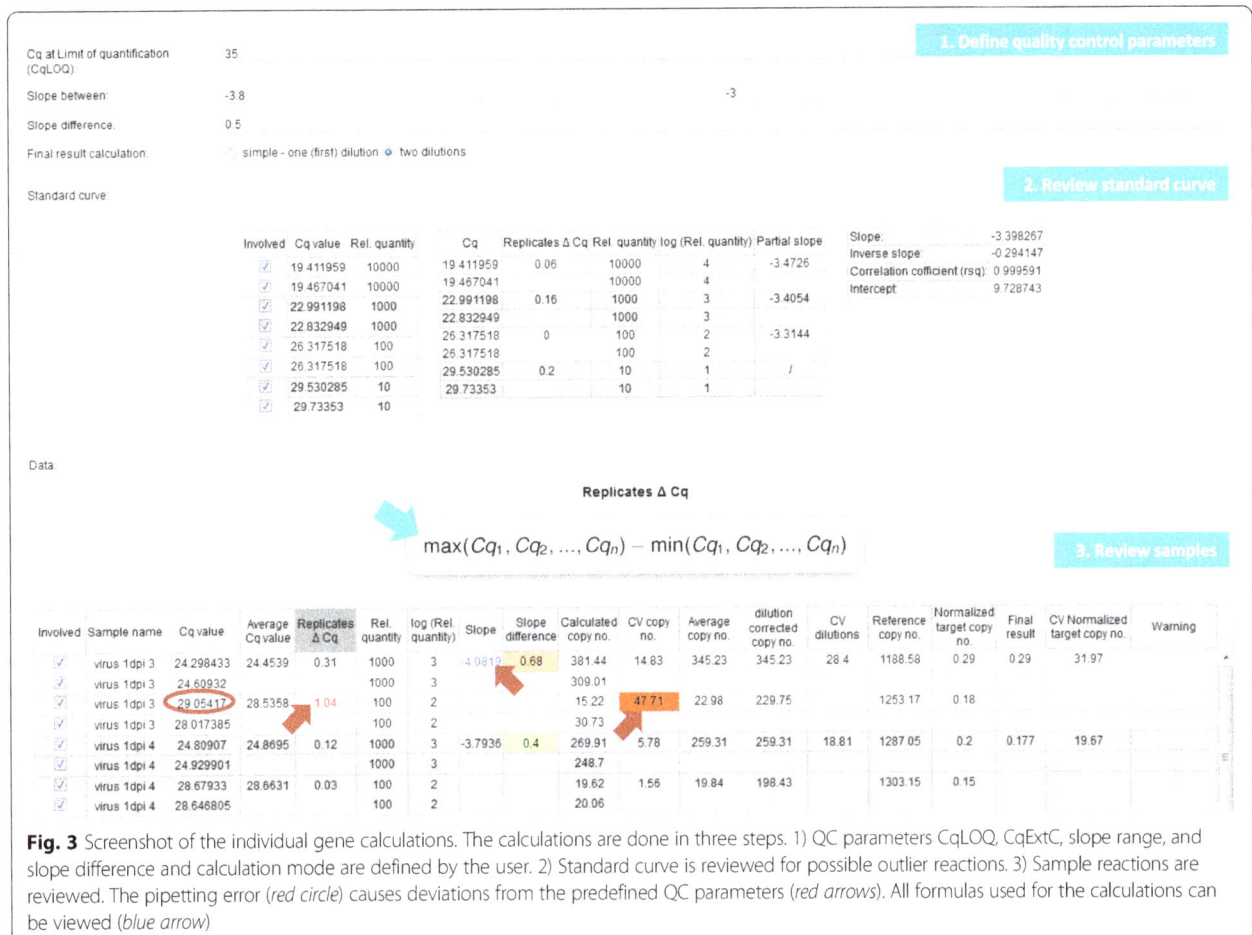

Fig. 3 Screenshot of the individual gene calculations. The calculations are done in three steps. 1) QC parameters CqLOQ, CqExtC, slope range, and slope difference and calculation mode are defined by the user. 2) Standard curve is reviewed for possible outlier reactions. 3) Sample reactions are reviewed. The pipetting error (*red circle*) causes deviations from the predefined QC parameters (*red arrows*). All formulas used for the calculations can be viewed (*blue arrow*)

or RNA) extraction procedure is defined by the user ensuring that only good quality data is used for calculations. By default, the C_{qExtC} is set to 34, therefore rarely affecting the quantification. Based on the assumption that the reference genes are highly expressed, the users can, however, lower this threshold to identify outlier samples. If all the reference genes fail this criterion, the target gene final result is not calculated (Fig. 4).

Limit of quantification (LOQ). For each target gene assay, the C_q at the LOQ (C_{qLOQ}), specifying the lower limit of the quantification of the assay is defined by the user, either based on previous in-house validation data (if available) or estimated from the experiment's standard curve. On the other hand, the LOQ can be recognized by quantGenius as high variability (CV > 30) between the replicates' copy numbers, arising from pipetting stochasticity, assuming that the true pipetting errors have previously been manually removed. In the simple calculation, the final result for samples below LOQ is imputed based on the C_{qLOQ} and all sample reference gene data (Fig. 4a, Additional file 2, Equation 24). In the two-dilution calculation, the LOQ QC step is performed in two steps: a) if the first dilution (less diluted

reactions) is under LOQ, the final result is calculated as in the simple calculation (described above), b) if only the second dilution (more diluted reactions) is under LOQ, the first dilution is used for the calculation of the final result (Fig. 4b).

Limit of detection (LOD). If all reactions of the sample for a target gene are missing C_q values, indicating that the target DNA levels in the sample are under the LOD of the assay, then the final result is imputed based on the C_{qLOQ} copy numbers of all sample reference gene data (Fig. 4, Additional file 2, Equation 25). The final result is, therefore, a very small number (lower than the LOQ-imputed value but not zero) which makes further data analyses possible without additional data imputation. LOD imputation is performed only for target genes, as the reference genes must be present well above the LOQ.

Individual sample amplification efficiency control. This QC step is implemented only in the case of two-dilution calculation and is used to identify outlier samples with apparently inappropriate amplification efficiencies as compared to the one of the standard curve [6]. If the individual sample slope (Additional file 2, Equation 9) falls

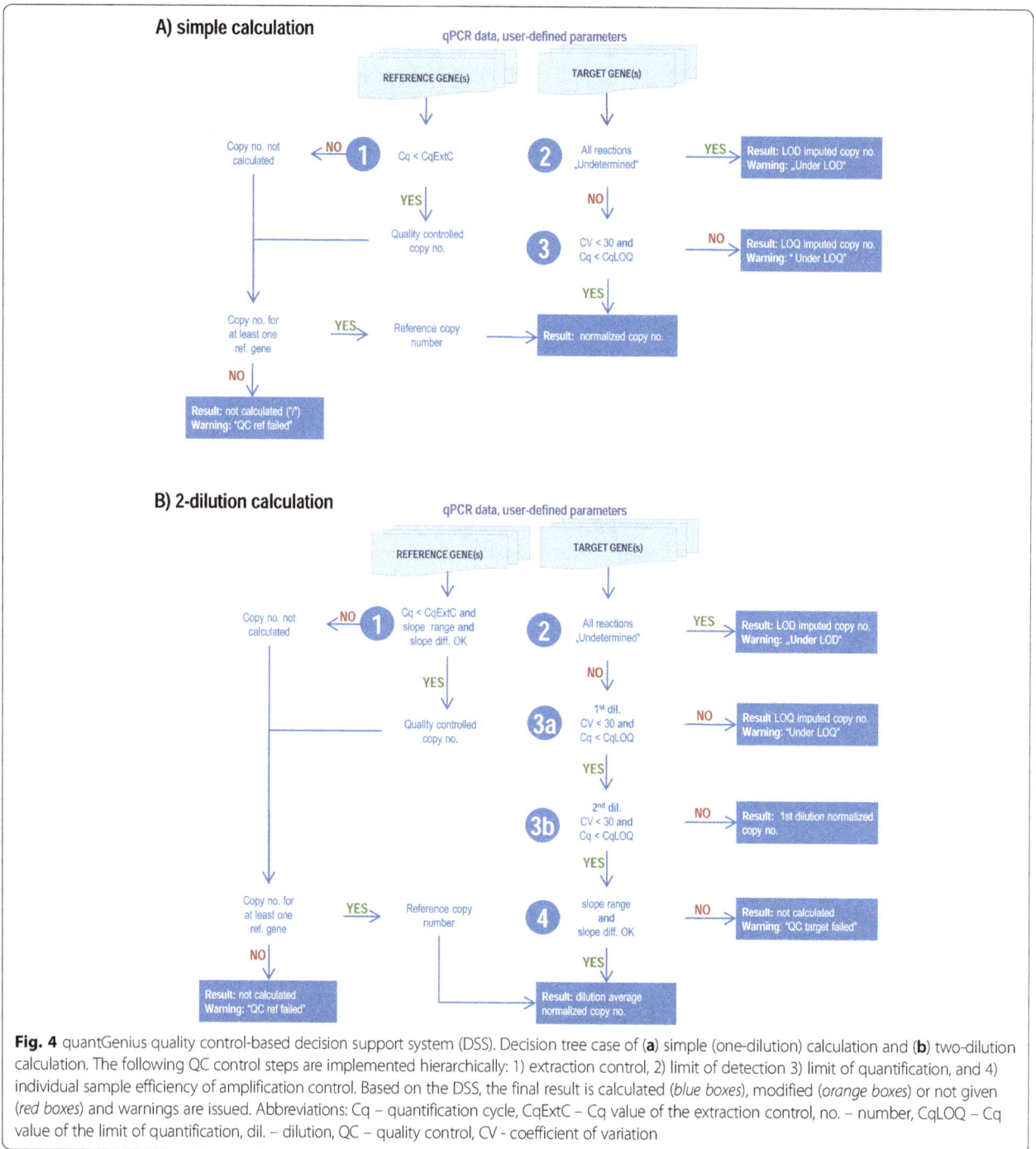

Fig. 4 quantGenius quality control-based decision support system (DSS). Decision tree case of (**a**) simple (one-dilution) calculation and (**b**) two-dilution calculation. The following QC control steps are implemented hierarchically: 1) extraction control, 2) limit of detection 3) limit of quantification, and 4) individual sample efficiency of amplification control. Based on the DSS, the final result is calculated (*blue boxes*), modified (*orange boxes*) or not given (*red boxes*) and warnings are issued. Abbreviations: Cq – quantification cycle, CqExtC – Cq value of the extraction control, no. – number, CqLOQ – Cq value of the limit of quantification, dil. – dilution, QC – quality control, CV - coefficient of variation

out of the pre-defined slope range or its difference from the standard curve's slope (Additional file 2, Equation 10) is bigger than the pre-defined maximum slope difference, the reference or target gene copy numbers are not calculated for this sample (Fig. 4b).

Export enabling further data analysis
All the data, imported sample names, quantities and C_q values, intermediate calculations and QC parameters as

well final results are available for the export from quantGenius to allow for further analysis and visualization in third-party software tools. All the data per individual gene can be exported in Excel (.xls) format (see example in Additional file 4). On the other hand, final results for all the target genes in the experiment can be also exported in a form of a sample-gene matrix in tab-delimited.txt or.xls formats. In the latter, the results are complemented with the QC warnings, so the user can distinguish between values, calculated directly from the sample data or the imputed values.

Comparison of features with advance qPCR analysis software tools

The quantification approach and crucial QC features of quantGenius were compared with similarly advanced software tools for qPCR data analysis: REST [14], one of the first software tools for qPCR analysis, two popular commercial packages qBASE+ (Biogazelle NV, [10]) and GenEx (MultiD Analyses AB) as well as an open source tool DAG Expression [15], one of the rare tools that uses standard curve based quantification (Table 1). It is important to note that the compared software tools have additional features that are not included in quantGenius, such as qualitative QC parameters (positive and negative controls, control of genomic DNA removal etc.) or further steps in the data analysis pipeline such as statistical analysis, graph plotting etc. These features were not included in quantGenius as it is focused on the quantification aspect of the qPCR data analysis pipeline.

Performance validation

The current version of the application was tested in-house for a year to detect and remove coding bugs. Further, we have analysed 50 experiments from different projects, where 40 were set on 384-well plates and 10 on the Fluidigm 48.48 Dynamic Arrays. Quantification and QC were performed in parallel in quantGenius and Microsoft Excel using preformatted formulas. A subset of the comparison is shown in the Additional files 4 and 5, respectively. Using both approaches, all the intermediate and final copy numbers, as well as those of the calculated QC parameters, were identical.

Use cases showing the importance of the quantGenius decision support system

To show the importance of proper QC in quantitative analyses we have reanalysed two datasets from different qPCR applications using quantGenius, a gene expression study and a genetically modified organism (GMO) quantification analysis.

For the gene expression use case, a subset of qPCR data from our previously published experiment [16], analysing two target genes in the response of potato to virus inoculation. The raw data (C_q values) and basic experimental details are available in Additional file 6, while the experimental details are available in the original publication [16]. Three quantification approaches were compared:

a) quantGenius two-dilution quantification with the default QC settings
b) standard curve quantification approach without any QC-DSS
c) commonly used $\Delta\Delta C_q$ approach [17], using only one dilution of the samples

The relative copy numbers obtained in the three approaches are presented in Fig. 5 and Additional file 7. The overall results of the methods correlate highly ($r > 0.99$) for both target genes. Nevertheless, the power of quantGenius is shown in the case of individual samples with low gene expression values and sub-optimal amplification efficiencies.

The expression of the PR-1b gene was near the LOQ in the mock-inoculated samples (demonstrated as C_q values near C_{qLOQ} and high inter-replicate CVs), which resulted in high copy number variation (CV > 50) between different quantification approaches (Fig. 5, top panel, a arrows, Additional file 7). With quantGenius, the copy number values below LOQ are imputed with a small value number that is in the range of values calculated for other samples near LOQ. The user is alerted with a warning and will take this into account when interpreting the results. On the other hand, in the samples where only more diluted reactions were under the LOQ, only the less diluted reactions were used for the quantification.

In both target genes, there were cases of inhibition of amplification in individual samples, resulting in outlier

Table 1 Comparison of selected features of quantGenius and other software tools

Analysis tool/Feature	Quantification method	Multiple reference genes	Quality control factor					
			Replicates	Extraction control	LOQ	LOD	Sample efficiency	gDNA
quantGenius	Std.curve	+	+	+	+	+	+	-
GenEx	ΔΔCq-E/Std.curve	+	+	-	-	+	-	+
qBase+	ΔΔCq-E	+	+	+	-	-	-	-
REST	ΔΔCq-E	+	+	-	-	-	-	-
DAG Expression	Std.curve	+	+	-	-	-	-	-

Quantification method, use of multiple reference genes for normalization and implementation of QC factors in quantification are compared. Std.curve – standard curve, ΔΔCq-E – efficiency corrected ΔΔCq method, replicates – replicate variability, extraction control – extraction efficiency, LOQ/LOD imputation – identification and imputation of copy numbers that are under LOQ or LOD, respectively, sample efficiency – individual sample efficiency estimate, gDNA – gDNA contamination correction

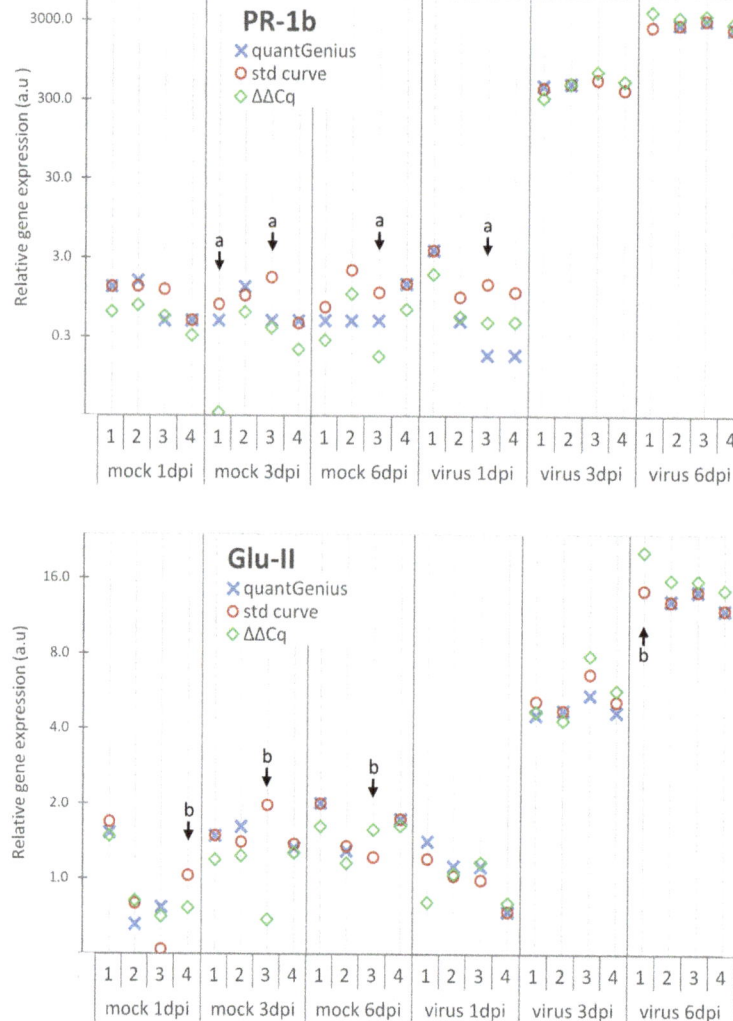

Fig. 5 Importance of implemented QC-DSS as shown in the gene expression use case. Expression of two target genes (*PR1-b, upper panel* and *Glu-II, lower panel*) was analysed in mock- and virus-inoculated potato plants at one, three and six days post infection (dpi). *EF*-1 and *COX* were used for normalization [16], (Additional file 6). Relative expression values obtained by quantGenius (*cross*) are compared to the ones obtained using standard curve quantification without QC performed (std curve, *circle*) and the $\Delta\Delta C_q$ quantification approach ($\Delta\Delta C_q$, *diamond*). To get comparable values in the three approaches, the results of each approach were normalized to one of the samples (virus 3dpi 2) and then scaled to the average expression of the first experimental group (mock 1dpi). *a arrows* - examples of samples with C_q values near LOQ showing high variability among the quantification approaches used. *b arrows* - examples of outlier samples with an efficiency problem detected in either the target or the reference gene where results are not calculated in quantGenius

results, which are especially evident in the Glu-II gene results (Fig. 5, b arrows). In these cases, quantGenius does not calculate the final result and thus again increases the reliability of the outputs of the quantification.

The second dataset is from the GMO diagnostics, where the quantity the GMO (Round-up-ready soybean, RRS) in the samples is quantified as a ratio of the transgene and reference gene (soybean lectin) copy numbers, both calculated from the standard curve of the reference material with known GMO content [3]. In the presented example, strong inhibition for the both, the reference gene and the transgene assays in both of the analysed

DNA extractions from the same sample was observed which resulted in more diluted DNA reactions having lower C_q values than less diluted ones (Additional file 8). Without the QC, the calculated % of the GMO would have been ranging from 56 to 1090%, depending on the DNA isolation and dilution used. On the other hand, in the quantGenius workflow, the results for this sample are not given, primarily because of unacceptable efficiency of the reference gene (see decision tree in Fig. 4). For this sample, the DNA isolation and qPCR analysis were repeated and it then passed QC and the GMO content was determined to be 33% (data not shown).

Discussion

The paper presents a web application for quantification of nucleic acids, integrating unique QC-based DSS (Fig. 4), built based on the acknowledged qPCR standards [1, 3, 18] which ensure that only high-quality data is used for biological interpretation. Most qPCR data analysis tools (partly reviewed in [13]) have been designed with a simple experimental design. Individual QC steps that are implemented in quantGenius are also included in other software tools (Table 1). None, however, to our knowledge, uses individual sample efficiency estimates as a QC step. Moreover, the application was built to be simple and intuitive and offers full flexibility for different experimental setups. Although the same calculations, including QC, can be done in spreadsheet software such as Microsoft Excel, the use of quantGenius does not require manual interventions for either QC or data preparation for other analysis tools. Combined with the import data consistency check-up, the use of quantGenius greatly reduces the risk of human errors when handling the data. The QC steps implemented in the DSS are the ones critical for quantification, whereas the users need to perform initial (qualitative) QC steps, such as checking fluorescence curves, qDNA contamination, the efficiency of reverse transcription, non-template or other controls, prior to importing the data to quantGenius.

quantGenius is based on quantification using a standard curve [2]. Although this approach is more robust and gives the user the biggest flexibility in the cases of suboptimal samples and/or assays [3, 6, 18] and also eliminates the need for additional interplate calibration if the same standard curve is used on all plates [3], it is implemented only in some qPCR data analysis tools (Table 1, [13]). It was previously shown that with ideal samples and assays, the results of more commonly used $\Delta\Delta Ct$ and the standard curve approach are identical [15], as was also confirmed by the presented case study, where the correlation of the quantGenius and $\Delta\Delta C_q$ results was really high (Fig. 5, Additional file 7).

In quantGenius, normalization to several validated reference genes is enabled, as it is considered the gold standard for most of the experimental setups and is also recommended by the MIQE guidelines [1, 19]. Still, the selection and validation of the reference genes should be performed beforehand by specialized tools (GeNorm, NormFinder, BestKeeper; reviewed in [12], GrayNorm [20]).

Lower copy numbers of the reference gene can indicate problems with DNA/RNA isolation or reverse transcription yields [21], leading to unreliable quantification of target genes. The extraction control implemented in quantGenius eliminates such samples from further analyses (Fig. 4).

Depending on the biological system studied, the targets in individual samples may not be detected (are under the LOD). Moreover, low amounts of DNA in the qPCR reaction can increase the measurement uncertainty due to the high variability of quantity estimations caused by the occurrence of stochastic effects, therefore only the reactions above the LOQ can be accurately quantified [5]. The reactions where Cq values are not determined are treated differently in different analysis approaches: they are either excluded from downstream analysis, which makes further calculation impossible and can lead to unnecessary information loss or even false interpretation. Alternatively, these reactions are assigned a maximum obtained C_q value which leads to biased inference or they are imputed using different statistical models [22]. In the quantGenius data analysis scheme (Fig. 4), the values below LOD and LOQ are imputed taking into account the target gene copy numbers at the LOQ and average reference gene copy numbers, resembling the background correction implemented in high-throughput gene expression analysis methods [23]. Therefore, the imputed values are comparable but appropriately lower than the ones within quantification range of the assay where the LOD imputed values are lower than the LOQ imputed ones. In this way, the user can easily spot the imputed values when inspecting the resulting output matrix and take appropriate caution when interpreting such results as was shown in our gene expression use case (Fig. 5, Additional file 7). Nevertheless, in cases, where the target DNA is truly absent (e.g absence of microorganism or transgene), the LOD imputation may result in false "positive" result and in these cases the exported data matrix without the imputed values should be used for interpretation of the results.

Low reproducibility of the C_q values from technical replicates can be an indication of an unstable assay, a pipetting error or stochastic effects due to the low amount of DNA in the reaction [24]. The latter is implemented in quantGenius, as an indication of below LOQ target DNA amounts [6] which allows for robust analysis.

There is currently no consensus on how sample specific PCR efficiencies should be calculated and used for robust quantification. Although the individual sample amplification efficiencies determined from the amplification curves increase the random error of qPCR quantifications [24], the individual sample efficiency determination has a great value for outlier detection [25]. However, as the reaction efficiency is both sample and assay dependent [6], use of RNA spike-ins is not the best option for individual sample efficiency. Therefore, quantGenius workflow includes a simple control of PCR efficiency in individual samples by comparing the C_q values of two dilutions of the same sample to identify outlier samples with suboptimal efficiencies. As quantification is in those cases not accurate, no result is given for those samples. The presented approach is associated with slightly higher cost of wet-lab analysis (caa 15% higher cost for chemicals), but on the other hand it greatly increases the quality and reliability of the data, especially in samples where the presence of inhibitors is expected, such as plant samples, food and feed samples, environmental samples, microorganisms grown in complex media etc. [6, 11]. This kind of outlier samples were also observed in our gene expression dataset (Fig. 5, Additional file 7) and in the GMO quantification use case (Additional file 8). The default limits of acceptable individual sample efficiencies are quite loose, allowing for reliable detection of two-fold copy number differences. The stringency of this QC parameter can be modified depending on the application which will result in change of the quantification measurement uncertainty [3, 4]. However, in matrices free of inhibitors (e.g., cell cultures extracts, plasmid DNA), a simple (one-dilution) approach, which is also available in quantGenius, can be used safely.

To promote quantGenius use within the scientific community, the application was is registered in the ELIXIR Tools and Data Services Registry (https://bio.tools) [26]. Future improvements are envisaged to automate data import, which is especially beneficial for the analysis of data generated by high-throughput platforms. Moreover, the connection of the application database to other databases (such as gene, assay or experimental data) will contribute to data management following the FAIR Data Principles [27].

Conclusions

As opposed to black box solutions, quantGenius was designed by biologists with ease of use, flexibility and transparency in mind. It is an intuitive and easy to use tool for qPCR data organization, analysis and decision support in various qPCR applications. The integration of QC-based DSS makes it unique and enables researchers to spend more time for interpreting the biology behind the results than analysing the data.

Additional files

Additional file 1: Schema (ER diagram) of the database and data organization in quantGenius.

Additional file 2: Equations used in quantGenius workflow.

Additional file 3: qPCR data import screenshot.

Additional file 4: Example of the export of results and intermediate values for an individual gene.

Additional file 5: quantGenius performance validation: calculations in Excel.

Additional file 6 Experimental data of the gene expression use case.

Additional file 7: Comparison of gene expression values calculated by quantGenius and other standard methods.

Additional file 8: GMO quantification use case.

Abbreviations
Cq: Quantification cycle; C_{qExtC}: Cq value of the extraction control; C_{qLOQ}: Cq value of the limit of quantification; CV: Coefficient of variation; DSS: Decision support system; GMO: Genetically modified organism; LOD: Limit of quantification; LOQ: Limit of detection; QC: Quality control; qPCR: Quantitative PCR; RRS: Round-up-ready soybean

Acknowledgements
We would like to acknowledge Henrik Krnec and dr. Živa Ramšak for their assistance with the application deployment and Tina Demšar for help with the GMO quantification use case data.

Funding
This work was supported by the Slovenian Research Agency [P4-0165, N4-002].

Authors' contributions
ŠB, AR, KS, MP, MPN and KG designed and tested the web application, MS did the programming. ŠB and KG wrote the manuscript. All authors read and approved the final manuscript.

Competing interests
The authors declare that they have no competing interests.

References
1. Bustin SA, Benes V, Garson JA, Hellemans J, Huggett J, Kubista M, et al. The MIQE guidelines: minimum information for publication of quantitative real-time PCR experiments. Clin Chem. 2009;55:611–22.
2. Bustin S. Absolute quantification of mRNA using real-time reverse transcription polymerase chain reaction assays. J Mol Endocrinol. 2000;25:169–93.
3. Žel J, Milavec M, Morisset D, Plan D, Van den Eede G, Gruden K. How to reliably test for GMOs (SpringerBriefs in food, health, and nutrition). New York: Springer; 2012.
4. Nolan T, Huggett J, Sanchez E. Good practice guide for the application of quantitative PCR (qPCR). LGC. 2013. http://www.gene-quantification.de/national-measurement-system-qpcr-guide.pdf. Accessed 19 May 2017.
5. Karlen Y, McNair A, Perseguers S, Mazza C, Mermod N. Statistical significance of quantitative PCR. BMC Bioinformatics. 2007;8:131.
6. Cankar K, Stebih D, Dreo T, Zel J, Gruden K. Critical points of DNA quantification by real-time PCR-effects of DNA extraction method and sample matrix on quantification of genetically modified organisms. BMC Biotechnol. 2006;6:37.
7. Svec D, Tichopad A, Novosadova V, Pfaffl MW, Kubista M. How good is a PCR efficiency estimate: Recommendations for precise and robust qPCR efficiency assessments. Biomol Detect Quantif. 2015;3:9–16.
8. Trypsteen W, De Neve J, Bosman K, Nijhuis M, Thas O, Vandekerckhove L, et al. Robust regression methods for real-time PCR. Anal Biochem. 2015;480:34–6.
9. Ruijter JM, Pfaffl MW, Zhao S, Spiess AN, Boggy G, Blom J, et al. Evaluation of qPCR curve analysis methods for reliable biomarker discovery: Bias, resolution, precision, and implications. Methods. 2013;59:32–46.
10. Hellemans J, Mortier G, De Paepe A, Speleman F, Vandesompele J. qBase relative quantification framework and software for management and automated analysis of real-time quantitative PCR data. Genome Biol. 2007;8:R19.
11. Pérez LM, Fittipaldi M, Adrados B, Morató J, Codony F. Error estimation in environmental DNA targets quantification due to PCR efficiencies differences between real samples and standards. Folia Microbiol (Praha). 2013;58:657–62.
12. Kozera B, Rapacz M. Reference genes in real-time PCR. J Appl Genet. 2013;54:391–406.
13. Pabinger S, Rödiger S, Kriegner A, Vierlinger K, Weinhäusel A. A survey of tools for the analysis of quantitative PCR (qPCR) data. Biomol Detect Quantif. 2014;1:23–33.

14. Pfaffl MW, Horgan GW, Dempfle L. Relative expression software tool (REST) for group-wise comparison and statistical analysis of relative expression results in real-time PCR. Nucleic Acids Res. 2002;30:e36.

15. Ballester M, Cordón R, Folch JM. DAG expression: high-throughput gene expression analysis of real-time PCR data using standard curves for relative quantification. PLoS One. 2013;8:e80385.

16. Baebler Š, Witek K, Petek M, Stare K, Tušek-Znidaric M, Pompe-Novak M, et al. Salicylic acid is an indispensable component of the Ny-1 resistance-gene-mediated response against Potato virus Y infection in potato. J Exp Bot. 2014;65:1095–109.

17. Schmittgen TD, Livak KJ. Analyzing real-time PCR data by the comparative C(T) method. Nat Protoc. 2008;3:1101–8.

18. BS EN ISO 21570. Foodstuffs—Methods of analysis for the detection of genetically modified organisms and derived products—Quantitative nucleic acid based methods. British Standards. 2006.

19. Vandesompele J, De Preter K, Pattyn F, Poppe B, Van Roy N, De Paepe A, et al. Accurate normalization of real-time quantitative RT-PCR data by geometric averaging of multiple internal control genes. Genome Biol. 2002; 3:research0034.

20. Remans T, Keunen E, Bex GJ, Smeets K, Vangronsveld J, Cuypers A. Reliable gene expression analysis by reverse transcription-quantitative PCR: reporting and minimizing the uncertainty in data accuracy. Plant Cell. 2014;26:3829–37.

21. Bustin S, Dhillon HS, Kirvell S, Greenwood C, Parker M, Shipley GL, et al. Variability of the reverse transcription step: practical implications. Clin Chem. 2015;61:202–12.

22. McCall MN, McMurray HR, Land H, Almudevar A. On non-detects in qPCR data. Bioinformatics. 2014;30:2310–6.

23. Ritchie ME, Phipson B, Wu D, Hu Y, Law CW, Shi W, et al. limma powers differential expression analyses for RNA-sequencing and microarray studies. Nucleic Acids Res. 2015;43:e47.

24. Nordgård O, Kvaløy JT, Farmen RK, Heikkilä R. Error propagation in relative real-time reverse transcription polymerase chain reaction quantification models: the balance between accuracy and precision. Anal Biochem. 2006;356:182–93.

25. Bar T, Ståhlberg A, Muszta A, Kubista M. Kinetic Outlier Detection (KOD) in real-time PCR. Nucleic Acids Res. 2003;31:e105.

26. Ison J, Rapacki K, Ménager H, Kalaš M, Rydza E, Chmura P, et al. Tools and data services registry: a community effort to document bioinformatics resources. Nucleic Acids Res. 2016;44:D38–47.

27. Wilkinson MD, Dumontier M, Aalbersberg IJ, Appleton G, Axton M, Baak A, et al. The FAIR Guiding Principles for scientific data management and stewardship. Sci Data. 2016;3:160018.

18

AGeNNT: annotation of enzyme families by means of refined neighborhood networks

Florian Kandlinger[1,2], Maximilian G. Plach[1] and Rainer Merkl[1*]

Abstract

Background: Large enzyme families may contain functionally diverse members that give rise to clusters in a sequence similarity network (SSN). In prokaryotes, the genome neighborhood of a gene-product is indicative of its function and thus, a genome neighborhood network (GNN) deduced for an SSN provides strong clues to the specific function of enzymes constituting the different clusters. The Enzyme Function Initiative (http://enzymefunction.org/) offers services that compute SSNs and GNNs.

Results: We have implemented AGeNNT that utilizes these services, albeit with datasets purged with respect to unspecific protein functions and overrepresented species. AGeNNT generates refined GNNs (rGNNs) that consist of cluster-nodes representing the sequences under study and Pfam-nodes representing enzyme functions encoded in the respective neighborhoods. For cluster-nodes, AGeNNT summarizes the phylogenetic relationships of the contributing species and a statistic indicates how unique nodes and GNs are within this rGNN. Pfam-nodes are annotated with additional features like GO terms describing protein function. For edges, the coverage is given, which is the relative number of neighborhoods containing the considered enzyme function (Pfam-node). AGeNNT is available at https://github.com/kandlinf/agennt.

Conclusions: An rGNN is easier to interpret than a conventional GNN, which commonly contains proteins without enzymatic function and overly specific neighborhoods due to phylogenetic bias. The implemented filter routines and the statistic allow the user to identify those neighborhoods that are most indicative of a specific metabolic capacity. Thus, AGeNNT facilitates to distinguish and annotate functionally different members of enzyme families.

Keywords: Sequence similarity network, SSN, Genome neighborhood network, GNN, Genome content, Enzyme function, Homology-free annotation

Background

A common method for annotating a protein is homology-based transfer of function by means of sequence comparison. Possible matches are organized in databases like InterPro [1] or Pfam [2] and the usage of such databases simplifies the assignment of protein function due to the comprehensive characterization of their entries. InterPro comprises signatures from more than ten repositories. Pfam entries subsume sequences and functions of individual protein domains which are accessed by their Pfam-ID.

However, the level of misannotation in some databases can exceed 80%, if sequence similarity is the only measure to assign function [3]. Reliability increases with the integration of orthogonal methods

like genome neighborhoods (GNs). In prokaryotic genomes, genes are organized in operons and commonly, the corresponding gene-products have related functions like the enzymes that catalyze subsequent steps of a metabolic pathway. Thus, GN algorithms utilize the fact that short distances between genes allow for the prediction of a functional coupling of their products, if the GN is conserved across many phylogenetically diverse species [4]. Such GNs are particularly useful to characterize elements of large enzyme superfamilies. More than a third of them are functionally diverse, *i. e.*, homologous members catalyze reactions with different EC numbers [5]. If these homologs are part of different operons, the GNs of isofunctional enzymes must be similar and a GN comparison must discriminate functionally different enzymes.

To identify putatively isofunctional enzymes, one can compare their sequences pairwise and cluster them, if

* Correspondence: Rainer.Merkl@ur.de
[1]Institute of Biophysics and Physical Biochemistry, University of Regensburg, D-93040 Regensburg, Germany
Full list of author information is available at the end of the article

sequence similarity exceeds a superfamily-specific threshold Th. However, depending on the Th value, clusters may break or regroup. Thus, sequence similarity networks (SSNs) that represent sequences as nodes and their pairwise similarity as weighted edges are a more flexible concept to model subtle sequence relationships that may interlink sequence clusters [6]. It follows that the set of GNs deduced for all nodes of an SSN also form networks, commonly named genome neighborhood networks (GNNs).

In order to facilitate the prediction of specific enzyme functions, the Enzyme Function Initiative (EFI, http://enzymefunction.org/) is developing and disseminating high throughput *in silico* methods and offers services that compute for a given set of protein sequences SSNs and GNNs. In this context, the GN of a given gene product is represented by a set of Pfam-IDs, listing all protein domains found in the adjacent $\pm nb$ neighbors, where nb is chosen by the user. To date, these are the only GNNs deduced from a large number of genomes. We have implemented AGeNNT that Automatically Generates refined Neighborhood NeTworks. AGeNNT utilizes the EFI services but processes in- and output in order to create function-oriented, intuitive and easy to interpret GNNs.

Methods

Computing the thresholds *A-Th* and *S-Th*

The two alternative thresholds A-Th and S-Th computed by AGeNNT are based on the analysis of two values that result if a chosen threshold (Th) is used to eliminate edges of an SSN: According to terminology introduced earlier [7], $Nn(Th)$ is the resulting number of nodes interconnected by edges and $SE(Th)$ is the resulting number of edges. Both network parameters can be combined to a ratio value $Nsv(Th)$:

$$Nsv(Th) = \frac{SE(Th)}{Nn(Th)} \qquad (1)$$

Applying a low Th-value induces the predominant elimination of edges that connect nodes belonging to *different* protein families, because their pairwise sequence similarity is low. Consequently, the increase of Th decreases $SE(Th)$ but not $Nn(Th)$, as these nodes are still connected to other members of the same family. Thus, the ratio $Nsv(Th)$ becomes smaller until the specific Th value is reached that also induces the elimination of edges within sequence clusters. For this Th value and larger ones, isolated nodes will arise that increase $Nsv(Th)$. By step-wise incrementing Th_t to Th_{t+1}, AGeNNT determines the lowest threshold inducing the raise of $Nsv(Th)$; this is the A-Th value computed according to [7].

We suggest an alternative, smoothed threshold named S-Th that is based on the relative changes $relNn(Th)$ and $relSE(Th)$:

$$relNn(Th) = \frac{Nn(Th)}{Nn(Th_{min})}, \ relSE(Th) = \frac{SE(Th)}{SE(Th_{min})} \qquad (2)$$

Here, Th_{min} is the smallest pairwise similarity score occurring within the considered SSN. We define S-Th as the lowest value Th_t for which the gain of isolated nodes is higher than the loss of eliminated edges:

$$S\text{-}Th = \underset{Th_t}{\mathrm{argmin}} \big((relNn(Th_t) - relNn(Th_{t+1})) > \\ (relSE(Th_t) - relSE(Th_{t+1})) \big) \qquad (3)$$

Computing a measure for the uniqueness of Pfam-nodes and cluster-nodes

Inspired by the work of A. Kalinka [8], we utilized the hypergeometric distribution in order to assess the uniqueness of network elements. A GNN is a graph consisting of cluster-node c_j (representing n sequence-nodes s_i from the corresponding SSN), Pfam-nodes p_l, and edges $e_i^l = (s_i, p_l)$ interconnecting a sequence-node and a Pfam-node.

For a given GNN, let $N = |\{s_i\}|$ be the total number of sequence-nodes. Let $M_l = |\{e_i^l\}|$ be the number of sequence-nodes interconnected to a specific p_l and let $k_l^j = |\{s_i \in c_j | \exists e_i^l\}|$ be the number of sequence-nodes s_i belonging to c_j and connected to a certain p_l. Using the hypergeometric distribution, one can compute the probability that k_l^j out of the n sequence-nodes have p_l in their GN:

$$P_{p_l}^{c_j}\left(X = k_l^j\right) = \frac{\binom{M_l}{k_l^j}\binom{N - M_l}{n - k_l^j}}{\binom{N}{n}} \qquad (4)$$

As a measure of the "uniqueness" of an edge (cluster-node c_j, Pfam-node p_l), AGeNNT lists the following value:

$$unique(p_l, c_j) = -\log_{10}\left(P_{p_l}^{c_j}(k_l^j)\right) \qquad (5)$$

As an approximation, we assume independence of the occurrence of the different Pfam-nodes p_l that belong to the GN of cluster-node c_j. Thus, we determine the uniqueness of a GN for a certain cluster c_j according to:

$$unique(c_j) = \sum_{p_l \in pfam(c_j)} -\log_{10}\left(P_{p_l}^{c_j}(k_l^j)\right) \qquad (6)$$

Here, $pfam(c_j)$ is the set of Pfam-nodes interconnected to sequence-nodes belonging to c_j. A high uniqueness value of a cluster-node arises, if its GN consists of many, exclusively linked Pfam-nodes. Analogously, AGeNNT computes the uniqueness of Pfam-nodes:

$$unique(p_l) = \sum_{c_j \in cluster(p_l)} -\log_{10}\left(P_{p_l}^{c_j}\left(k_l^j\right)\right) \qquad (7)$$

Here, $cluster(p_l)$ is the set of cluster-nodes interconnected to the Pfam-node p_l.

Eliminating subspecies

In SSNs deduced from InterPro families, sequences are annotated with a unique number (taxid) indicating their phylogenetic origin. Thus, for these well characterized datasets, AGeNNT can correct for strongly overrepresented species. To do so, AGeNNT utilizes a list (ftp://ftp.ncbi.nlm.nih.gov/pub/taxonomy/taxcat.zip) generated by the NCBI that indicates for each taxid whether it belongs to a species or a subspecies. At wish, AGeNNT eliminates all sequences originating from the genomes of subspecies prior to the computation and analysis of an rGNN.

User-defined whitelist

The user can specify a whitelist consisting of a text file containing one Pfam-ID per line. Pfams that are not part of this whitelist will be skipped during the process of generating a filtered GNN. The built-in whitelist consists of 7176 Pfams-IDs whose description contains a reference to an enzymatic function like an EC number or the terms "enzyme" or "catalytic".

Listing the phylogenetic origin of sequences

Based on taxids, AGeNNT determines for each cluster-node the normalized frequencies with which the different phyla contribute sequences. These numbers are listed as *PhylumStat* values.

Download and installation of AGeNNT

AGeNNT is a standalone Java application that can be used on many computer platforms. However, for the analysis of large networks (especially SSNs), we recommend a powerful CPU and at least 32 GB main memory. After download, AGeNNT can be started without any additional configuration by executing a command script. See the README file to be found after installation in the ..\agennt directory. For the generation of neighborhood networks and their visualization, the program Cytoscape is needed. It can be downloaded from http://www.cytoscape.org/. During the first call of Cytoscape within AGeNNT, the user has to specify the location of the Cytoscape executable. Afterwards, AGeNNT starts Cytoscape without further assistance.

Results

Function of AGeNNT

AGeNNT requires an SSN computed by the EFI-EST service (http://efi.igb.illinois.edu/efi-est/) as input. The computation of an SSN is detailed in [9] and demonstrated in Additional file 1. EFI-EST provides a set of output files containing networks of different granularity. If the network has less than 10 M edges, a full network is provided. Additionally, representative node (rep-node) networks are offered. In a rep-node k network, sequences that share at least k% identical resides are collected in the same sequence node. Thus, the number of nodes and edges as well as phylogenetic bias can be reduced by analyzing a rep-node 95 or a rep-node 80 network.

Due to their comprehensive annotation, we recommend the usage of SSNs deduced from InterPro families and the BLAST E-value cut-off 1E-5 in order to create most comprehensive networks. The results shown below are based on InterPro version 58 or 60. The outcome of the subsequently applied network clustering algorithm depends on the threshold Th applied to the edge weight distribution of the SSN. AGeNNT utilizes a previously proposed heuristic [7] and an in-house method to compute two alternatives, named A-Th and S-Th. After the user has selected a threshold, AGeNNT eliminates from the initial SSN all edges not reaching Th and initiates the computation of a GNN. To start this EFI service, the user has to specify the size of the genome neighborhood (nb between ±3 and ±10 genes), the co-occurrence lower limit (1 to 100%) and an email address. The co-occurrence lower limit is the fraction of sequences from the given sequence cluster that possess a gene-product encoding the functionality of a Pfam-node p_l within their nb neighborhood.

After the completion of the EFI service, AGeNNT downloads the GNN to the user's computer and offers the visualization of the EFI GNN, a colored SSN, and the refined GNN (rGNN) by means of Cytoscape [10]. A GNN interlinks cluster-nodes (representing a sequence cluster of the corresponding SSN) and Pfam-nodes (representing the corresponding GNs). Each GN consists of those Pfam-nodes p_l that reach the specified co-occurrence lower limit. Similar to the sequence-centered version of the EFI GNNs, the "hubs" of the network are the cluster-nodes that are linked to "spokes" that are the Pfam-nodes. Due to this representation, it is easy to determine and to compare the GNs of different sequence clusters; compare Fig. 1a. In comparison with EFI-GNNs, rGNNs benefit from the following features that decisively support the annotation of enzymes:

- AGeNNT assists the user in specifying the threshold *th* required to filter the initial SSN prior to GNN generation; see Formula (3).

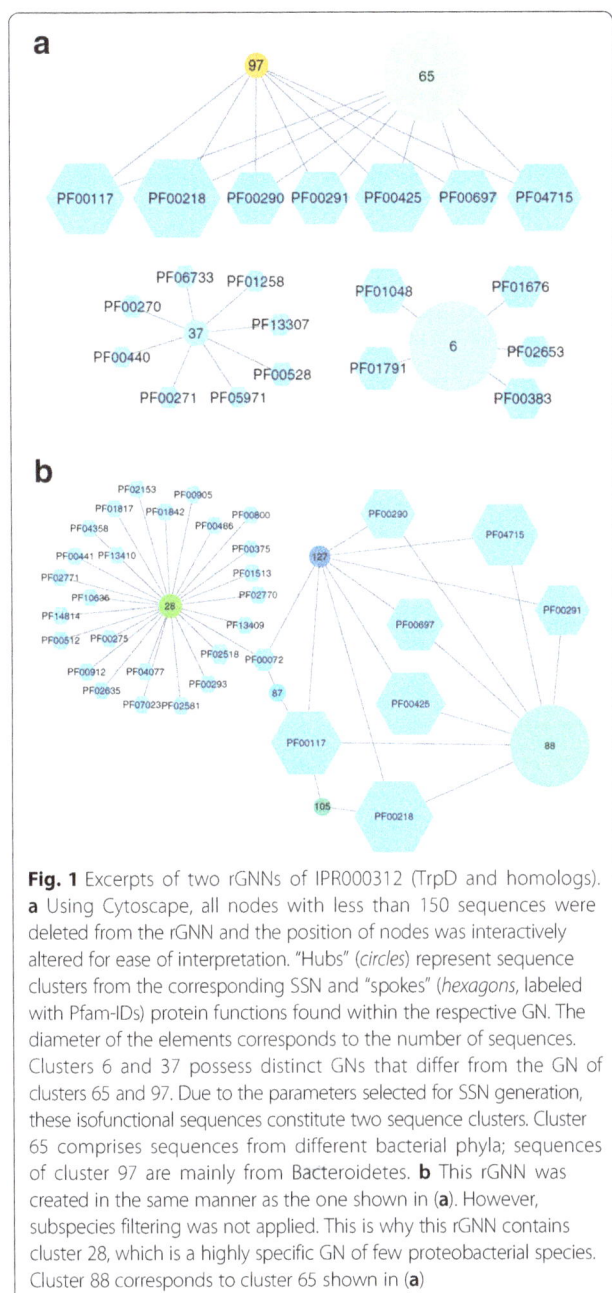

Fig. 1 Excerpts of two rGNNs of IPR000312 (TrpD and homologs). **a** Using Cytoscape, all nodes with less than 150 sequences were deleted from the rGNN and the position of nodes was interactively altered for ease of interpretation. "Hubs" (*circles*) represent sequence clusters from the corresponding SSN and "spokes" (*hexagons*, labeled with Pfam-IDs) protein functions found within the respective GN. The diameter of the elements corresponds to the number of sequences. Clusters 6 and 37 possess distinct GNs that differ from the GN of clusters 65 and 97. Due to the parameters selected for SSN generation, these isofunctional sequences constitute two sequence clusters. Cluster 65 comprises sequences from different bacterial phyla; sequences of cluster 97 are mainly from Bacteroidetes. **b** This rGNN was created in the same manner as the one shown in (**a**). However, subspecies filtering was not applied. This is why this rGNN contains cluster 28, which is a highly specific GN of few proteobacterial species. Cluster 88 corresponds to cluster 65 shown in (**a**)

- To reduce graph complexity, AGeNNT optionally eliminates gene neighbors without enzymatic function based on an editable "whitelist" containing Pfam-IDs, which represent enzymatic functions.
- Each Pfam-node is additionally annotated with GO terms that specify the molecular function and the metabolic process in which the enzyme is involved.
- On demand, AGeNNT reduces phylogenetic bias. Then, each species can only contribute one sequence to each cluster node and the sequences from related subspecies are eliminated. To further characterize the phylogenetic distribution of the enzymes, AGeNNT

determines the normalized frequencies of all phyla contributing to each cluster-node.

- The size of the nodes corresponds to the number of clustered sequences (*SeqCount*). Thus, dominant as well as more special GNs can be discriminated easily.
- For each edge, interconnecting a cluster node and a Pfam node, the *Coverage* is given, which is the relative number of genomes supporting this link of a cluster-node and an enzyme function (Pfam-node) encoded in the considered neighborhood.
- In order to assist the user in comparing GNs, AGeNNT computes statistics to assess the "uniqueness" of edges, Pfams, and GNs; see Formulae (4) to (7).

The usage of AGeNNT is detailed in Additional file 1. In the following, we motivate the computation of a novel threshold, show typical results and a novel application of rGNNs, and illustrate the predictive power of rGNNs by analyzing some retrospective test cases. The processed rGNNs of all test cases can be downloaded from our GitHub repository; see release tab.

S-Th, a more adequate threshold for the analysis of dense SSNs

The aim of SSN generation is to distribute sequence-nodes among cluster-nodes that represent each a certain protein function, *i. e.* an isofunctional protein family. In order to make possible the clustering by means of graph analysis techniques, the edges interconnecting the sequence-nodes of an EFI SSN are labeled with scores indicating the pairwise similarity determined by BLAST [11]. It turned out that the performance of network clustering algorithms depends on thresholding the edges prior to clustering [7] and for each network a proper threshold Th has to be found. If chosen correctly, Th separates inter-family edges from intra-family edges [7]. For the assessment of their threshold (A-Th value), the authors had analyzed four sequence sets consisting at most of 1308 sequences, each representing the full diversity of a protein family [7]. Due to the low pairwise similarity of many sequences, the corresponding SSNs were sparse networks, which is no longer the case for current datasets. Nowadays, most datasets representing protein families contain large numbers of sequences, which are often highly similar to each other. Thus, the resulting SSNs are dense networks, as the number of edges is usually at least ten fold higher than the number of nodes. To illustrate this circumstance, Table 1 lists characteristic values of the five InterPro families analyzed below. As a consequence, the A-Th value is often chosen too high for a representative analysis. Therefore, we suggest an alternative, smoothed threshold named S-Th (Formula (3)), which eliminates more edges; see Table 1.

Table 1 Characteristics of InterPro families and resulting SSNs

InterPro family	# seq	Rep-node 100				Rep-node 80			
		# nodes	# edges	A-Th	S-Th	# nodes	# edges	A-Th	S-Th
IPR000312	21,626	17,712	71,376,439	190	107	5446	6,789,466	195	97
IPR004651	10,868	8521	36,297,962	140	101	1830	1,672,923	104	88
IPR023016	9463	7428	27,581,256	144	78	1920	1,842,066	114	67
IPR015890	29,878	14,614	96,362,962	259	96	8388	31,307,224	259	86
IPR007115	10,421	7848	12,511,920	70	40	2901	1,609,102	57	34

The first column gives the name of the InterPro family and the second one the number of sequences belonging to this dataset. The four columns entitled Rep-node 100 and Rep-node 80, respectively, list the number of nodes and edges of the corresponding SSN and the thresholds A-Th and S-Th. For the generation of the dataset, the BLAST E-value cut-off 1E-5 was used

An rGNN of the glycosyl transferase family

IPR000312 is named glycosyl transferase family 3 and subsumes all enzymes that transfer a phosphorylated ribose substrate. The family includes anthranilate phosphoribosyltransferases (TrpD, EC 2.4.2.18) and thymidine phosphorylases (DeoA, EC 2.4.2.4). To create Fig. 1a, we chose the proposed S-Th value of 97, eliminated subspecies and created an rGNN by collecting ±10 neighbors at a co-occurrence value of 20%. For visualization, we used Cytoscape and eliminated all nodes representing < 150 sequences. For all representations of rGNNs shown below, Cytoscape's organic layout was applied initially and the position of nodes was rearranged interactively for ease of interpretation.

The comparison of the neighborhoods makes clear that the homologs constituting IPR000312 possess at least three different enzymatic functions: The GN of clusters 97 (362 sequences) and 65 (3858 sequences) comprises PF00117, PF00218, PF00290, PF00291, PF00425, PF00697, and PF04715, which represent gene products of the canonical tryptophan operon surrounding *trp*D. For the edges interconnecting cluster-node 97 and Pfam-nodes, all *Coverage* values are > 0.72, for cluster 65 the values are in the range between 0.4 and 0.6. These numbers indicate a lower GN conservation for the physiologically diverse TrpD sequences which are mainly from Proteobacteria and Firmicutes clustered in node 65, in contrast to the GNs of the physiologically more related TrpD sequences of cluster 97, which are mainly from Bacteroidetes. The GN of cluster 6 (3834 sequences) represents a typical GN of nucleoside phosphorylases (DeoA). The 452 sequences of cluster 37 are annotated as glycosyl transferases. However, the GN contains two domains of a DNA unwinding helicase (PF00271, PF06733) suggesting DNA interaction for these 452 sequences. Indeed, one representative, YbiB, has recently been characterized as a DNA-binding protein [12]. Altogether, this retrospective test case demonstrates the predictive power of GNNs, because the GNs are indicative of putative functions for all three clusters of homologs.

The number of Pfam-nodes belonging to these GNs is relatively small, because the elimination of sequences from subspecies removed any species-specific conservation of gene arrangements. Figure 1b is part of an rGNN created with the same parameters albeit lacking subspecies elimination. This rGNN contains the cluster-node 28 representing 661 sequences mostly from Proteobacteria. The corresponding SSN makes clear that these sequences are combined in not more than 60 sequence nodes and the three most populated ones represent 228, 173, and 73 sequences originating predominantly from subspecies. Due to their close phylogenetic relationship, their genome neighborhood is highly conserved. This is why the GN of cluster 28 contains 25 additional Pfam nodes that occur as neighbors in the genomes of few proteobacteria. The uniqueness value of cluster-node 28 is 6667, which also indicates a highly specific combination of enzyme functions.

This example illustrates that the elimination of phylogenetic bias helps to avoid GNs whose content is overly specific due to a close phylogenetic relationship. Such GNs can be misleading, if the ±10 neighborhoods contain enzymes from different operons which commonly show no strong functional coupling. On the other hand, the GN of cluster 88 indicates the robustness of rGNN topology: As expected, this dominating GN represents the tryptophan operon. It consists of PF00117 (TrpG), PF00218 (TrpC), PF00290 (TrpA), PF00291 (TrpB), PF00425 (TrpE, catalytic domain), PF00697 (TrpF), and PF04715 (TrpE, N-terminal region). This set is identical to the GNs of clusters 65 and 97 of Fig. 1a. The numbering of the clusters is different, because the subspecies filter alters the composition of the sequence sets. Figure 1a is lacking a GN corresponding to cluster 28 of Fig. 1b, because after subspecies elimination, the respective sequence nodes contained less than 150 sequences each and were eliminated.

In order to further demonstrate the robustness of rGNNs, especially with respect to the parameters chosen for their generation, we created 27 rGNNs based on the rep-node 100, rep-node 80, and rep-node 60 files of

IPR000312 and varied in a systematic manner the neighborhood (±3, ±6, and ±10) and the co-occurrence (10, 20, and 30%) values. The corresponding Figures S1–S3 can be found in Additional file 2. To allow for a straightforward comparison of the networks, the same color code was used for all plots and all graphs were oriented the same way. A comparison of the networks makes clear that – as expected – the complexity is highest for the rep-node 100 network analyzed with a ±10 neighborhood and a 10% co-occurrence. For each rep-node k dataset, the complexity of GNs decreases for a smaller neighborhood and a higher co-occurrence value. Moreover, complexity goes down with the decrease of the k value. However, as can be seen, characteristic Pfam-nodes like (PF00117, PF00218), (PF00383, PF01791), and (PF00271, PF06733), which are important for functional assignment of cluster-nodes representing TrpD, DeoA, or YbiB–like enzymes are present in all networks.

In summary, this systematic analysis of one InterPro family testifies to the robustness of the SSN/rGNN approach, because the assignment of three specific enzyme functions is possible for a broad range of parameter combinations.

The GNN of a monofunctional enzyme has a simple topology

*his*F is encoded in the histidine operon of Bacteria and its gene product catalyzes a cyclization reaction during the sixth step of histidine biosynthesis. For bacterial HisF, no further function has been described. In order to illustrate how adjacent operons affect the composition of GNs, we computed an SSN for the InterPro family IPR004651 with a BLAST E-value of 1E-5 by means of the EFI-EST service. We applied the threshold S-Th, which was 97, and eliminated subspecies. An rGNN was created by collecting ±10 neighbors at a co-occurrence value of 20%. The number of sequences belonging to different nodes varied from 1 to 3574. To eliminate low populated nodes, all nodes with a *SeqCount* value ≤ 25 were eliminated. Figure 2 shows the resulting rGNN; interestingly, it contains not more than six cluster-nodes.

Cluster node 26 (uniqueness 43) represents 3574 sequences from different phyla. The corresponding GN consists of PF00117 (glutamine amidotransferase class I), PF00475 (imidazoleglycerol-phosphate dehydratase), PF00977 (histidine biosynthesis protein), PF01502 (phosphoribosyl-AMP cyclohydrolase) and PF01503 (phosphoribosyl-ATP pyrophosphohydrolase). These enzymes constitute the core of the canonical histidine operon and the two smaller clusters 6 (uniqueness 1345, 1040 mainly proteobacterial sequences) and 16 (uniqueness 1054, 813 bacterial sequences with a large fraction of Actinobacteria) possess in their specific GNs these Pfam-nodes as well. Clusters 6 and 16 share in their GNs

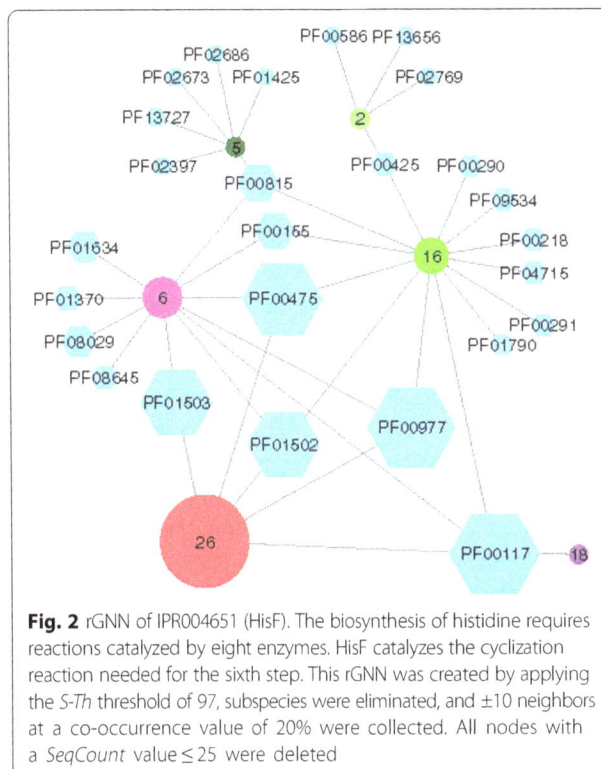

Fig. 2 rGNN of IPR004651 (HisF). The biosynthesis of histidine requires reactions catalyzed by eight enzymes. HisF catalyzes the cyclization reaction needed for the sixth step. This rGNN was created by applying the *S-Th* threshold of 97, subspecies were eliminated, and ±10 neighbors at a co-occurrence value of 20% were collected. All nodes with a *SeqCount* value ≤ 25 were deleted

PF00155 (aminotransferase class I and II) and PF00815 (histidinol dehydrogenase), which are also involved in histidine biosynthesis, but are seemingly less conserved in bacterial operons. Two Pfam-nodes specific for cluster 6, namely PF08029 and PF01634, are domains of the ATP phosphoribosyltransferase, which is also related to histidine biosynthesis. Most likely, the remaining two enzymes PF08645 (polynucleotide kinase 3 phosphatase) and PF01370 (NAD dependent epimerase/dehydratase) are not involved in histidine biosynthesis but are colocated.

Interestingly, five of the seven enzyme functions specific for cluster 16 are related to tryptophan biosynthesis. These are PF00218 (indole-3-glycerol phosphate synthase, TrpC), PF00290 (tryptophan synthase alpha chain, TrpA), PF0291 (pyridoxal-phosphate dependent enzyme, TrpB), PF00425 (TrpE, catalytic domain), and PF04715 (Trp E, N-terminal region). Inspecting the HisF neighborhood by using the *Genome Browser* of the BioCyc server [13] shows that the histidine and tryptophan operons of Mycobacteria and other Actinobacteria are directly adjacent in their genomes. This is why the GN of cluster 6 contains functions from two operons. Cluster 2 (uniqueness 634) represents 76 sequences dominantly from Euryarchaeota, cluster 5 (uniqueness 1093) contains 45 sequences from Spirochaetes, and cluster 18 (uniqueness 8) 44 sequences mainly from Proteobacteria. Without deeper analysis, these marginally populated GNs are difficult to interpret. In contrast, the composition of the three dominating GNs shows that the rGNN of a mono-functional enzyme has a relatively simple topology.

Correlating enzymatic function of HisA and PriA and the localization of their genes

For two actinobacterial species, namely *Streptomyces coelicolor* and *Mycobacterium tuberculosis*, the existence of an enzyme named PriA has been reported [14]. With respect to sequence and structure, PriA is highly similar to HisA, which catalyzes the isomerization of an aminoaldose in histidine biosynthesis. Interestingly, Actinobacteria lack a *trp*F gene and it turned out that PriA is a bi-functional homolog of HisA, which adopts the roles of HisA in histidine and of TrpF in tryptophan biosynthesis. The evolution of PriA is unclear; it has been suggested that its bi-functionality is an evolutionary response to the loss of the *trp*F gene and that a narrowing down of the PriA specificity in certain Actinobacteria occurred after the horizontal gene transfer of a whole *trp* operon [15]. An SSN of the HisA/PriA superfamily (IPR023016) made clear that *his*A genes are present in all major phylogenetic groups and that the occurrence of annotated *pri*A genes is indeed restricted to the Actinobacteria. Moreover, by characterizing ancestral HisA enzymes, it was made plausible that HisA has been a bi-functional enzyme for at least 2 billion years, most likely without any evolutionary pressure [16]. The latter results are strong evidence for the assumption that PriA is a typical HisA successor due to the bi-functionality of the ancestral HisA enzymes.

If one can show that the GN of bi-functional PriA enzymes is highly similar to the GN of typical HisA enzymes, one can further confirm that PriA is more a typical than an exceptional HisA homolog. Figure 3 shows the rGNN determined for IPR023016. Using the EFI-SSN service and an *E*-value of 1E-5 as BLAST cut-off, an SSN was created. The *S-Th* value of 75 and elimination of subspecies was chosen prior to the generation of the rGNN. For highest sensitivity, a ±10 neighborhood was analyzed; however, in order to eliminate enzymes belonging to the adjacent tryptophan operon (compare Fig. 2), a co-occurrence of 25% was used to compute the rGNN. Using Cytoscape, all cluster-nodes representing ≤ 20 sequences were deleted. Besides few outliers, all HisA homologs from Actinobacteria belong to sequence-cluster 1 that represents 579 sequences. Cluster-nodes 4 (821 sequences) and 42 (1946 sequences) comprise sequences from different bacterial phyla. The GNs of these three clusters overlap to a great extent and consist of enzymes involved in histidine biosynthesis. Most importantly, the GN of the actinobacterial HisA homologs does not contain enzyme functions not found in the other GNs of bacterial HisA enzymes. This finding makes clear that with respect to functional coupling, the actinobacterial PriA enzymes do not differ from other HisA homologs.

The function of chorismate-utilizing enzymes can be deduced from their specific GNs

The sequences of the C-terminal domain of chorismate-utilizing enzymes constitute IPR015890. Among these

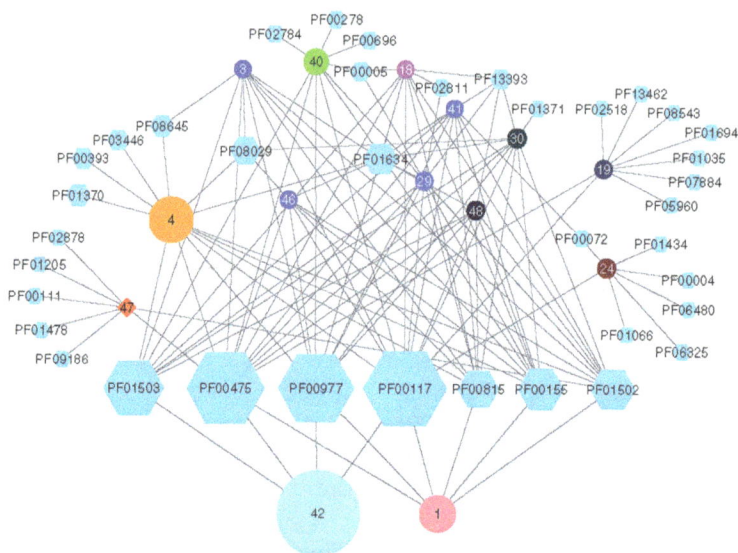

Fig. 3 rGNN of IPR023016 (HisA/PriA). PriA is a bifunctional enzyme that adopts the roles of both HisA in histidine and of TrpF in tryptophan biosynthesis. Cluster-node 1 contains the actinobacterial HisA homologs (PriA) and cluster-nodes 4 and 42 homologs from different Bacteria. The GNs of all three cluster-nodes overlap to a great extent and consist of enzymes involved in histidine biosynthesis. The Pfams belonging to the overlap of GNs are arranged in a line

enzymes are anthranilate synthases (AS), aminodeoxy-chorismate synthases (ADCS), isochorismate synthases (ICS), and salicylate synthases (SS). AS and ADCS catalyze mechanistically related reactions using ammonia as a nucleophile in tryptophan and folate biosynthesis, respectively. Although ICS and SS are highly similar to AS and ADCS with respect to sequence and structure, they utilize water instead of ammonia as a nucleophile and are part of secondary metabolic biosynthetic pathways leading to iron-chelating siderophores (e. g. enterobactin, yersiniabactin, and mycobactin) and electron-transport compounds (menaquinone).

It has recently been shown that only two amino acid substitutions in AS are sufficient to generate a bi-functional enzyme that forms isochorismate as efficiently as does a native ICS while retaining AS activity [17]. Thus, the comparison of sequences in this enzyme family may be misleading and the determination of function from sequence homology may be erroneous. However, as these enzymes are part of different metabolic pathways, their GNs should be indicative of their predominant function.

The rGNN of this enzyme family (Fig. 4) was computed using the EFI-SSN service and an E-value of 1E-5 as BLAST cut-off. The S-Th value of 93 and elimination of subspecies was selected prior to the generation of the rGNN with a ±10 neighborhood and a co-occurrence of 20%. Using Cytoscape, all cluster-nodes representing ≤ 100 sequences were eliminated and the organic layout was utilized for visualization.

Dominating is cluster-node 38, which represents 8191 AS and ADCS sequences from several major bacterial and archaeal phyla. Its GN consists of only one Pfam-node, PF00117 (glutamine amidotransferase class I) that subsumes the glutaminases TrpG and PabA, which are part of the heterodimeric AS and ADCS complexes and deliver the ammonia required for the AS and ADCS reactions. Most likely, the occurrence of AS and ADCS enzymes in many, phylogenetically less related species and the variability of the genome neighborhood is the reason that this GN consists of not more than the gluta-minase subunits. PF00117 also belongs to the GN of cluster-node 59 which additionally contains 16 other Pfam-nodes. The respective sequences are almost exclusively of

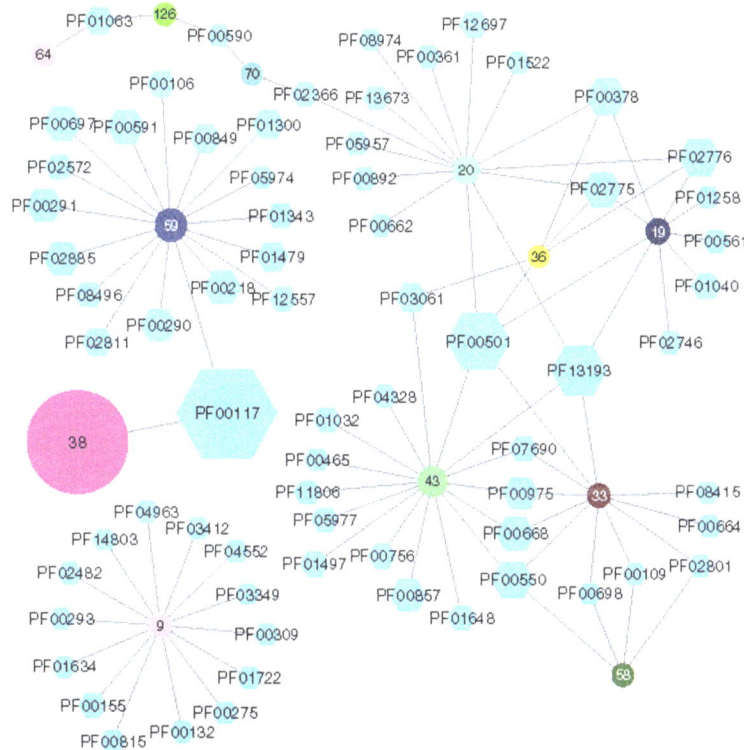

Fig. 4 rGNN of IPR015890 (chorismate-utilizing enzymes, C-terminal domain). The family of chorismate-utilizing enzymes comprises several homologous but functionally diverse enzymes from different primary and secondary metabolic pathways. The enzyme AS from tryptophan biosynthesis is represented by cluster-nodes 38 and 59, whose GN represents the typical tryptophan operon. The GN of ADCS from folate biosynthesis (cluster-node 64) is less conserved and consists of one Pfam-node, representing the enzyme catalyzing the subsequent step of folate biosynthesis. The homologous ICS and SS enzymes are part of different biosynthesis pathways that lead to iron-chelating siderophores as well as electron transport compounds. The rGNN greatly assists in separating these enzymes into different isofunctional groups: The GNs make clear that cluster-nodes 20, 36, and 19 represent MenF-type ICS, cluster-node 43 represents EntC-type ICS, and cluster-node 33 represents SS enzymes

proteobacterial origin and this close relationship is the reason that this GN contains other enzymes of the *trp* operon, namely PF00218 (TrpC), PF00697 (TrpF), and PF00290 (TrpA). The GNs of other ADCS also seems to be highly diverse, the main ADCS cluster, cluster-node 64, only contains PF01063 (amidotransferase class IV) which represents PabC, the enzyme that catalyzes the subsequent step in folate biosynthesis.

Cluster-node 9 is special due to a unique GN consisting of 14 Pfams. The sequences comprising this cluster are almost exclusively from *Acinetobacter* and *Psychrobacter* species and are annotated as either AS or ADCS. However, the GN does not contain a single enzyme typical for tryptophan or folate operons but instead contains several DNA- or ATP-binding proteins, hydrolases, and a histidinol dehydrogenase. Presumably, the sequences of cluster-node 9 represent homologous copies of AS or ADCS that are part of a different functional context in *Acinetobacter* and *Psychrobacter* and they may be interesting candidates for elucidating enzymatic function.

As mentioned above, IPR015890 also contains ICS and SS. These two enzymes are highly similar in sequence and structure and their respective reactions differ only in the different processing of their common product. Thus, false annotations are very common for ICS and SS. Moreover, several isozymes of ICS are known that catalyze the same reaction in different metabolic contexts. For example, the ICS EntC is part of the biosynthesis of the siderophore enterobactin, whereas the ICS MenF is part of the biosynthesis of menaquinone. The rGNN can help to reliably assign the ICS sequence of this enzyme family to the different isotypes. For example, the cluster-nodes 20, 36, and 19 represent ICS sequences from Proteobacteria, Firmicutes, and Bacteroidetes, respectively, and share five Pfams in their GNs (PF00378, PF02775, PF02776, PF03061, and PF00501). However, the GN of cluster-node 19 exclusively contains PF01040, which represents the UbiA prenyltransferase family. One homolog of this family, MenA, is part of the biosynthesis of menaquinone [18]. Along the same lines, the GN of cluster-node 20 exclusively contains PF00662, which represents a family of NADH-ubiquinone oxidoreductases, some of which also accept menaquinone as an electron acceptor [19].

Cluster-node 43 represents the other isotype of ICS, EntC. Its GN contains PF00975 (thioesterase domain) and PF00668 (condensation domain), which are part of the non-ribosomal peptide-synthetases that catalyze the formation of enterobactin-type siderophores [20]. Moreover, this GN contains PF00857 (EntB), the enzyme that catalyzes the step following the EntC reaction in the biosynthesis of enterobactin. Cluster-nodes 43 and 33 have in common four Pfams in their respective GNs (PF00975, PF00550, PF07690, and PF00668). However, the GN of cluster-node 33 does not contain PF00857 (EntB), which supports the annotation of these sequences as SS. In contrast to ICS, these enzymes directly convert chorismate to salicylate and not via an additional step (catalyzed by EntB) to 2,3-dihydro-2,3-dihydroxybenzoate.

In summary, the analysis of the InterPro family of chorismate-utilizing enzymes illustrates how GNs in combination with the functionality of AGeNNT can help to reliably dissect large families of homologous enzymes with diverse functions and different metabolic contexts.

rGNN analysis immediately reveals at least three different functions of PTPS enzymes

The InterPro entry IPR007115 is named "6-pyruvoyl tetrahydropterin synthase/QueD family protein" and contains 10,421 sequences that possess 18 different domain architectures [1]. 6-pyruvoyl tetrahydropterin synthase (PTPS) catalyzes the conversion of dihydroneopterin triphosphate to 6-pyruvoyl tetrahydropterin, which is the second of three enzymatic steps in the synthesis of tetrahydrobiopterin from GTP [21]. The enzyme QueD, which contributes to the biosynthesis of queuosine, a hypermodified base in the wobble position of some tRNAs in bacteria and eukaryotes [22], is also part of this functionally diverse superfamily [23]. In total, at least six PTPS homologs with different enzymatic functions named PTPS-I – PTPS-VI have been described to date, which are also often misannotated and hard to discern via sequence similarity alone [24, 25]. Moreover, several bacterial species possess more than one PTPS homolog and a standard GNN approach did not allow the accurate and precise identification of the different enzymatic functions [25]. We expected a better performance of an rGNN analysis due to the following reasons:

1) AGeNNT's built-in filters reduce noise and eliminate Pfam-nodes less relevant for a functional assignment.
2) AGeNNT's representation of GNs allows the detection of clearly distinct GNs but also function-related overlaps between clade-specific GNs.

To begin with, we computed an rGNN for IPR007115 based on the rep-node 80 file and used our standard parameters, i. e., the BLAST *E*-value cut-off 1E-5, a ±10 neighborhood, and a co-occurrence of 20%. Additionally, we applied the *S-Th* value of 34 and eliminated subspecies. By utilizing Cytoscape's *Select* command, we generated the network named rGNN_7115_150 that contains all cluster- and Pfam-nodes representing at least 150 sequences; see Fig. 5.

Dominant elements of rGNN_7115_150 are three cluster-nodes (62, 71, and 65) that possess distinct GNs. The GN of cluster 71 (uniqueness 182, 1183 sequences)

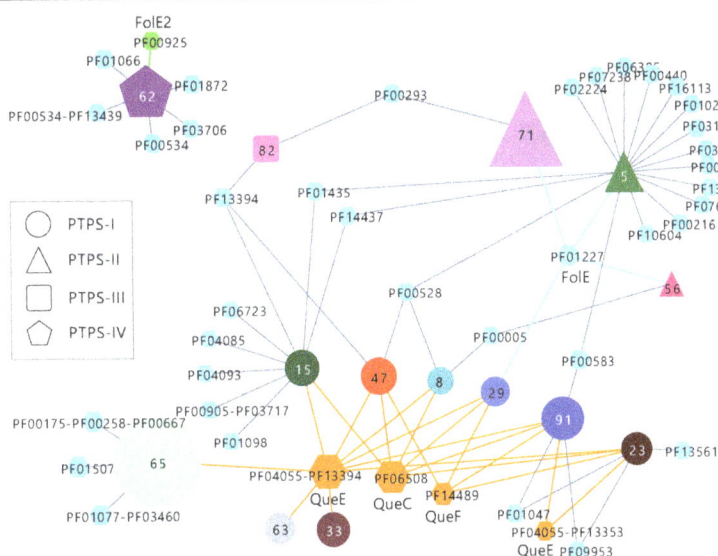

Fig. 5 rGNN of IPR007115 (6-pyruvoyl tetrahydropterin synthase/QueD family protein). This rGNN (named rGNN_7115_150) contains all nodes that represent more than 150 sequences. Dominating are cluster-nodes 65 (PTPS-I enzymes), 71 (PTPS-II enzymes), and 62 (PTPS-IV enzymes). Dominating Pfam-nodes are related to queosine biosynthesis (QueC, QueE, and QueF). The GNs of PTPS-II enzymes (cluster 71) and of PTPS-III enzymes consist of not more than two Pfam-nodes. The GNs of PTPS-IV enzymes contain FolE2; FolE belongs to the GN of cluster-nodes 5, 56, and 71

contains not more than two Pfam-nodes, namely PF01227 (GTP cyclohydrolase I; including the folate biosynthetic enzyme FolE) and PF00293 (nudix family proteins that hydrolyze a wide range of organic pyrophosphates). Interestingly, cluster-node 62 (uniqueness 1629, 561 sequences) also contains a GTP cyclohydrolase I, namely PF00925 (GTP cyclohydrolase I; includes the folate biosynthetic enzyme FolE2), but also PF01872 (RibD C-terminal domain) which is involved in riboflavin biosynthesis [1]. Cluster 65 (uniqueness 1046) represents 1463 sequences mainly from Proteobacteria. The most prominent Pfam-node of the corresponding GN is PF04055-PF13394 (radical SAM superfamily-4Fe-4S single cluster domain that includes the queosine biosynthesis enzyme QueE).

This Pfam-node occurs in nine GNs and the corresponding cluster-nodes (65, 15, 47, 8, 29, 91, 23, 63, and 33) represent a total of 4065 sequences. The GNs of these clusters also contain PF06508 (QueC), PF14489 (QueF) and PF4055-PF13353 (QueE), all of which contribute to queosine biosynthesis. Thus, the abundance of sequences and the density of the network make clear that QueD (PTPS-I) functionality is most prevalent in this InterPro family. As a first result, we postulate at least three different enzyme functions for PTPS homologs; the analysis of a conventional GNN did not allow such a classification [25].

In order to illustrate a more profound analysis of a complex case, we generated a second, more detailed network named rGNN_7115_30 that contains all cluster- and

Pfam-nodes representing at least 30 sequences; see Additional file 2: Figure S4. Additionally, we analyzed sequence motifs, which are indicative of the functionality of PTPS homologs [24, 25] and can easily be deduced for cluster-nodes; see the protocol given in the legend of Additional file 2: Figure S4. As expected, the concerted analysis of known motifs and GNs allows a more precise specification of PTPS functions: For example, the clusters predicted as being involved in queosine biosynthesis showed the motif CxxxHGH that is typical for PTPS-I functionality [25]. All sequences of cluster-node 71 contain the PTPS-II typical sequence motif CxxxxxHGH [25]. In contrast, the sequences of cluster-nodes 5 and 56 contain the motifs CxxxxHGH and CxxxHGY, thus we termed them PTPS-II-like. The sequences of cluster-node 82 share the motif ExxHGH indicative of PTPS-III functionality and those of cluster-node 62 preferentially contain the PTPS-IV-typical sequence signature FGPAQ [25].

Importantly, our rGNN approach identifies FolE2 as element of the PTPS-IV neighborhood, which has not been recognized previously, possibly due to the low number of analyzed sequences. Moreover, our analysis of more than 10,000 sequences suggests a highly variable GN of PTPS-II sequences (cluster-node 71); the coverage of the edge interconnecting cluster-node 71 and PF01227 is not higher than 0.13. A more detailed analysis of rGNN_7115_30 can be found in the legend of Additional file 2: Figure S4.

Taken together, by analyzing this InterPro family of PTPS homologs, we have demonstrated that the visualization of GNs as interwoven networks helps to corroborate similarities and differences of GNs. By eliminating smaller GNs, the user can create a "bird eye's view" on enzyme interactions and identify those ones that are conserved in many, phylogenetically distinct genomes. These results nicely supplement findings deduced from the specific analysis of individual genomes and are additionally robust against the effect of "genomic hitchhiking". Most likely, some microbial neighborhoods are merely due to the expression level and not to the functional theme of a given neighbourhood [26], which makes some individual cases enigmatic.

Discussion

Potentials and limitations of SSNs, GNNs, and rGNNs

The exponential growth of sequence data demands the development of robust and simple to use techniques to support the experimental biologist in analyzing and restructuring functionally diverse enzyme superfamilies. SSNs are an easy to use alternative to multiple sequence alignments (MSAs) and phylogenetic trees. For large and diverse sequence sets, it is difficult to construct a reliable MSA. This in turn impedes the identification of sequence motifs that are specific for the different functions of homologs and also the computation of a phylogenetic tree. On the other hand, the two-dimensional distances used in SSNs represent much of the information underlying phylogenetic trees [27]. Moreover, when analyzing and annotating large superfamilies, the main goal is not to create the optimal representation of sequence similarity, but to allow the user the visualization of many protein attributes that are orthogonal to sequence similarity and represent derived information. As has been demonstrated, by mapping these features onto SSNs by means of interactive software like Cytoscape, the informed user can rapidly develop hypotheses about the function of family members [16, 27–33].

The synergistic use of SSNs and GNNs further assists the user in assigning enzyme function, because the genome neighbors are expected to be functionally related to the enzymes under study. Along this line, experiments confirmed that the majority of the 2333 enzymes of the proline racemase superfamily catalyze only three known reactions. Thus, by using GNNs without additional information, the function of > 85% of the family members could be predicted [34].

We made plausible that the simultaneous analysis of many phylogenetically unrelated genomes reduces the risk of creating overly specific GNs containing functionally unrelated enzymes. To the best of our knowledge, the EFI services are the only algorithms that deduce and combine GNs for a large number of homologous protein sequences. In contrast, alternatives like the *Multi-Genome Browser* of *BioCyc* [13] or PSAT [35] are focusing on the analysis of few genes or are restricted to co-expressed genes [36] or mammalian genomes, like G-Nest [37]. Thus, the strength of EFI GNNs is the possibility to deduce GNs from many genomes and to combine them. Moreover, these GNNs are an ideal representation of the commonalities and differences of GNs found for homologous proteins.

AGeNNT provides additional guidance in functional annotation. However, the user has to anticipate several kinds of bias that may affect the composition of the resulting rGNNs:

1) In contrast to initial implementations of GNs [4], the EFI GNNs comprise all protein functions encoded by neighboring genes irrespective of their orientation and their regulation. Thus, a GN may contain functionally unrelated gene-products belonging to different operons. This risk is higher, if a large neighborhood value was chosen for GNN compilation and if the operon under study consists of only few genes.

2) If some species (and subspecies) are overrepresented in a cluster-node, their species-specific GN can dominate the GN determined for the whole cluster-node.

3) The composition of the individual GNs depends on the clusters constituting the SSN.

A combination of enzyme functions from several operons to the same GN (problem 1) is more likely for cluster-nodes representing closely related species but unlikely for less related ones, because the genome location of operons is not conserved on the grand scale. By inspecting the *PhylumStat* value computed by AGeNNT for each cluster-node, the user can deduce the phylogenetic relationship of the corresponding species.

Eliminating phylogenetic bias (problem 2) is difficult; however, the subspecies filter offered by AGeNNT and the usage of rep-node networks enables the user to purge a fair amount of this bias. Additionally, high uniqueness values are indicative of highly specific GNs related to a certain evolutionary environment or taxon. If accompanied by a broad phylogenetic spread shown by *PhylumStat*, small uniqueness values signal GNs that are found in many, less related taxa.

In order to assess the robustness of the findings deduced by means of rGNNs (problem 3), we recommend to utilize several parameter combinations and to compare the outcome. For example, if the parameters chosen for SSN generation give rise to a large "super cluster" consisting of several clusters $c_1 .. c_n$, the GN deduced from the corresponding rGNN might be indicative of only one cluster c_j but not for all of them.

Moreover, if a genome contains two copies of a gene, the GNN is a composite of two genome neighborhoods. To identify such cases, the user can control the *SeqCount* and *Coverage* values to be found in the *Edge Table* of Cytoscape's *Table Panel*. These values give for all edges that interconnect a cluster-node and a Pfam-node the absolute and relative number of genomes supporting this link. If these values are relatively small, the corresponding enzyme function occurs only in a fraction of clustered GNs.

Conclusions

For monofunctional enzymes found in many phyla, the GN is conserved to a great extent and after filtering, the rGNN is of low complexity because adjacent operons vary in their composition. rGNNs of high complexity are indicative of homologous enzymes possessing different functions. The representation of these GNs generated by AGeNNT assists the user in discriminating highly specific genome content and a more common functional coupling of enzymes deduced from a large number of phylogenetically less related species. The rich annotation of cluster-nodes supports the user in structuring large protein families and in deducing putative functions based on the differing metabolic contexts.

Abbreviations
EFI: Enzyme Function Initiative; GN: Genome neighborhood; GNN: Genome neighborhood network; NCBI: National Center for Biotechnology Information; Pfam: Protein family; rGNN: refined genome neighborhood network; SSN: Sequence similarity network; Th: Threshold

Acknowledgements
We thank John Gerlt and Jörg Keller for continued support and Thomas Kinateder for the analysis of two InterPro families.

Funding
No funding was received for this study.

Authors' contributions
FK designed, implemented and validated the software. RM and MGP analyzed rGNNs and wrote the manuscript. All authors read and approved the paper.

Competing interests
The authors declare that they have no competing interests.

Software and datasets
AGeNNT, the processed rGNNs used to create Figs. 1, 2, 3, 4 and 5, the input and the final results of the tutorial presented in Additional file 1 can be downloaded from https://github.com/kandlinf/agennt.

Author details
[1]Institute of Biophysics and Physical Biochemistry, University of Regensburg, D-93040 Regensburg, Germany. [2]Faculty of Mathematics and Computer Science, University of Hagen, D-58084 Hagen, Germany.

References

1. Mitchell A, Chang HY, Daugherty L, Fraser M, Hunter S, Lopez R, McAnulla C, McMenamin C, Nuka G, Pesseat S, Sangrador-Vegas A, Scheremetjew M, Rato C, Yong SY, Bateman A, Punta M, Attwood TK, Sigrist CJ, Redaschi N, Rivoire C, Xenarios I, Kahn D, Guyot D, Bork P, Letunic I, Gough J, Oates M, Haft D, Huang H, Natale DA, et al. The InterPro protein families database: the classification resource after 15 years. Nucleic Acids Res. 2015;43:D213–21.
2. Finn RD, Coggill P, Eberhardt RY, Eddy SR, Mistry J, Mitchell AL, Potter SC, Punta M, Qureshi M, Sangrador-Vegas A, Salazar GA, Tate J, Bateman A. The Pfam protein families database: towards a more sustainable future. Nucleic Acids Res. 2016;44:D279–85.
3. Schnoes AM, Brown SD, Dodevski I, Babbitt PC. Annotation error in public databases: misannotation of molecular function in enzyme superfamilies. PLoS Comput Biol. 2009;5:e1000605.
4. Overbeek R, Fonstein M, D'Souza M, Pusch GD, Maltsev N. The use of gene clusters to infer functional coupling. Proc Natl Acad Sci U S A. 1999;96: 2896–901.
5. Almonacid DE, Babbitt PC. Toward mechanistic classification of enzyme functions. Curr Opin Chem Biol. 2011;15:435–42.
6. Gerlt JA, Bouvier JT, Davidson DB, Imker HJ, Sadkhin B, Slater DR, Whalen KL. Enzyme Function Initiative-Enzyme Similarity Tool (EFI-EST): A web tool for generating protein sequence similarity networks. Biochim Biophys Acta. 2015;1854:1019–37.
7. Apeltsin L, Morris JH, Babbitt PC, Ferrin TE. Improving the quality of protein similarity network clustering algorithms using the network edge weight distribution. Bioinformatics. 2011;27:326–33.
8. Kalinka AT. The probability of drawing intersections: extending the hypergeometric distribution. arXiv preprint arXiv:13050717. 2013.
9. Gerlt JA. Tools and strategies for discovering novel enzymes and metabolic pathways. Perspect Sci. 2016;9:24–32.
10. Smoot ME, Ono K, Ruscheinski J, Wang PL, Ideker T. Cytoscape 2.8: new features for data integration and network visualization. Bioinformatics. 2011; 27:431–2.
11. Altschul SF, Madden TL, Schaffer AA, Zhang J, Zhang Z, Miller W, Lipman DJ. Gapped BLAST and PSI-BLAST: a new generation of protein database search programs. Nucleic Acids Res. 1997;25:3389–402.
12. Schneider D, Kaiser W, Stutz C, Holinski A, Mayans O, Babinger P. YbiB from *Escherichia coli*, the defining member of the novel TrpD2 family of prokaryotic DNA-binding proteins. J Biol Chem. 2015;290:19527–39.
13. Caspi R, Altman T, Billington R, Dreher K, Foerster H, Fulcher CA, Holland TA, Keseler IM, Kothari A, Kubo A, Krummenacker M, Latendresse M, Mueller LA, Ong Q, Paley S, Subhraveti P, Weaver DS, Weerasinghe D, Zhang P, Karp PD. The MetaCyc database of metabolic pathways and enzymes and the BioCyc collection of pathway/genome databases. Nucleic Acids Res. 2014;42:D459–71.
14. Barona-Gómez F, Hodgson DA. Occurrence of a putative ancient-like isomerase involved in histidine and tryptophan biosynthesis. EMBO Rep. 2003;4:296–300.
15. Noda-García L, Camacho-Zarco AR, Medina-Ruíz S, Gaytán P, Carrillo-Tripp M, Fülöp V, Barona-Gómez F. Evolution of substrate specificity in a recipient's enzyme following horizontal gene transfer. Mol Biol Evol. 2013;30:2024–34.
16. Plach MG, Reisinger B, Sterner R, Merkl R. Long-term persistence of bi-functionality contributes to the robustness of microbial life through exaptation. PLoS Genet. 2016;12:e1005836.
17. Plach MG, Löffler P, Merkl R, Sterner R. Conversion of anthranilate synthase into isochorismate synthase: implications for the evolution of chorismate-utilizing enzymes. Angew Chem Int Ed. 2015;54:11270–4.
18. Meganathan R. Biosynthesis of menaquinone (vitamin K2) and ubiquinone (coenzyme Q): a perspective on enzymatic mechanisms. Vitam Horm. 2001; 61:173–218.
19. Friedrich T. The NADH:ubiquinone oxidoreductase (complex I) from *Escherichia coli*. Biochim Biophys Acta. 1998;1364:134–46.
20. Raymond KN, Dertz EA, Kim SS. Enterobactin: an archetype for microbial iron transport. Proc Natl Acad Sci U S A. 2003;100:3584–8.
21. Nar H, Huber R, Heizmann CW, Thony B, Burgisser D. Three-dimensional structure of 6-pyruvoyl tetrahydropterin synthase, an enzyme involved in tetrahydrobiopterin biosynthesis. EMBO J. 1994;13:1255–62.
22. Iwata-Reuyl D. Biosynthesis of the 7-deazaguanosine hypermodified nucleosides of transfer RNA. Bioorg Chem. 2003;31:24–43.

23. Reader JS, Metzgar D, Schimmel P, de Crécy-Lagard V. Identification of four genes necessary for biosynthesis of the modified nucleoside queuosine. J Biol Chem. 2004;279:6280–5.

24. Phillips G, Grochowski LL, Bonnett S, Xu H, Bailly M, Blaby-Haas C, El Yacoubi B, Iwata-Reuyl D, White RH, de Crécy-Lagard V. Functional promiscuity of the COG0720 family. ACS Chem Biol. 2012;7:197–209.

25. Zallot R, Harrison KJ, Kolaczkowski B, de Crécy-Lagard V. Functional annotations of paralogs: a blessing and a curse. Life (Basel). 2016;6:39.

26. Rogozin IB, Makarova KS, Murvai J, Czabarka E, Wolf YI, Tatusov RL, Szekely LA, Koonin EV. Connected gene neighborhoods in prokaryotic genomes. Nucleic Acids Res. 2002;30:2212–23.

27. Atkinson HJ, Morris JH, Ferrin TE, Babbitt PC. Using sequence similarity networks for visualization of relationships across diverse protein superfamilies. PLoS One. 2009;4:e4345.

28. Dai X, Mashiguchi K, Chen Q, Kasahara H, Kamiya Y, Ojha S, DuBois J, Ballou D, Zhao Y. The biochemical mechanism of auxin biosynthesis by an arabidopsis YUCCA flavin-containing monooxygenase. J Biol Chem. 2013; 288:1448–57.

29. Bearne SL. The interdigitating loop of the enolase superfamily as a specificity binding determinant or 'flying buttress'. Biochim Biophys Acta. 2017;1865:619–30.

30. Jia B, Jia X, Kim KH, Jeon CO. Integrative view of 2-oxoglutarate/Fe(II)-dependent oxygenase diversity and functions in bacteria. Biochim Biophys Acta. 2017;1861:323–34.

31. Jia B, Jia X, Hyun Kim K, Ji Pu Z, Kang MS, Ok Jeon C. Evolutionary, computational, and biochemical studies of the salicylaldehyde dehydrogenases in the naphthalene degradation pathway. Sci Rep. 2017;7:43489.

32. Zhang X, Carter MS, Vetting MW, San Francisco B, Zhao S, Al-Obaidi NF, Solbiati JO, Thiaville JJ, de Crécy-Lagard V, Jacobson MP, Almo SC, Gerlt JA. Assignment of function to a domain of unknown function: DUF1537 is a new kinase family in catabolic pathways for acid sugars. Proc Natl Acad Sci U S A. 2016;113:E4161–9.

33. Huang H, Carter MS, Vetting MW, Al-Obaidi N, Patskovsky Y, Almo SC, Gerlt JA. A general strategy for the discovery of metabolic pathways: d-threitol, l-threitol, and erythritol utilization in *Mycobacterium smegmatis*. J Am Chem Soc. 2015;137:14570–3.

34. Zhao S, Sakai A, Zhang X, Vetting MW, Kumar R, Hillerich B, San Francisco B, Solbiati J, Steves A, Brown S, Akiva E, Barber A, Seidel RD, Babbitt PC, Almo SC, Gerlt JA, Jacobson MP. Prediction and characterization of enzymatic activities guided by sequence similarity and genome neighborhood networks. elife. 2014;3:e03275.

35. Fong C, Rohmer L, Radey M, Wasnick M, Brittnacher MJ. PSAT: a web tool to compare genomic neighborhoods of multiple prokaryotic genomes. BMC Bioinformatics. 2008;9:170.

36. Faria J, Davis J, Edirisinghe J, Taylor R, Weisenhorn P, Olson R, Stevens R, Rocha M, Rocha I, Best A, DeJongh M, Tintle M, Parelo B, Overbeek R, Henry C. Computing and applying atomic regulons to understand gene expression and regulation. Front Microbiol. 2016;7:1819.

37. Lemay DG, Martin WF, Hinrichs AS, Rijnkels M, German JB, Korf I, Pollard KS. G-NEST: a gene neighborhood scoring tool to identify co-conserved, co-expressed genes. BMC Bioinformatics. 2012;13:253.

Permissions

All chapters in this book were first published in BMC BIOINFORMATICS, by BioMed Central; hereby published with permission under the Creative Commons Attribution License or equivalent. Every chapter published in this book has been scrutinized by our experts. Their significance has been extensively debated. The topics covered herein carry significant findings which will fuel the growth of the discipline. They may even be implemented as practical applications or may be referred to as a beginning point for another development.

The contributors of this book come from diverse backgrounds, making this book a truly international effort. This book will bring forth new frontiers with its revolutionizing research information and detailed analysis of the nascent developments around the world.

We would like to thank all the contributing authors for lending their expertise to make the book truly unique. They have played a crucial role in the development of this book. Without their invaluable contributions this book wouldn't have been possible. They have made vital efforts to compile up to date information on the varied aspects of this subject to make this book a valuable addition to the collection of many professionals and students.

This book was conceptualized with the vision of imparting up-to-date information and advanced data in this field. To ensure the same, a matchless editorial board was set up. Every individual on the board went through rigorous rounds of assessment to prove their worth. After which they invested a large part of their time researching and compiling the most relevant data for our readers.

The editorial board has been involved in producing this book since its inception. They have spent rigorous hours researching and exploring the diverse topics which have resulted in the successful publishing of this book. They have passed on their knowledge of decades through this book. To expedite this challenging task, the publisher supported the team at every step. A small team of assistant editors was also appointed to further simplify the editing procedure and attain best results for the readers.

Apart from the editorial board, the designing team has also invested a significant amount of their time in understanding the subject and creating the most relevant covers. They scrutinized every image to scout for the most suitable representation of the subject and create an appropriate cover for the book.

The publishing team has been an ardent support to the editorial, designing and production team. Their endless efforts to recruit the best for this project, has resulted in the accomplishment of this book. They are a veteran in the field of academics and their pool of knowledge is as vast as their experience in printing. Their expertise and guidance has proved useful at every step. Their uncompromising quality standards have made this book an exceptional effort. Their encouragement from time to time has been an inspiration for everyone.

The publisher and the editorial board hope that this book will prove to be a valuable piece of knowledge for researchers, students, practitioners and scholars across the globe.

List of Contributors

Simona De Summa, Rosamaria Pinto and Stefania Tommasi
IRCCS-Istituto Tumori "Giovanni Paolo II", Molecular Genetics Laboratory, viale Orazio Flacco, 65, 70124 Bari, Italy

Giovanni Malerba, Antonio Mori and Vladan Mijatovic
Department of Neuroscience, Biomedicine and Movement Sciences, Section of Biology and Genetics, University of Verona, Strada Le Grazie 8, 37135 Verona, Italy

Anna Kusnezowa and Lars I. Leichert
Institute of Biochemistry and Pathobiochemistry – Microbial Biochemistry, Ruhr University Bochum, Universitätsstr. 150, 44780 Bochum, Germany

Pierre Borgnat and Benjamin Audit
Univ Lyon, Ens de Lyon, Univ Claude Bernard Lyon 1, CNRS, Laboratoire de Physique, F-69342 Lyon, France

Rasha E. Boulos
Univ Lyon, Ens de Lyon, Univ Claude Bernard Lyon 1, CNRS, Laboratoire de Physique, F-69342 Lyon, France

Nicolas Tremblay
Univ Lyon, Ens de Lyon, Univ Claude Bernard Lyon 1, CNRS, Laboratoire de Physique, F-69342 Lyon, France
CNRS, GIPSA-lab, Grenoble, France

Alain Arneodo
Univ Lyon, Ens de Lyon, Univ Claude Bernard Lyon 1, CNRS, Laboratoire de Physique, F-69342 Lyon, France
LOMA, Université de Bordeaux, CNRS, UMR 5798, 51 Cours de le Libération, 33405 Talence, France

Tanchanok Wisitponchai and Chatchai Tayapiwatana
Division of Clinical Immunology, Department of Medical Technology, Faculty of Associated Medical Sciences, Chiang Mai University, Chiang Mai 50200, Thailand
Center of Biomolecular Therapy and Diagnostic, Faculty of Associated Medical Sciences, Chiang Mai University, Chiang Mai 50200, Thailand

Kuntida Kitidee
Center of Biomolecular Therapy and Diagnostic, Faculty of Associated Medical Sciences, Chiang Mai University, Chiang Mai 50200, Thailand
Center for Research and Innovation, Faculty of Medical Technology, Mahidol University, Bangkok 10700, Thailand

Watshara Shoombuatong
Center of Data Mining and Biomedical Informatics, Faculty of Medical Technology, Mahidol University, Bangkok 10700, Thailand

Vannajan Sanghiran Lee
Thailand Center of Excellence in Physics, Commission on Higher Education, Bangkok 10400, Thailand
Department of Chemistry, Faculty of Science, University of Malaya, Kuala Lumpur 50603, Malaysia

Miles Aron, Richard Browning and Eleanor Stride
Department of Engineering Science, Institute of Biomedical Engineering, University of Oxford, Oxford OX3 7DQ, UK

Dario Carugo
Department of Engineering Science, Institute of Biomedical Engineering, University of Oxford, Oxford OX3 7DQ, UK
Faculty of Engineering and The Environment, University of Southampton, Southampton SO17 1BJ, UK

Erdinc Sezgin and Christian Eggeling
MRC Human Immunology Unit, Weatherall Institute of Molecular Medicine, University of Oxford, Headley Way, Oxford OX3 9DS, UK

Jorge Bernardino de la Serna
MRC Human Immunology Unit, Weatherall Institute of Molecular Medicine, University of Oxford, Headley Way, Oxford OX3 9DS, UK

Research Complex at Harwell, Central Laser Facility, Rutherford Appleton Laboratory, Science and Technology Facilities Council, Harwell-Oxford OX11 0FA, UK

Zhenzhou Wang and Haixing Li
State Key Laboratory for Robotics, Shenyang Institute of Automation, Chinese Academy of Sciences, Shenyang, China

Xiang Gao
Department of Public Health Sciences, Loyola University Chicago Health Sciences Division, Maywood, IL 60153, USA

Huaiying Lin and Kashi Revanna
Department of Public Health Sciences, Loyola University Chicago Health Sciences Division, Maywood, IL 60153, USA
Center for Biomedical Informatics, Loyola University Chicago Health Sciences Division, Maywood, IL 60153, USA

Qunfeng Dong
Department of Public Health Sciences, Loyola University Chicago Health Sciences Division, Maywood, IL 60153, USA
Center for Biomedical Informatics, Loyola University Chicago Health Sciences Division, Maywood, IL 60153, USA
Bioinformatics Program, Loyola University Chicago Lake Shore Campus, Chicago, IL 60660, USA
Department of Computer Science, Loyola University Chicago Water Tower Campus, Chicago, IL 60611, USA

Jiun-Hong Chen
Department of Life Science, National Taiwan University, Taipei 106, Taiwan

You-Yu Lin
Department of Life Science, National Taiwan University, Taipei 106, Taiwan
Graduate Institute of Clinical Medicine, National Taiwan University, Taipei 100, Taiwan

Chia-Hung Hsieh
Department of Forestry and Nature Conservation, Chinese Culture University, Taipei 111, Taiwan

Xuemei Lu
Laboratory of Disease Genomics and Individualized Medicine, Beijing Institute of Genomics, the Chinese Academy of Sciences, Beijing 100101, China

Jia-Horng Kao and Pei-Jer Chen
Graduate Institute of Clinical Medicine, National Taiwan University, Taipei 100, Taiwan

Ding-Shinn Chen
Graduate Institute of Clinical Medicine, National Taiwan University, Taipei 100, Taiwan
Genomics Research Center, Academia Sinica, Taipei 115, Taiwan

Hurng-Yi Wang
Graduate Institute of Clinical Medicine, National Taiwan University, Taipei 100, Taiwan
Institute of Ecology and Evolutionary Biology, National Taiwan University, Taipei 106, Taiwan
Research Center for Developmental Biology and Regenerative Medicine, National Taiwan University, Taipei 100, Taiwan

Surabhi Maheshwari
Department of Biological Sciences, Louisiana State University, Baton Rouge, LA, USA

Michal Brylinski
Department of Biological Sciences, Louisiana State University, Baton Rouge, LA, USA
Center for Computation & Technology, Louisiana State University, Baton Rouge, LA, USA

Sarvesh Nikumbh
Computational Biology & Applied Algorithmics, Max Planck Institute for Informatics, Saarland Informatics Campus, Building E1.4, D-66123 Saarbruecken, Germany

Nico Pfeifer
Computational Biology & Applied Algorithmics, Max Planck Institute for Informatics, Saarland Informatics Campus, Building E1.4, D-66123 Saarbruecken, Germany
Department of Computer Science, University of Tübingen, Sand 14, D-72076 Tübingen, German

Qing Wei and Ishita K. Khan
Department of Computer Science, Purdue University, West Lafayette, IN 47907, USA

Ziyun Ding
Department of Biological Science, Purdue University, West Lafayette, IN 47907, USA

Daisuke Kihara
Department of Biological Science, Purdue University, West Lafayette, IN 47907, USA

Department of Computer Science, Purdue University, West Lafayette, IN 47907, USA

Satwica Yerneni
Division of Biomedical Statistics and Informatics, Mayo Clinic, Rochester, MN 55905, USA

Boris Shabash and Kay C. Wiese
School of Computing Science, Simon Fraser University, 8888 University Drive, Burnaby, BC, Canada

P. K. Busk, B. Pilgaard, M. J. Lezyk, A. S. Meyer and L. Lange
Department of Chemical and Biochemical Engineering, Technical University of Denmark, Søltofts Plads, Building 229, 2800 Kgs. Lyngby, Denmark

Lingxiao Lan
School of Software Engineering, Beijing Jiaotong University, 3 Shangyuan Residence, Haidian District, 100044 Beijing, China

Ergude Bao
School of Software Engineering, Beijing Jiaotong University, 3 Shangyuan Residence, Haidian District, 100044 Beijing, China
Department of Botany and Plant Sciences, University of California, Riverside, 900 University Ave., 92521 Riverside, CA, USA

Dmitry Y. Mozzherin
University of Illinois, Illinois Natural History Survey, Species File Group, 1816 South Oak St., Champaign, IL, 61820, USA

Alexander A. Myltsev
IP Myltsev, Kaslinskaya St.Chelyabinsk, 454084, Russia

David J. Patterson
University of Sydney, Sydney, Australia

Akhilesh Kaushal, Hongmei Zhang, Wilfried J. J. Karmaus and Meredith Ray
Division of Epidemiology, Biostatistics, and Environmental Health, University of Memphis, Memphis 38152, TN, USA

Mylin A. Torres
Winship Cancer Institute, Emory University, 1365 Clifton Rd. NE, Atlanta 30322, GA, USA
Department of Radiation Oncology, Emory University School of Medicine, 1365 Clifton Rd. NE, Atlanta 30322, GA, USA

Alicia K. Smith
Winship Cancer Institute, Emory University, 1365 Clifton Rd. NE, Atlanta 30322, GA, USA
Department of Psychiatry and Behavioral Sciences, Emory University School of Medicine, 101 Woodruff Circle, Suite 4000, Atlanta 30322, GA, USA

Shu-Li Wang
National Institute of Environmental Health Sciences, National Health Research Institutes, Miaoli, Taiwan

Špela Baebler, Miha Svalina, Marko Petek, Katja Stare, Ana Rotter, Maruša Pompe-Novak and Kristina Gruden
Department of Biotechnology and Systems Biology, National Institute of Biology, Ljubljana 1000, SI, Slovenia

Maximilian G. Plach and Rainer Merkl
Institute of Biophysics and Physical Biochemistry, University of Regensburg, D-93040 Regensburg, Germany

Florian Kandlinger
Institute of Biophysics and Physical Biochemistry, University of Regensburg, D-93040 Regensburg, Germany
Faculty of Mathematics and Computer Science, University of Hagen, D-58084 Hagen, Germany

Index

www.ingramcontent.com/pod-product-compliance
Lightning Source LLC
Chambersburg PA
CBHW082047190326
41458CB00010B/3479